New Horizons for Second-Order Cybernetics

K&E Series on Knots and Everything — Vol. 60

New Horizons for Second-Order Cybernetics

Editors

Alexander Riegler
Vrije Universiteit Brussel, Belgium

Karl H Müller
International Academy for Systems and Cybernetic Sciences, Austria

Stuart A Umpleby
George Washington University, USA

NEW JERSEY • LONDON • SINGAPORE • BEIJING • SHANGHAI • HONG KONG • TAIPEI • CHENNAI • TOKYO

Published by

World Scientific Publishing Co. Pte. Ltd.
5 Toh Tuck Link, Singapore 596224
USA office: 27 Warren Street, Suite 401-402, Hackensack, NJ 07601
UK office: 57 Shelton Street, Covent Garden, London WC2H 9HE

Library of Congress Cataloging-in-Publication Data
Names: Riegler, Alexander, editor.
Title: New horizons for second-order cybernetics / edited by: Alexander Riegler, Vrije Universiteit Brussel, Belgium, Karl H. Müller, International Academy for Systems and Cybernetic Sciences, Austria, Stuart A. Umpleby, George Washington University, USA.
Description: New Jersey : World Scientific, [2018] | Series: Series on knots and everything ; volume 60 | Includes bibliographical references.
Identifiers: LCCN 2017027904| ISBN 9789813226258 (hardcover : alk. paper) | ISBN 9813226250 (hardcover : alk. paper)
Subjects: LCSH: Cybernetics--Research. | Cybernetics--Philosophy. | System theory.
Classification: LCC Q310 .N475 2018 | DDC 003/.5--dc23
LC record available at https://lccn.loc.gov/2017027904

British Library Cataloguing-in-Publication Data
A catalogue record for this book is available from the British Library.

Cover photo by Emilie Ceriez

Printed in Singapore

Dedicated to

Ranulph Glanville (1946–2014)

for his enormous efforts in promoting second-order cybernetics

Contents

Part II: Reflecting on the Perspectives for a Fivefold Agenda of Second-Order Cybernetics

Epilogue

Acknowledgement of Sources

All texts in Part I as well as the chapter "Mapping the Varieties of Second-Order Cybernetics" by Karl Müller and Alexander Riegler were previously published in 2016 in the journal *Constructivist Foundations* volume 11, number 3, http://constructivist.info/11/3 .

Prologue

A Brief History of (Second-Order) Cybernetics

Louis H. Kauffman & Stuart A. Umpleby

1. The field of cybernetics had its origins in a series of conferences sponsored by the Josiah Macy Jr. Foundation from 1946 to 1953. The title of the conferences was "Circular Causal and Feedback Mechanisms in Biological and Social Systems." After Norbert Wiener (1948) published his book, *Cybernetics; or, Control and Communication in the Animal and the Machine*, the Macy conferences were called the conferences on cybernetics (Pias 2003). During World War II scientists had worked on a wide range of problems – shipping supplies safely and efficiently across the North Atlantic, building radar-guided anti-aircraft guns, coding and decoding messages, conducting psychological warfare operations, and maintaining families when the men were away and women were working in factories. Scientists who had worked on projects during the war wanted to discuss their experiences and what they had learned. Wiener had expanded the discussion considerably and included other mathematicians and scientists in the conversation such as the Mexican researcher, physician and physiologist Arturo Rosenblueth. Gregory Bateson and Margaret Mead, two anthropologists, felt that in the course of solving these problems interesting ideas, in particular circular causality, had been developed by engineers and mathematicians. They hoped that the conferences would introduce these ideas to social scientists. Together with the conference chairman, Warren McCulloch, they worked to coach the participants in the meetings on how to talk with one another despite very different disciplinary frames of reference (Kline 2015).

2. Circular causality has long been a difficult topic for scientists to deal with. One of the informal fallacies is a caution to writers to avoid circular reasoning. Statistical methods, which are used to establish the confidence we have in causal relations, usually assume linear causal relations. However, circular causality is essential in any regulatory process. A thermostat regulating the heat in a room, a driver steering a car on a road, or a manager working to maintain the profitability of a firm are all engaged in a circular process. In each case the regulator affects the system being regulated, observes the results of actions and then formulates another course of action. Note that this sequence of observation and formulation is not only circular, it is more simultaneous than sequential.

3. Circular causality is essential in biology. Biological organisms survive due to the process of homeostasis. The body consists of many circular processes. We become hungry, so we eat. We become thirsty, so we drink. The body is satisfied for a time, but then the cycle repeats. The iris in the eye regulates the amount of light entering the eye. When the body becomes too hot, it sweats, and evaporation cools it. Biological survival is possible due to a large number of circular causal processes (Cannon 1939). Similarly, the work of

McCulloch and Pitts (1943) engaged for the first time models of cognition based on circularly interconnected models of logical neurons.

4. Just as biological systems depend on circular processes, social systems do as well. In commercial transactions and in families communication is fundamental. Each party seeks to influence the decisions and actions of others. The existence of complex organizations such as business firms requires many ongoing circular causal processes (Beer 1972). A business firm must continually recruit, hire, and train workers as the company grows and as current workers retire. Finding and working with customers and suppliers entails communication back and forth. Employees continually monitor and modify the internal processes in a firm – advertising, production, purchasing materials and maintaining equipment.

5. Governments also require many kinds of feedback, a circular process (Deutsch 1963). Ideally governments serve the interests of citizens and citizens control the operations of government through voting, lobbying, oversight by the press, and occasional law suits. The decisions of lower courts can be reviewed by higher courts and even a constitution can be changed by amendment. Given the vital role that circularity plays in biological and social systems, it is surprising that so much of science focuses on linear causal relations. Probably this happens because scientists seek certainty in their knowledge. They want to know what confidence they should have in a particular result of research. But statistical confidence intervals work with measurements (e.g., standard deviation) and linear causal relations. Circular causal relations are more challenging and usually rely on comparing results from a model with time series data.

6. Looking at circular causal processes has proven to be quite fruitful. For example, Humberto Maturana, Francisco Varela and Ricardo Uribe created the theory of autopoiesis, or self-production, to explain living systems (Maturana & Varela 1980). They noted that a living system has parts that engage in processes that result in new parts engaging in similar processes. There is some variation in the parts and processes produced, which enables evolution.

7. Because cyberneticians were interested in cognition, a biological process, they were interested in the role of the observer in scientific research. Although scientists sought for many years to exclude the observer in an effort to be unbiased and objective, cyberneticians noted that every statement made is made by an observer to an observer. That is, the observer has purposes within a social context and a history that includes national culture and academic training. Hence, observations independent of the characteristics of the observer are not physically possible. *Second-order cybernetics* has been an effort to incorporate this realization into cybernetics. The early work in cybernetics focused on the design of control devices. Second-order cybernetics was an effort to apply the same ideas to the observer or designer of control devices, hence to cognition (Foerster 2003).

8. A further development in cybernetics has been to reconsider social systems as collections of purposeful systems – individuals and organizations. Much of social science research has treated social systems as collections of interacting variables. This is possible only if one makes certain assumptions about the elements of social systems, for example the Efficient Market Hypothesis assumes that economic actors are rational, self-interested profit maximizers who all have the same information and complete information. Recently more attention has been given to the often improbable assumptions that scientists make in constructing their models. One assumption that has been carried over from the natural sciences to the social sciences is that theories do not alter the system observed. Although we

assume that physical objects do not change their behavior when scientific theories change, social systems do change their behavior depending on which theory is guiding actions, for example the theories of Adam Smith, Karl Marx, John Maynard Keynes, or Milton Friedman.

9. The latest developments in cybernetics have been theories of *reflexivity*. Vladimir Lefebvre (1982) suggests that there are two systems of ethical cognition depending on whether one believes the end justifies the means or the end does not justify the means. George Soros (1987) has pointed out that people in societies, including scientists, not only observe, they also participate. The fact that the elements of social systems are both observing and participating greatly increases uncertainty about future events within society and explains the fallibility of our predictions. Heinz von Foerster proposed that since our knowledge of the social world is limited by our experiences, we need other people, whose experiences are different from ours, to support or challenge our perceptions and conclusions. Karl Müller (Müller & Riegler 2014) has suggested that meta research involves a kind of reentry or reconsideration of previous findings. Louis Kauffman (2016 and this volume) has described science as a search for invariances in our contextual descriptions and the production/observation of objects through these invariances.

10. The contributions in this book illustrate that cybernetics is an important contribution to contemporary science. Physics and chemistry provide a theory of the material domain by explaining matter and energy processes. Cybernetics offers a theory of less tangible phenomena by explaining processes of communication and control (Umpleby, 2007). The cybernetics domain is different because both observers and theories influence what happens in social systems, and it is through social systems that all living science occurs. The contributions in this book suggest several possible future directions for cybernetics. They describe how science is changing and propose a unified point of view for classical and second-order science (Umpleby 2014).

References

Beer S. (1972) Brain of the firm. Herder and Herder, New York.

Cannon W. B. (1939) The wisdom of the body. W.W. Norton, New York.

Deutsch K. W. (1963) The nerves of government: Models of political communication and control. Free Press of Glencoe, London.

Foerster H. von (2003) Understanding understanding. Springer, New York.

Kauffman L. H. (2016) Cybernetics, reflexivity and second-order science. Constructivist Foundations 11(3): 489–497. http://constructivist.info/11/3/489

Kline R. (2015) The cybernetics moment: Or why we call our age the information age. Johns Hopkins University Press, Baltimore.

Lefebvre V. (1982) Algebra of conscience: A comparative analysis of Western and Soviet ethical systems. Reidel, Boston MA.

Maturana H. R. & Varela F. J. (1980) Autopoiesis and cognition: The realization of the living. Reidel, Boston MA.

McCulloch W. & Pitts W. (1943) A logical calculus of the ideas immanent in nervous activity. Bulletin of Mathematical Biophysics 5: 115–133.

Müller K. H. & Riegler A. (2014) Second-order science: A vast and largely unexplored science frontier. Constructivist Foundations 10(1): 7–15. http://constructivist.info/10/1/007

Pias C. (2003) Cybernetics: The Macy conferences 1946–1953. Diaphanes, Zurich.

Soros G. (1987) The alchemy of finance: Reading the mind of the market. Simon and Schuster, New York.

Umpleby S. (2007) Physical relationships among matter, energy and information. Systems Research and Behavioral Science 24(3): 369-372. http://blogs.gwu.edu/umpleby/recent-papers/

Umpleby S. A. (2014) Second order science: Logic, strategies, methods. Constructivist Foundations 10(1): 16-23. http://constructivist.info/10/1/016

Wiener N. (1948) Cybernetics: Or control and communication in the animal and the machine. MIT Press, New York.

Mapping the Varieties of Second-Order Cybernetics

Karl H. Müller & Alexander Riegler

Introduction

1. In his impressive study on the cybernetic moment, Ronald Kline (2015) claims that the move from first-order to second-order cybernetics was a dead end that did not produce long-lasting impacts for other disciplines. As such, it did not leave any significant traces. Similarly, Robert Martin (2015) maintains that second-order cybernetics (SOC) has failed to affect its scientific neighboring disciplines. These two general assessments on the internal and external weaknesses of second-order cybernetics form the central hypothesis we want to address in this editorial:

Kline-Martin-Hypothesis: As a research program, second-order cybernetics was

a. insufficiently developed,
b. has had no sustainable consequences for other scientific disciplines in the past, and
c. will remain mostly irrelevant in the future as well.

Thus, second-order cybernetics will fail miserably to meet the constructivist challenges (as defined by Riegler 2005, 2015), and researchers should bid their farewell to radical constructivism altogether (as claimed by Schmidt 2003) as an increasingly marginalized side-stream endeavor (see the contributions in Quale & Riegler 2010).

2. One of the goals of this editorial is to collect enough argumentative material to either support or reject each of the three parts of the Kline-Martin-hypothesis.

3. Another goal of this editorial and of this volume on second-order cybernetics is to continue and end an experiment that started with the special issue of *Constructivist Foundations* on second-order science (Riegler & Müller 2014). While the majority of the contributions in that issue saw second-order science (SOS) as mainly linked with the inclusion of an observer, a smaller part, including the editors, viewed second-order science as a genuinely new research domain operating with building blocks from first-order science in an analytical and reflexive manner and as significantly different and independent from inclusions or exclusions of observers (Müller & Riegler 2014).

4. In contrast to the 2014 issue, we now want to explore what has become of second-cybernetics (SOC), which emerged in the late 1960s by adding reflexivity to the first-order cybernetics of observed systems, which made it a "cybernetics of observing systems" (Foerster 1974). This reflexive turn in cybernetics deserves a closer look and should be evaluated using the following questions:

- What are the similarities and differences in the current approaches to SOC?
- What are the relations between contemporary SOC and other scientific disciplines?
- Given its age, are there new perspectives for contemporary cybernetics available, transcending the original SOC?
- What are the relations between SOC and second-order science and what could be possible functions and roles of cybernetics for second-order science?

5. With the answers to these questions, we expect to be able to offer several maps on:

- The current frameworks of SOC
- The significant differences and/or similarities between SOC and the domains of second-order science.

Reconstructing SOC

6. Let us start with a historical journey to construct a map of SOC that transforms an implicit order on its scope and dimensions into an explicit one. The starting point for this journey is Heinz von Foerster,[1] who on the occasion of his 90th birthday gave an interview in which he presented a selection of his five most important publications:[2]

a. *Das Gedächtnis. Eine quantenmechanische Untersuchung* [Memory. A Quantum-Mechanical Investigation] (Foerster 1948)
b. "Some Remarks on Changing Populations" (Foerster 1959)
c. "Doomsday: Friday, November 13, AD 2026" (Foerster, Mora & Amiot 1960)
d. "Objects: Tokens for Eigenbehaviors" (Foerster 1976)
e. "Ethics and Second-Order Cybernetics" (Foerster 1992)

7. At first, von Foerster's selection is quite surprising because only two of them – "Objects" and "Ethics," are included in his representational collection of papers, *Understanding Understanding* (Foerster 2003). We could be tempted to dismiss his peculiar choice and attribute it to his age or to other constraining factors. But on closer reflection, it becomes clear that his selection offers a new view on the scope and the main characteristics of SOC, which have been mostly ignored until recently.

8. Von Foerster's five most important publications can be separated into 3 categories:

Category 1 comprises only his unique monograph on memory. Von Foerster selected it because this publication helped him to move from post-war Austria, where he had no hopes for an academic career, to the core of one of the most productive and energetic science circles around Warren McCulloch's Macy conferences and to the American university and research system.

1. It must be emphasized that approaches similar to Heinz von Foerster's perspective on second-order cybernetics were built by Humberto Maturana (1970), Maturana & Francisco Varela (1980, 1987), Gordon Pask (1975a, 1975b, 1976) and others, but were not included in this editorial due to its focus on the contemporary varieties of second-order cybernetics.
2. "90 Jahre Heinz von Foerster – Die praktische Bedeutung seiner wichtigsten Arbeiten. [90 years Heinz von Foerster – The practical relevance of his most important publications]," DVD directed by Maria Pruckner, 2001. Malik Management Zentrum, St. Gallen, Switzerland.

Category 2 contains two articles on population dynamics. The first one discusses the dynamics of leucocytes, the second the long-term evolution of human populations. Both articles are based on a particular hyperbolic equation, which was later characterized as the "von Foerster-equation," to which von Foerster himself attributed quite some significance.

Category 3 includes von Foerster's articles on objects and ethics. While the former seems well justified, choosing the latter appears to be rather ill-founded because it seems to ignore many more obvious choices such as the articles on self-organizing systems and their environments (Foerster 1960), on computation in neural nets (Foerster 1967) and on constructing a reality (Foerster 1973). In particular, the last one has been reprinted many times.

9. The publications in the first two categories seem justified because the first publication fulfilled a vital function for von Foerster himself, and the papers in the second category offer a general diffusion formalism for his views on describing the dynamics of populations, which can be applied to self-organizing systems with very heterogeneous compositions, ranging from leucocytes to humans.

10. The third category, however, remains an enigma because the two articles it contains cover very different ground. The paper "Ethics and Second-Order Cybernetics" is based on a lecture that was held without using any formalisms. The article on objects, however, offers a highly formal account of the necessary emergence of eigenforms in recursively closed systems.

11. Fortunately, we can offer several arguments showing that the two articles in the third category must be considered as the foundations of SOC in particular and of a new perspective on the general methodology of science.

12. In order to find the particular points of relevance of the first of these two articles, we need to look to another publication from the same period, i.e., "Through the Eyes of the Other" (Foerster 1991). In both articles, von Foerster develops a general epistemic distinction on interacting with one's environment (or world) in two fundamentally different ways. This distinction was further elaborated during a week-long conversation in 1997.[3] There, von Foerster asked a question that gives rise to this fundamental epistemic differentiation:

> Am I an observer who stands outside and looks in as God-Heinz or am I part of the world, a fellow player, a fellow being? (Foerster 2014: 128)[4]

13. This epistemic split led to a specific terminology for these two very different epistemic modes, i.e., the endo-mode and the exo-mode (Müller 2016a).[5] His distinction of the two epistemic modes is also relevant for cybernetics itself as he links it to the distinction between first- and second-order cybernetics. This distinction…

---

3. This conversation formed the basis of Foerster (1997), which was translated to English in 2014.
4. Von Foerster developed a very intriguing list of the characteristics of these two fundamentally different epistemic modes towards one's world or environment; see Umpleby's contribution to this volume, in which he offers a detailed account of this epistemic split between "being apart" or "being a part" (For more details, see Müller 2016a: 161–186).
5. A similar distinction can be found in Otto Rössler's 1992 book on endo-physics. However, our distinction between exo and endo differs significantly from Rössler's and others (such as Atmanspacher & Dalenoort 1994) who assume a two-level structure of reality.

> is nothing else but a paraphrase of […] the two fundamentally different epistemological, even ethical, positions, where one considers oneself, on the one hand, as an independent observer who watches the world go by; or on the other hand, as a particular actor in the circularity of human relations. (Foerster 2003: 303)

14. So according to von Foerster, the following equivalences can be established:

- First-order cybernetics ≡ cybernetics in an exo-mode (from without) ≡ cybernetics of systems observed
- Second-order cybernetics ≡ cybernetics in an endo-mode (from within) ≡ cybernetics of observing systems.

15. In our interpretation, this split into epistemic modes transcends cybernetics itself and can be generalized in the following way across science.

- Endo-mode ≡ exploring the world from within ≡ endo-science
- Exo-mode ≡ exploring the world from without ≡ exo-science.

16. Unfortunately, von Foerster's shift of first- and second-order cybernetics to different epistemic modes was never fully appreciated by the cybernetic (nor the radical constructivist) community. In any case, we are convinced that the importance von Foerster attributed to this shift was the main reason why von Foerster selected the ethics paper as one of his five most important and influential publications. Although first-order cybernetics focuses on technology and goal-directed machines and SOC on living or observing systems, the scientific disciplines they are dealing with are not the key difference between them. It is rather the epistemic mode, exo or endo, that distinguishes them.

17. Aside from these few publications, von Foerster never produced a detailed account of the endo-mode as a new general methodology for science. He only insisted that SOC should and must be organized with a shift to the endo-mode in which the I of observers or researchers was to be an intrinsic element of the research process itself, which requires profound methodological changes. This becomes very clear in his statement, "I am the observed relation between myself and observing myself" (Foerster 2003: 257), where "I" invites a host of additional notions such as self-reference, self-observing, self-reflexivity and the like, as well as a long history of paradoxes based on the notions of I, self and reference, such as the paradoxes by Epimenides, Russell or Nelson and Grelling.

18. The embeddedness of SOC within a specific epistemic mode offers a new perspective, which we will apply to the contributions in this volume by grouping them into three categories.

Category 1: Here, SOC is considered an instance of building a new and general methodology of science that is not linked exclusively to cybernetics but that can be utilized across all scientific disciplines and for a variety of applied disciplines, and that can be extended to cover the arts as well. The articles by Louis Kauffman and Stuart Umpleby fall into this category, and our editorial provides supplementary material.

Category 2: The articles focus on instances of working with and in SOC within special scientific disciplines and domains. They comprise Diana Gasparyan's article, which links SOC with consciousness studies, and Bernard Scott's, which uses SOC to provide foundations for psychology.

	Exo-mode	Endo-mode	Authors in this volume
General methodology of science	Established methodology	SOC_E as constructing a new general methodology or a special methodology for the social sciences	Kauffman, Umpleby
Theories, models, research programs, foundations for scientific disciplines	First-order cybernetics with a focus on theories and models of control and communication	SOC_E with a focus on transforming research programs and entire scientific disciplines	Scott, Gasparyan
Applications in science and the arts	"Applied" cybernetics with a focus on technological applications	SOC_E with a focus on reflexive practices in "applied" disciplines, including the arts	Sweeting, Scholte

Table 1. A two-dimensional map of the varieties of second-order$_E$ cybernetics in this book.

Category 3: SOC is used in applied disciplines and artistic fields. This category includes Ben Sweeting's article, which combines SOC with design research, and Tom Scholte's, which applies SOC to studies in theater performances.

19. Based on this categorization, we will use the notation of *second-order$_E$ cybernetics* (SOC_E) to denote the type of cybernetics that is linked to the specific epistemological endo-mode *E* of exploring and investigating the world from within. Table 1 summarizes our argument and offers a coherent perspective on its scope and dimensions (see also Müller 2016a, 2016b).

20. The concept of SOC_E is important because it points to a more general background that explains why the shift to an endo-mode and to cybernetics as a cybernetics "from within" was of such great importance to von Foerster that he expected the latter two publications in his top-five list to become the igniting spark for a scientific revolution.

21. For von Foerster, such a revolution was clearly needed. For him, the "flawless" but "sterile" paths of the established scientific method rest on the pillar of objectivity and become "counter-productive in contemplating any evolutionary process, be it the growing up of an individual, or a society in transition" (Foerster 2003: 204). By contrast, by giving up "objectivity," the new general methodology of science is not based on pure or arbitrary subjectivity, as short-sighted critics may assume, but on

- a recursively closed research organization in a triadic network[6] of researcher(s),
- a domain of investigation, and, finally,

6. Triadic networks were repeatedly emphasized by von Foerster. A paradigmatic example is the triadic network of a chicken, a rooster and an egg, which are linked through generative relations where "you cannot say who was first and you cannot say who was last" (Foerster 2003: 284). Triadic networks in science can be built between (1) actors or researchers, (2) an environment or domain of investigation and (3) a common language, grammar, rule system or, more generally, a knowledge base. For the importance of generative relations within such a triadic research network, see Müller (2007).

- a scientific knowledge base that includes, among other things, theories, models, rule-systems, research designs, etc.

22. This special type of research organization can and must lead to eigenforms of variable forms and nature (see von Foerster's article on "objects"). These eigenforms emerge as a recursive consensus formed among all three nodes in the triadic research network involved.

23. In his contribution to this volume, Kauffman presents a detailed account of such an endo-methodology for science that covers all the necessary elements in research processes, including observations, experiments and data-collections. It also describes these elements in a significantly different way than in the established science methodology, which stands on objectivity, truth and, as its underlying epistemology, hypothetical realism. What has been presented by us as the epistemic mode from within is analyzed by Kauffman as a reflexive domain in which a community of researchers also become actors. While we will not replicate all the details of Kauffman's article here, we want to point out two highly relevant features of his perspective on endo-methodology.

Feature 1

24. Contrary to the established understanding of physics that considers physical experiments as a way to confirm objective, "real" or "true" features of the world, Kauffman offers a radically different account of physical experiments. This account is based on the Church-Curry fixed point theorem and states that for any reflexive domain D and any component C in D there is an element X in D such that X becomes a fixed point for C: $CX=X$ and X is, thus, the eigenform for C (see also Kauffman 2005, 2009). For physical experiments E that are conducted by a single researcher or a group of researchers R, there is a necessary element X so that $E(R, X)=X$. An interesting shift occurs here from the realist account of experiments as a search for objective features of nature and reality to the composition of researchers conducting physical experiments. In this endo-perspective, physical experiments can be reproduced again and again with identical results X because these experiments E become invariant under different compositions of R. Different researchers R with different age and gender, with different regional and cultural backgrounds, with different research designs, etc. will end in the eigenform-configuration of $E(R, X)=X$. Obviously, experiments can fail under different compositions of R, they can be disconfirmed and yield a new X, which can be rejected again, etc. But the main emphasis in an endo-methodology lies on the researchers themselves and their changing compositions over time. $E(R, X)=X$ states that a special outcome X becomes invariant over time with respect to the possible configurations of researchers R.

Feature 2

25. The endo-methodology offers a stringent interpretation of several seemingly strange ideas formulated by von Foerster on self-writing theories such as the following proposition:

> The laws of physics, the so-called 'laws of nature,' can be described by us. The laws of brain functions – or even more generally – the laws of biology, must be written in such a way that the writing of these laws can be deducted from them, i.e., they have to write themselves. (Foerster 2003: 231)

26. Additionally, von Foerster points to the special status of any theory of the brain, which must be written by a brain:

> It is clear that if the brain sciences do not want to degenerate into a physics or chemistry of living – or having once lived – tissue they must develop a theory of the brain: $T(B)$. But, of course, this theory must be written by a brain: $B(T)$. This means that this theory must be constructed in a way as to write itself: $T(B(T))$. (Foerster 2003: 195)

27. In reflexive domains, the theories T of the brain, of biology, or of nature, can actually write themselves, which can be described by the same formalism for eigenforms: $T(R, X)=X$.

28. It was one of the unexpected surprises of editing this volume that Kauffman presented such a concise perspective for a SOC-inspired endo-methodology, and that Umpleby presented an outline of such an endo-methodology for the social sciences together with a rich agenda to be pursued in the future.[7]

29. We can now formulate our mappings of the articles in this volume and their specific cognitive family resemblances within the framework of SOC_E as cybernetics from within or as endo-cybernetics as a comprehensive research agenda with five large-scale domains that corresponds with the contributions in this volume.[8]

Agenda I: Building an alternative general scientific methodology

30. This line of investigation was forcefully pursued by Kauffman. His article contains the special building blocks of such a general methodology that potentially covers a variety of disciplines, including quantum physics, the life sciences, the cognitive sciences, the social sciences and the technical sciences. Therefore, we suggest that it would be very productive to broaden and diversify Kauffman's approach into a larger survey publication (or "handbook" if you will) that offers a contemporary account of the status of endo-research across various scientific disciplines and focuses also on the interactions between theories and their diffusion across society and nature.

Agenda II: Building specialized endo-methodologies for different scientific disciplines

31. The second agenda is reflected in Umpleby's article, which offers the appropriate elements of an endo-methodology for the main areas of Umpleby's work in management science, sociology and social policy. At present, scientific disciplines such as management science or sociology use methodologies that are still framed in an exo-mode and it becomes, thus, a challenging and comprehensive task to transform these exo-methodologies into suitable endo-methodologies. A state of the art summary of an exo-methodology of empirical social research such as that by Jürgen Bortz and Nicola Döring (2015) can be

---

7. Many crucial in-depth explorations of the endo-methodology are still missing, such as a study of observer effects across academic disciplines, or the actual differences in outcomes between research in endo- and exo-mode. However, the current volume has already assembled some of the ingredients that will help to undertake these investigations.
8. To our knowledge, Glanville was the only scholar in recent years who worked simultaneously on all five agendas of SOC_E while interacting with heterogeneous groups in Europe, Australia, China and the US.

challenged by a comprehensive endo-methodology that gives rise to additional debates on the comparative advantages and disadvantages or on the differences in the actual empirical outcomes of these exo- and endo-approaches. Additionally, the second agenda should also undertake in-depth comparative studies between the endo-mode, other non-traditional methodologies and the exo-mode with respect to their research organization, their dialogical and consensus-building capacities, their ethical commitments, their types of interventions and, finally, their empirical outcomes.In his contribution, Umpleby makes the strong point that the view from within can give rise to a much broader range of problem solutions and can also lead to a more active role of researchers in changing their environments, not only interpreting it.

Agenda III: Offering foundational frameworks as well as reframing and contextualizing research problems across all scientific disciplines

32. The third agenda is exemplified by the respective articles of Scott and Gasparyan. The former presents a foundational outline for psychology, using a well-known framework from SOC_E, namely Pask's conversation theory (Pask 1975a, 1975b, 1976). The latter offers a new perspective on reframing notorious problems in consciousness studies with the help of building blocks from SOC_E. Recently, Michael Lissack (2016a, 2016b) has emphasized the importance of contextualizations of debates, controversies and models within science, which could become yet another task for this agenda. It enables an in-depth analysis of researchers, their cultural and cognitive backgrounds, their interaction patterns and their lines of conflicts.

Agenda IV: Creating reflective circular practices within applied disciplines

33. The fourth agenda addresses a large number of applied disciplines, including design, architecture, education science (Scott 2014), public administration and the large number of therapeutic and consulting domains. Here, SOC_E can become a constant source of inspiration for the re-organization of research activities and practical work in the general format of recursively closed reflexive circles, as shown in Sweeting's article.

Agenda V: Building special circular reflexive approaches for special niches within artistic domains

34. The fifth agenda carries SOC_E into those artistic domains in which the organization of recursively closed reflexive circles and eigenforms can lead to new innovative art. While this agenda may be restricted to certain niches in the artistic domain, Scholte's paper shows that it can at least be applied to the world of theaters and theater performances.

Summary

35. Cybernetics is a transdisciplinary scientific discipline that brought goals and circularity into the domain of scientific investigations. Initially, this was accomplished in the 1940s and 1950s through notions such as feedback mechanisms, feed-forward, purpose and control. Between the late 1960s and 1970s, cybernetics expanded to promote circular research designs and reflexive investigations that include the researchers or the observers as a necessary component. So far, this second-order$_E$ cybernetics has fallen short of reach-

ing its full potential. Despite promising attempts such as that by Bruce Clarke (2009), the agenda of SOC_E has not been significantly pursued by a large number of researchers, and a large potential still awaits its use, with highly innovative gains not only for cybernetics itself but for all of science.

Towards a map of second-order$_E$ cybernetics and of second-order science

36. The differentiation of cybernetics as cybernetics from without or first-order cybernetics and cybernetics from within or SOC_E raises the question as to the relations between SOC_E and second-order science. The immediate response would most probably locate SOC_E as just another instance of second-order science or of science from within.

37. Although this initial response suggests itself, we will propose quite a different answer that may appear counter-intuitive and even paradoxical at first. We propose that SOC_E as cybernetics from within or as endo-cybernetics is not included in the domains of second-order science. The reason for this complete split is grounded in a basic difference between two strictly separate contexts for the concept of second-order itself, which is already hinted at in von Foerster's *Cybernetics of Cybernetics* (Foerster 1974).

38. Firstly, the cover of *Cybernetics of Cybernetics* presents cybernetics situated at two different mutually linked levels:

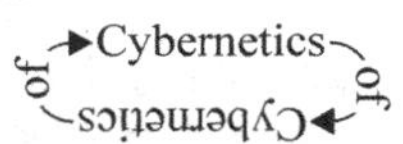

39. This format of SOC not only follows Margaret Mead's article on cybernetics of cybernetics (1968), it is further extended by von Foerster to the control of control and to the communication of communication (Foerster 1974: III), and generalized by him through Lars Löfgren's (1979, 1992) notion of autological concepts (Foerster 1984), i.e., concepts that can be applied to themselves such as understanding understanding, control of control, learning about learning, functions of functions, geometry of geometry, model of models (see also Müller 2007).

40. In Müller & Riegler (2014) and Müller (2016a), this first line of second-order concepts has recently been advanced and systematized to differentiate between first-order science and second-order science, separated by their different levels and linked through a dense set of bottom-up and top-down links: any first-order discipline, due to its enormous research output, can be linked to a potential second-order discipline, studying these outputs in comparative detail. Examples are first-order sociology and second-order sociology, first-order psychology and second-order psychology, first-order economics and second-order economics, first-order physics and second-order physics or first-order cybernetics and second-order cybernetics.

41. Secondly, page 1 of *Cybernetics of Cybernetics* presents the well-known definitions of first-order cybernetics as the cybernetics of observed systems, and of second-order cybernetics as the cybernetics of observing (systems). This distinction based on observed and observing was further elaborated by von Foerster in the paper on objects (Foerster 1976) and the related paper "In the Eyes of the Other" (Foerster 1991), emphasizing the new epistemic mode of exploring the world from within.

	Exo-mode ("from without")	Endo-mode ("from within")
First-order science	Established science accounts, using the still dominant general exo-methodology (e.g., first-order cybernetics)	First person approaches (I/we); participatory research designs (e.g., action research), using the new, general endo-methodology of consensus formations ("eigenforms"); (e.g., SOC_E)
Second-order science	Meta-studies, systematic literature reviews, second-order studies, etc. in the established general exo-methodology	Specifications of goals of researchers; first person approaches (I/we), operating in a new endo-methodology with a high degree of transparency and reproducibility in the pursuit of second-order$_L$ investigations

Table 2. A two-dimensional map on second-order science and SOC.

42. This line of thinking was further strengthened and generalized in the course of this volume and was summarized as the mode from within, as the endo-mode and as the mode that transforms the operations in disciplinary research and leads to new disciplines, this time in the form of endo-sociology, as opposed to established exo-sociology, to endo-psychology in contrast to exo-psychology, to endo-economics versus exo-economics, to endo-physics against exo-physics or to endo-cybernetics alias SOC_E or to cybernetics from within.

43. Thus, SOC can refer to two significantly different areas of inquiry, i.e., as cybernetics from within, i.e., as endo-cybernetics or as cybernetics at the second-order science level. Accordingly, we would like to introduce the concepts of

- second-order$_E$
- second-order$_L$

where second-order$_E$ is linked with a particular epistemic mode, i.e., with the endo-mode, and second-order$_L$ is linked with a particular level in science, i.e., with the second-order level and, thus, with second-order science. Second-order$_E$ cybernetics is the focus of this volume, in which second-order$_L$ cybernetics is only marginally present. This editorial, by contrast, could be considered as a proto-study in second-order$_L$ cybernetics on the varieties of second-order$_E$ approaches.

44. Table 2 presents a map comparing the varieties of first- and second-order science with the two epistemic modes, from without and from within. It shows that second-order$_E$ cybernetics and first-order cybernetics or cybernetics from without become different cybernetic approaches for exploring the world while first-order science and second-order science can be pursued, in principle, in both modes.

45. The mapping, with its focus on the varieties of SOC in the contexts of first-order science and second-order science, can be further generalized to cover all scientific disciplines. This can be accomplished by a 2×2 matrix between types or levels of science and the two epistemic modes.

- Established disciplines, such as sociology, psychology, economics and physics, can be organized as research traditions for exploring the world from without, in an

	Exo-mode ("from without")	Endo-mode ("from within")
First-order science	Established accounts using the once dominant exo-methodology (e.g., established empirical) social research, most of economics, physics, etc.)	First person approaches (I/we); participatory research designs (e.g., action research), using the new, general endo-methodology of consensus formations ("eigenforms"); (e.g., endo-sociology, endo-psychology, endo-economics, endo-physics, etc.)
Second-order science	Second-order studies in *X* (sociology, psychology, economics, physics, etc.), using an established exo-methodology	Second-order studies in *X* (sociology, psychology, economics, physics, etc.), operating in a new endo-methodology with a high degree of transparency, interactivity and reproducibility

Table 3. A two-dimensional map for any established scientific discipline such as sociology, psychology, economics, physics.

exo-mode, or, alternatively, from within, in an endo-mode, as in the instances of exo-sociology and endo-sociology or exo-physics and endo-physics.

- Likewise, second-order sociology, second-order psychology, second-order economics or second-order physics as disciplines for reflecting on the explorations of their corresponding first-order disciplines can be organized in an established exo-mode or in an endo-mode as well (e.g., second-order exo-sociology, second-order endo-sociology).

46. Table 3 offers a map for scientific disciplines and the potential for research in four different disciplinary areas, located on two different levels with two epistemic modes for each level.

47. Through Tables 1 to 3, we can also go back to the first stage in our experiment, where we explored the scope and dimensions of second-order science (Müller & Riegler 2014). We had to leave this first stage with the inconclusive result of a high degree of heterogeneity and diversity with respect to the status of second-order science. However, we are now able to present the final map in Table 4. It provides an overview of all the target articles and their locations within the new landscapes of SOS_L and SOC_E.

48. Table 4 suggests that we have conducted an experiment with the authors and ourselves on two key concepts for radical constructivism, namely on second-order science and second-order cybernetics, with the surprising result that almost all of the articles in both issues deal with one of the five agendas of SOC_E whereas only our editorial, a single contribution for the issue on second-order science, i.e., Völker & Scholl (2014), and the concluding part of Kauffman's article in this volume focus on second-$order_L$ science.

49. With Table 4, we can conclude our experiment because we have arrived at a fresh perspective on the scope and the differences between two key concepts of second-order science and second-order cybernetics. Tables 1 to 4 reveal at least four surprising aspects.

a. The most surprising result for us was the dual definition of SOC, which was apparently developed in two significantly different versions during its conception between 1968 and 1975.

Second-order science in Riegler & Müller (2014)	
Second-order$_L$ science:	
Editorial	Basic outline of second-order science
Völker & Scholl	Focus on meta-analyses
Kauffman (concluding parts)	Eigenforms of eigenforms
Second-order$_E$ cybernetics:	
Umpleby	Agenda II
Füllsack	Agenda I
Allrøe & Noe	Agenda III
Vörös	Agenda III
Aufenvenne et. al.	Agenda III
Second-order cybernetics in this volume	
Editorial	Agenda I–V (as general targets)
Glanville	Agenda I–V (as research outputs)
Kauffman	Agenda I
Umpleby	Agenda II
Scott	Agenda III
Gasparyan	Agenda III
Sweeting	Agenda IV
Scholte	Agenda V

Table 4. Mapping the various contributions in Riegler & Müller (2014) and in the current volume.

- SOC_E was based on the initial definition by von Foerster as the cybernetics of observing systems, but remained in an under-critical state and was never developed into a full research program.
- SOC_L was linked to Mead and her self-referential desire to study cybernetic domains such as the American Society for Cybernetics with tools and methods from cybernetics. However, it has never been followed by a stream of publications on cybernetic studies of cybernetic studies nor by any other self-referential investigations.

b. As a consequence, SOS_L, following a level differentiation, and SOC_E in its endo-mode must and should be arranged in different science landscapes, in which SOS_L occupies the domains of the second-order level while SOC_E retains its status as a first-order transdiscipline but shifts to the epistemic mode from within.

c. The four maps in this editorial offer several so far largely unexplored research spaces with a high innovation potential for both SOS_L and SOC_E. These open frontiers can and must be investigated, due to the two strictly independent dimensions of levels and epistemic modes, each for themselves. Advances and breakthroughs in SOS_L will leave SOC_E unaffected and vice versa.

d. Finally, both SOC_E and SOS_L also become drivers towards higher levels of reflexivity in science. Second-order$_L$ science becomes reflexive with respect to first-order science and second-order$_E$ cybernetics operates in reflexive domains and in circles of actions and reflections, and turns reflexive with respect to researchers and their domain of investigation.

50. Tables 1 to 4 offer a novel way for recombining second-order$_{E,L}$ dimensions into a new architecture of science and into a rich and diversified research agenda for second-order$_L$ science on the one hand and second-order$_E$ cybernetics on the other (see also Müller 2016a).

The Kline-Martin-Hypothesis revisited

51. Based on what we have presented so far, the Kline-Martin hypothesis, which we introduced at the beginning, can be accepted in two of its three parts, but must be rejected in its third part:

Confirmed: During recent decades, SOC has been developed insufficiently as a research program.

Confirmed: SOC has exhibited no or very limited sustainable consequences for other scientific disciplines. In essence, SOC has become limited to the activities of small circles of socio-cyberneticians and second-order cyberneticians.

Rejected: SOC will remain mostly irrelevant in the future. This assumption must be rejected and this rejection is not based on the uncertain status of predictions in general, but on two significantly different arguments:

a. SOC_E as presented by the authors in this volume and summarized in this editorial has an innovative agenda to pursue. To our knowledge, for the first time, SOC_E has been recreated here in a systematic way as a comprehensive and transdisciplinary research program.
b. SOC_E must be credited for its role in promoting reflexivity in science. The title of Umpleby's article "Second-Order Cybernetics as a Fundamental Revolution in Science" already points in this direction and this argument will become even stronger in the future due to a strong movement towards higher levels of reflexivity in science.

52. So after almost seventy years of existence, cybernetics can finally pursue a rich research agenda that consists of classical or first-order cybernetics, SOC_E as cybernetics from within and the new post-cybernetics agenda as a second-order science discipline that reflects on reflections on reflections as in the instance of a second-order$_L$ study that is based on second-order$_L$ studies of lower order (Müller 2016b).

53. Moreover, the present issue with its shift to and its emphasis on a new general science methodology should put an end to the sterile philosophical battles between the proponents of radical constructivism and those of scientific realism. Second-order$_E$ cybernetics can move away from these battles into the core of science by offering general and special methodologies for research and new instruments for reframing and contextualizing research problems across all scientific disciplines.

Conclusion

54. As a concluding point, we would like to offer additional contextual information on the current shifts and transformations in science that have formed the background for our experiment with second-order science and second-order cybernetics.

55. In the rapidly growing role of reflexivity in science (Umpleby 2007, 2010), reflexivity can arise in three different contexts.

Context 1: Reflexivity can be characterized as a fundamental feature of actor-based systems or networks, as shown in the works of Vladimir Lefebvre (1977, 1982), George Soros (1994) and many others. These publications are presented in detail in Müller (2015a, 2015b).

Context 2: Reflexivity can also become a core ingredient for science and research, which was the topic of the collection of articles edited by Frederick Steier more than 25 years ago (Steier 1991), by anthropologists such as Clifford Geertz (1983, 2000) and by sociologists such as Margaret Archer (2007, 2012), Ulrich Beck (Beck, Giddens & Lash 1994) and Pierre Bourdieu (Bourdieu 1977, 1990; Bourdieu & Wacquant 1992).

Context 3: But reflexivity can also arise due to a new epistemic mode from within between researchers, their domains of investigation and their descriptive or explanatory accounts, which become relevant for all scientific disciplines, not just for the social sciences.

56. In our previous editorial on second-order science (Müller & Riegler 2015), we suggested that second-order science should be characterized as a "vast and largely unexplored science frontier" and as a new and significant reflexive component for the general science system itself, where second-order science becomes a reflexive scientific study of scientific inputs or outputs of first-order science. However, this revolution in reflexivity is just one of several elements that fundamentally transform the architecture of contemporary science. The revolution of reflexivity is accompanied by a revolution in complexity (Hollingsworth & Müller 2008) and by a revolution in information and communication technologies, which transform the scientific work spaces into digital work spaces and the scientific knowledge base into a digital knowledge base. The last two transformations in work spaces and knowledge bases can be summarized as a phase transition from it-science to bit-science following John Wheeler's dictum "it from bit" (Wheeler 1994) and Nicholas Negroponte's slogan "from atoms to bits" (Negroponte 1996), or as the transition from material books, libraries and journals to digital media, e-books and digital knowledge bases.

57. We would like to end our editorial by returning to von Foerster, who suggested that scientific research programs and research traditions do not disappear at a point when they no longer function properly, but rather at their peak. The Copernican revolution replaced the Ptolemaic system "at its height as to accuracy of its predictions" (Foerster 2003: 284). This point is also well-reflected in innovation theory, which claims that old technologies are replaced at their state of perfection (Utterback 1994). A second Copernican revolution is well under way and about to change science from its established forms and operations (Science I) to new contexts and configurations (Science II) where SOC_E could play a significant role.

Afterword: Ranulph Glanville's contributions to second-order cybernetics

58. This volume would not have been possible without the inspirations from Ranulph Glanville, who had acted as a driving force behind this volume since 2013. Tragically, he died on 20 December 2014. Since Glanville could not contribute to this volume with a target article, we will highlight his perspectives on SOC, which in our view constitutes a highly innovative yet barely appreciated research program for SOC.

59. Glanville's general framework is documented in three volumes with the common title "The Black B∞x" (Glanville 2009, 2012, 2014), from which we extract three key aspects that characterize his work (for further details, see Müller 2015c).

Aspect 1: Glanville's approach to SOC

60. Glanville viewed himself as being in a position that offers the necessary generalizations and abstractions of what his intellectual in-group within SOC produced: "My work might be thought of as a generalization of the work of others" (Glanville 2012: 192).

61. In another version, Glanville saw himself in the role of the principal designer, who produces the possibilities for others to pursue their cybernetic work:

> Many of us watch games. Some play, others umpire or referee. Still others govern games and make/remake the rules. There are some who create the field of play, mark and maintain it. The potential (Gibson's affordance) is not however, limited to games: others may use the ground in a completely unanticipated way, unintended by those who set up the ground. Behind all these is the person who creates the possibility of the blank field on which all this potential can be expected. That person is me, and that is my work: I create the unformed, empty field. (Glanville 2012: 35)

62. Still in another variation, he placed his work at the level $n+1$, under the assumption that level n was the highest and most general level explored by the rest of second-order cyberneticians:

> Cybernetics is often considered a meta-field. The Cybernetics of Cybernetics is, thus, a meta-meta-field. My work is, therefore, a meta-meta-meta-field. (Glanville 2012: 192)

63. Aside from being most general by necessity, Glanville described the central research question as follows:

> My major initial concern was to develop a set of concepts that might explain how, while we all observe and know differently, we behave as if we were observing the same 'thing.' What structure might support this? (Glanville 2012: 192).

64. Thus, the crucial research problem for Glanville was transcendental in nature because he was searching for the conditions of the possibility for observing, knowing, communicating, etc. and operated, therefore, on a very special level of abstraction.

65. Due to this unusual abstraction level, Glanville also abolished a separation that has become almost trivial and self-evident in the course of the long-term evolution of science, i.e., the separation between scientific problems on the one hand and problems in artistic or in applied domains such as education, architecture or design on the other hand. Glanville offered a general framework that is so much broader or wider than the normal scope of established research programs. His framework covers very heterogeneous territo-

ries and crosses easily the borders and boundaries between basic science, applied science, technology, design or arts.

Aspect 2: Glanville's circles of doing and reflecting

66. Volume 1 of Glanville's "Black B∞x" series, "Cybernetic Circles" (Glanville 2012), contains five major sections, namely cybernetics, objects, black box, distinction and variety. Volume 2, "Living in Cybernetic Circles" (Glanville 2014), is composed of five chapters: design, representation, knowing, education and others (the last chapter contains mainly tributes to Pask, von Foerster, Gerard de Zeeuw, Alfred Locker, Ernst von Glasersfeld and Richard Jung).

67. From the differentiation in the two volumes, one could conclude that Glanville established the theoretical foundations of his transcendental framework in volume 1 and applied it to areas such as design and education in the second volume. But this would miss the special form in which Glanville organized his theoretical or analytical production on the one hand and his practical work on the other hand. Whatever he did, Glanville arranged it in a permanent circle of theory and practice or doing and reflecting. In this, he followed Donald Schön's concept of the reflective practitioner, in which Schön defined reflection in a circular manner as an action that leads to new actions (Schön 1983, 1987, 1991). The action-reflection circle can be described as a circular organization between two contexts:

a. The context of production, where a problem solution is reached, an artistic object is created or a technological system is constructed.
b. The context of reflection, which, on the other hand, deals with the consequences, including ethical issues, further implications or modifications of the finished product, future developments, etc.

68. Glanville lived constantly within these production-reflection circles and shifted positions frequently, taking circularity seriously and as a stable form for theoretical or practical action.

Aspect 3: Glanville as reflexive author

69. In much of what has been written about second-order cybernetics and observers, the authors of those texts remain in the shadows, i.e., practically implicit. Here, Glanville differed radically from most of the usual observer-inclusive accounts. He was constantly an explicit element in the texts he produced and followed the model of a reflexive author. For decades, he worked as a pioneer, exploring the ramifications of observing from within in defining concepts, in developing models, in designing, in learning interactions, etc. In his publications, his co-presence as an author or observer is always explicit and never implicit.

70. These three aspects characterize Ranulph Glanville, who had vigorously promoted SOC since the late 1960s. Since he was a genuinely transdisciplinary scholar, he was also the "tragic hero" of SOC, as he pursued a highly complex research agenda covering architecture, communication theory, cybernetics, design and music. This, due to its scope, complexity and diversity, was too often hardly understood and barely recognized by his disciplinarily thinking peers.

Acknowledgements

We are grateful to Anthony Hodgson, Bernard Scott and Stuart Umpleby for their insightful criticism and comments.

References

For a continuously expanding list of Ranulph Glanville's publications, see http://cepa.info/author/glanville-r

Archer M. S. (2007) Making our way through the world. Human reflexivity and social mobility. University of Cambrdge Press, Cambridge.

Archer M. S. (2012) The reflexive imperative in late modernity. Cambridge University Press, Cambridge.

Atmanspacher H. & Dalenoort G. J. (eds.) (1994) Inside versus outside. Endo- and exo-concepts of observation and knowledge in physics, philosophy and cognitive science. Springer, Berlin.

Beck U., Giddens A. & Lash S. (1994) Reflexive modernization. Politics, tradition and aesthetics in the modern social order. Stanford University Press, Stanford.

Bortz J. & Döring N. (2015) Forschungsmethoden und Evaluation für Human- und Sozialwissenschaftler [Research methods and evaluation for humanities scholars and social scientists]. Fourth edition. Springer, Berlin.

Bourdieu P. (1977) Outline of a theory of practice. Cambridge University Press, Cambridge.

Bourdieu P. (1990) The logic of practice. Polity Press, Cambridge.

Bourdieu P. & Wacquant L. J. D. (1992) An invitation to reflexive sociology. Chicago University Press, Chicago.

Clarke B. (2009) Heinz von Foerster's demons: The emergence of second-order systems theory. In: Clarke B. & Hansen M. (eds.) Emergence and embodiment: New essays in second-order systems theory. Duke University Press, Durham: 34–61.

Foerster H. von (1948) Das Gedächtnis. Eine quantenmechanische Untersuchung. Franz Deuticke, Vienna.

Foerster H. von (1959) Some remarks on changing populations. In: Stohlman F. Jr. (ed.) The kinetics of cellular proliferation. Grune and Stratton, New York: 382–407.

Foerster H. von (1960) On self-organizing systems and their environments. Yovits M. C. & Cameron S. (eds.) Self-organizing systems. Pergamon Press, London: 31–50. Reprinted in: Foerster H. von (2003) Understanding understanding. Springer, New York: 1–19. http://cepa.info/1593

Foerster H. von (1967) Computation in neural nets. Currents in Modern Biology 1: 47–93. Reprinted in: Foerster H. von (2003) Understanding understanding. Springer, New York: 21–100.

Foerster H. von (1973) On constructing a reality. Preiser F. E. (ed.) Environmental design research. Volume 2. Dowden, Hutchinson & Ross, Stroudberg: 35–46. Reprinted in: Foerster H. von (2003) Understanding understanding. Springer, New York: 211–227. http://cepa.info/1278

Foerster H. von (ed.) (1974) Cybernetics of cybernetics. University of Illinois, Urbana.

Foerster H. von (1976) Objects: Tokens for (eigen-)behaviors. ASC Cybernetics Forum 8(3–4): 91–96. Reprinted in: Foerster H. von (2003) Understanding understanding. Springer, New York: 261–271. http://cepa.info/1270

Foerster H. von (1984) Principles of self-organization – In a socio-managerial context. In: Ulrich H. & Probst G. J. (eds.) Self-organization and management of social systems. Springer, Berlin: 2–24. http://cepa.info/1678

Foerster H. von (1991) Through the eyes of the other. In: Steier F. (ed.) Research and reflexivity. Sage Publications, London: 63–75. http://cepa.info/1729

Foerster H. von (1992) Ethics and second-order cybernetics. Cybernetics & Human Knowing 1(1): 9–20. Reprinted in: Foerster H. von (2003) Understanding understanding. Springer, New York: 287–304. http://cepa.info/1742

Foerster H. von (1997) Der Anfang von Himmel und Erde hat keinen Namen: Eine Selbsterschaffung in 7 Tagen. Edited by A. Müller & K. H. Müller. Döcker-Verlag, Vienna. English translation: Foerster H. von (2014).

Foerster H. von (2003) Understanding understanding: Essays on cybernetics and cognition. Springer, New York.

Foerster H. von (2014) The beginning of heaven and earth has no name: Seven days with second-order cybernetics. Fordham University Press, New York.

Foerster H. von, Mora P. M. & Amiot L. W. (1960) Doomsday: Friday, November 13 AD 2026. Science 132: 1291–1295. http://cepa.info/1596

Geertz C. (1983) Local knowledge: Further essays in interpretative anthropology. Basic Books, New York.

Geertz C. (2000) Available light: Anthropological reflections on philosophical topics. Princeton University Press, Princeton.

Glanville R. (2009) The black b∞x. Volume 3: 39 Steps. Edition echoraum, Vienna.

Glanville R. (2012) The black b∞x. Volume 1: Cybernetic circles. Edition echoraum, Vienna.

Glanville R. (2014) The black b∞x. Volume 2: Living in cybernetic circles. Edition echoraum, Vienna.

Hollingsworth J. R. & Müller K. H. (2008) Transforming socio-economics with a new epistemology. Socio-Economic Review 3(6): 395–426.

Kauffman L. H. (2005) EigenForm. Kybernetes 34(1/2): 129–150. http://cepa.info/1271

Kauffman L. H. (2009) Reflexivity and eigenform: The shape of process. Constructivist Foundations 4(3): 121–137. http://constructivist.info/4/3/121

Kline R. R. (2015) The cybernetics moment or why we call our age the information age. The Johns Hopkins University Press, Baltimore.

Lefebvre V. A. (1977) The structure of awareness: Toward a symbolic language of human reflexion. Sage, New York.

Lefebvre V. A. (1982) Algebra of conscience: A comparative analysis of Western and Soviet ethical systems. Reidel, Dordrecht.

Lissack M. (2016a) Second order science: Examining hidden presuppositions in the practice of science. Foundations of Science. First online: 18 January 2016.

Lissack M. (2016b) What second-order science reveals about scientific claims: Incommensurability, doubt, and a lack of explication. Foundations of Science. First online: 14 January 2016.

Löfgren L. (1979) Unfoldment of self–reference in logic and in computer science. In: Jensen F. & Mayoh B. & Moeller K. (eds.) Proceedings of the 5th Scandinavian logic symposium. Aalborg Press, Aalborg: 205–229.

Löfgren L. (1992) Complementarity in language: Toward a general understanding. In: Carvallo M. (ed.) Nature, cognition, and system II: Complementarity and beyond. Kluwer, Dordrecht: 113–153.

Martin R. J. (2015) Second-order cybernetics, radical constructivism and the biology of cognition. Cybernetics & Human Knowing 22(2–3):169–182.

Maturana H. R. (1970) Biology of cognition. Biological Computer Laboratory (BCL) Research Report BCL 9.0. University of Illinois, Urbana. Reprinted in: Maturana H. R. & Varela F. J. (1980) Autopoiesis and cognition. The realization of the living. Reidel, Dordrecht: 5–58. http://cepa.info/535

Maturana H. R. & Varela F. J. (1980) Autopoiesis and cognition: The realization of the living. Reidel, Dordecht.

Maturana H. R. & Varela F. J. (1987) The tree of knowledge: The biological roots of human understanding. Shambhala, Boston.

Mead M. (1968) Cybernetics of cybernetics. In: Foerster H. von, White J. D., Peterson L. J. & Russell J. K. (eds.) Purposive Systems. Spartan Books, New York: 1–11. http://cepa.info/2634

Müller K. H. (2007) The BCL: An unfinished revolution of an unfinished revolution. In: Müller A. & Müller K. H. (eds.) (2007) An unfinished revolution? Heinz von Foerster and the Biological Computer Laboratory | BCL, 1958–1976. Edition echoraum, Vienna: 407–466.

Müller K. H. (2015a) A silent revolution in reflexivity. Journal of Systemics, Cybernetics and Informatics 13(6): 70–81. http://cepa.info/2790

Müller K. H. (2015b) The multiple faces of reflexive research designs. Journal of Systemics, Cybernetics and Informatics 13(6): 87–98. http://cepa.info/2791

Müller K. H. (2015c) De profundis. Glanville's transcendental framework for second-order cybernetics. Cybernetics & Human Knowing 22(2–3): 27–47. http://cepa.info/2699

Müller K. H. (2016a) Second-order science. The revolution of scientific structures. Edition echoraum, Vienna.

Müller K. H. (2016b) Post-disciplinary cybernetics. The science of reflecting on reflections. Edition echoraum, Vienna.

Negroponte N. (1996) Being digital. Vintage Books, New York.

Pask G. (1975a) The cybernetics of human learning and performance. A guide to theory and research. Hutchinson Educational, London.

Pask G. (1975b) Conversation, cognition and learning. A cybernetic theory and methodology. Elsevier, New York.

Pask G. (1976) Conversation theory. Applications in education and epistemology. Elsevier, New York.

Quale A. & Riegler A. (eds.) (2010) Can radical constructivism become a mainstream endeavor? Constructivist Foundations 6(1). http://constructivist.info/6/1

Riegler A. (2005) Editorial: The constructivist challenge. Constructivist Foundations 1(1): 1–8. http://constructivist.info/1/1/001

Riegler A. (2015) What does the future hold for radical constructivism? In: Raskin J. D., Bridges S. K. & Kahn J. S. (eds.) Studies in meaning 5: Perturbing the status quo in constructivist psychology. Pace University Press, New York: 64–86. http://cepa.info/1285

Riegler A. & Müller K. H. (eds.) (2014) Special issue on second-order science. Constructivist Foundations 10(1). http://constructivist.info/10/1

Rössler O. E. (1992) Endophysics. Die Welt des inneren Beobachters. Merve, Berlin.

Schmidt S. J. (2003) Geschichten & Diskurse. Abschied vom Konstruktivismus. Rowohlt, Reinbek. English translation: Schmidt S. J. (2007) Histories & discourses. Rewriting constructivism. Imprint Academic, Exeter.

Schön D. A. (1983) The reflective practitioner: How professionals think in action. Temple Smith, London.

Schön D. A. (1987) Educating the reflective practitioner. Jossey-Bass, San Francisco.

Schön D. A. (ed.) (1991) The reflective turn: Case studies in and on educational practice. Teachers College, New York.

Scott B. (2014) Education for cybernetic enlightenment. Cybernetics & Human Knowing 21: 1–2: 199–205. http://cepa.info/1286

Soros G. (1994) The alchemy of finance: Reading the mind of the market. Second edition. John Wiley & Sons, New York.

Steier F. (ed.) (2001) Research and reflexivity. Sage Publications, London.

Umpleby S. A. (2007) Reflexivity in social systems: The theories of George Soros. Systems Research and Behavioral Science 24: 515–522. http://cepa.info/1280

Umpleby S. A. (2010) From complexity to reflexivity: The next step in the systems sciences. In: Trappl R. (ed.) Cybernetics and systems 2010. Austrian Society for Cybernetic Studies, Vienna: 281–286. http://cepa.info/891

Utterback J. M. (1994) Mastering the dynamics of innovation: How companies can seize opportunities in the face of technological change. Harvard Business School Press, Boston.
Völker J. & Scholl A. (2014) Do the media fail to represent reality? A constructivist and second-order critique of the research on environmental media coverage and its normative implications. Constructivist Foundations 10(1): 140–149. http://constructivist.info/10/1/140
Wheeler J. A. (1994) At home in the universe. American Institute of Physics, New York.

Part I:
Exploring Second-Order Cybernetics and its Fivefold Agenda

Second-Order Cybernetics as a Fundamental Revolution in Science

Stuart A. Umpleby

Introduction

1. Heinz von Foerster published in a variety of disciplines, including on the mechanism of memory (Foerster 1948), population growth (Foerster, Mora & Amiot 1960), self-organization (Foerster 1960), constructivist philosophy (Foerster 2003d), and ethics (Foerster 2003c). However, von Foerster also made a more fundamental contribution by showing how to expand the scientific enterprise with a new way of operating scientifically. My goal in this target article is to explain how von Foerster did that, and how that work has developed. This is about a scientific revolution not only in cybernetics but in the way science is done. It involves advances in the practice of science, and it leads to major shifts in our conception of the goals of science.

2. In order to make von Foerster's intended fundamental revolution for the science system more intelligible, I will proceed in five stages: first, by offering a short history of second-order cybernetics; second, by analysing the major contents of von Foerster's fundamental revolution in science; third, by describing an institute that acted for decades in accord with this new approach; fourth, by showing the advantages of this unfinished revolution; and fifth, by suggesting some new directions for research in second-order cybernetics.

Part 1: A short personal history of second-order cybernetics

3. After World War II there was excitement about the utility of applied science. One outgrowth of that excitement was the Josiah Macy Jr. Foundation conferences in New York City between 1946 and 1953. They were chaired by Warren McCulloch. Gregory Bateson and Margaret Mead were influential in starting them and continuing them. The title of those conferences was "Circular Causal and Feedback Mechanisms in Biological and Social Systems" (Pias 2003). After Norbert Wiener published his 1948 book *Cybernetics: or Control and Communication in the Animal and the Machine*, the conferees began to call the meetings the "Macy conferences on cybernetics."

4. As a student, when I first heard von Foerster talking about the role of the observer, it sounded to me very similar to what Thomas Kuhn said in *The Structure of Scientific Revolutions*, which was popular at the time. In a well-known passage, Kuhn writes that

> [...] the proponents of competing paradigms practice their trades in different worlds. One contains constrained bodies that fall slowly. The other, pendulums that repeat their motions again and again [...] Practicing in different worlds, the two groups of scientists see different things when they look from the same point in the same direction. Again, that is not to say they can see anything they please. Both are looking at the world, and what they look at

> has not changed. But in some areas they see different things, and they see them in different relations one to the other. That is why a law that cannot even be demonstrated to one group of scientists may occasionally seem intuitively obvious to another. Equally it is why before they can hope to communicate fully, one group or the other must experience the conversion that we have been calling a paradigm shift. (Kuhn 1962: 150)

5. However, when I showed this passage to von Foerster, he told me that was not what he meant by including the observer in science. After he published the article "On Constructing a Reality" (Foerster 2003d), which described a number of neurophysiological experiments, I realized what he had in mind. I thought von Foerster was proposing a third view of the scientific enterprise – not the normative view of Popper or the sociological view of Kuhn but a biological view that included awareness of how the brain functions. I felt von Foerster's idea offered an opportunity to make a scientific revolution (Umpleby 1974). See Table 1 for a comparison of the respective epistemologies of Popper, Kuhn, and von Foerster.[1]

6. There are many neurophysiological examples dating back at least to the 1960s and 1970s that support von Foerster's article, including: the well-known cocktail-party effect (one can focus on a single conversation among many conversations going on at a party); the physiological fact that even though images on the retina are bottom-up we still see things upright; the existence of the blind spot on the retina, which is not perceived as a gap since the brain fills in this space with the sensation that surrounds it; and Richard Held and Alan Hein's (1963) reports on an experiment with kittens showing that the brain constructs three dimensional space by coordinating signals from both muscles and eyes.

7. These experiments illustrate that the brain works in ways that we are not aware of. The brain seems to "construct a reality" based on sensory input. Since people have different experiences – language, home life, culture, religion, academic training, and job experiences – each person's "reality" is in some respects unique, though our knowledge of the physical and social world has many common features. To the usual philosophical critique of science, second-order cybernetics adds a scientific critique of philosophy

8. These and many more neurophysiological experiments provide the *biological* foundation of second-order cybernetics. The experiments are essential for second-order cybernetics because they show that "observations independent of the characteristics of the observer" are not physically possible. This is quite a different view from what is assumed in the usual methodology of science. As Humberto Maturana and von Foerster have pointed out, "Anything said is said *by* an observer" and "Anything said is said *to* an observer" (Foerster 2003a: 283) or "Everything said is said by an observer to another observer, who can be himself or herself" (Maturana 1978: 31). Our experiences are interpreted using conclusions we have drawn from earlier experiences (Maturana & Varela 1992). Francisco Varela and Wolf Singer (1987) provided empirical support for the prediction that only about 20% of neurones to the visual structure of the brain are from the retina, while at least 40% come from the visual cortex. Visual interpretations seem to be based more on past experiences than on present sensations.

---

1. This was first presented in my presentation "Unifying epistemologies by combining world, description and observer" at the annual meeting of the American Society for Cybernetics, Urbana, Illinois, 29 March–1 April 2007. An early description of the three points of view is given in a table, "Three Versions of Cybernetics," in Umpleby (1997).

Popper	Kuhn	von Foerster
A normative view of epistemology: how scientists should operate	A sociological view of epistemology: how groups of scientists operate	A biological view of epistemology: how the brain functions
Non-science vs. science	Steady progress vs. revolutions	Realism vs. constructivism
Solve the problem of induction: conjectures and refutations	Explain turmoil in original records vs. smooth progress in textbooks	Include the observer within the domain of science
How science as a picture of reality is tested and grows	How paradigms are developed and then replaced	How an individual constructs a "reality"
Scientific knowledge exists independent of human beings	Even data and experiments are interpreted	Ideas about knowledge should be rooted in neurophysiology
We can know what we know and do not know	Science is a community activity	If people accept this view, they will be more tolerant

Table 1. Three philosophical positions.

9. Reflecting on the observer and the importance of interpretation have long been themes in philosophy, the humanities in general, and occasionally in science. However, in the post World War II period in the United States, scientists sought to emphasize the objectivity of science and to condemn any suggestion of subjectivity as antiquated, inappropriate, uninformed, and wrong. Even in the definition of science, scientific methods were described as a way of eliminating observer bias.

10. In 1974, von Foerster used the term "second-order cybernetics" for the first time at a meeting of the American Society for Cybernetics in Philadelphia (Foerster 2003a). In the same year, the book *Cybernetics of Cybernetics* was published, which mentioned "second-order cybernetics" as the "cybernetics of observing systems," in contrast to first-order cybernetics as the "cybernetics of observed systems" (Foerster 1974: 1). By this time, there were a number of theoretical achievements in the field of cybernetics: McCulloch (1965) had defined and developed experimental epistemology; Bateson (1972) was doing work on schizophrenia and the double bind and had published *Steps to an Ecology of Mind*; Wiener's (1948) concept of a second industrial revolution was widely known; computers and robotics were advancing rapidly; there was some work being done by artists and composers (Reichardt 1968; Brün 2004); Maturana's concept of autopoiesis was attracting considerable attention as a way of explaining the organization and operation of living systems (Maturana 1975). Work on second-order cybernetics was just beginning when funding from government research grants ceased, due in part to the Mansfield Amendment.

11. In the early 1970s, there was turmoil on college campuses in the United States, and Congress wanted to cool the campuses. One of the causes of the turmoil was that much research on campuses was funded by the Department of Defense (DOD). At the same time, students were opposed to the Vietnam War. So Mike Mansfield, a liberal Democrat and Senate Majority Leader, proposed the Mansfield Amendment. It required that all DOD funding have a military mission. That meant that von Foerster, who was being funded by the Air Force Office of Scientific Research and the Office of Naval Research, had to declare the military mission of his research. He said the research had no military mission.

Consequently, he could no longer receive funding from the people who had been funding his research (Umpleby 2003). In contrast, people in the artificial intelligence (AI) labs at MIT, Stanford, and Carnegie Mellon, when asked about the military implications of AI, invented the concept of an electronic battlefield. That justification was very well received, both at the Pentagon and in Congress, because it held the promise of fewer American casualties in wars. The result was that the flow of federal money shifted from both cybernetics and AI to just AI and robotics.

12. The Biological Computer Laboratory (BCL) at the University of Illinois in Urbana-Champaign had been a leading center for cybernetics research since it was founded by von Foerster in 1958. There were efforts to find other funding for BCL, but they were not successful in achieving the previous level of support. So the laboratory was closed in 1975 when von Foerster retired and moved to California. Those interested in cybernetics then did not have a laboratory in the US dedicated to cybernetics research, since Warren McCulloch and Norbert Wiener, both at MIT, had died in the 1960s. In the late 1970s, to continue our conversations about cybernetics among the scholars who visited BCL, including Maturana, Varela, Lars Löfgren, Stafford Beer, Gordon Pask, Ranulph Glanville, and others, we moved the BCL network into cyberspace. This period was the early days of experiments with digital messaging. Murray Turoff had created the Electronic Information Exchange System (EIES) at New Jersey Institute of Technology in Newark. The National Science Foundation was funding several experimental trials on small research communities, using EIES. I received one of nine grants for these experimental trials. I invited the former BCL people to communicate with each other using this new medium (Umpleby 1979; Umpleby & Thomas 1983).

13. The American Society for Cybernetics (ASC), which was founded in 1964, had passed through a period of inactivity in the mid 1970s due to personality conflicts. The conflicts were resolved and ASC was revived by Barry Clemson, Doreen Steg, Larry Heilprin, and Klaus Krippendorff (Krippendorff & Clemson 2016). In addition to electronic messaging and collaboration via EIES, we resumed holding annual conferences in 1980 (Umpleby 2016). Usually, on the first day of these conferences, we held a tutorial for those new to the field. The tutorials covered both first- and second-order cybernetics. We felt we were beginning a scientific revolution in the field of cybernetics. The field had always had two orientations, which were becoming more well-defined (Corona & Thomas 2010). There were those who wanted to design electrical and mechanical equipment and those who wanted to understand human cognition. For the engineers, good work meant building something. For the biologists, philosophers, and social scientists, good work was a contribution to knowledge. These different goals led to some harsh exchanges and a further separation between the technical and philosophical branches of cybernetics.

14. We also conducted tutorials for the European Meetings on Cybernetics and Systems Research (EMCSR) in Vienna, Austria, the Dutch Systems Group in Amsterdam, and a few other conferences. Pask, Glanville, and Gerhard de Zeeuw organized symposia for the Vienna and Amsterdam conferences. Von Foerster, Maturana, Varela, and Ernst von Glasersfeld were invited as keynote speakers. We were working on the introduction of the notion of second-order cybernetics and related ideas. Several definitions of first and second-order cybernetics were created (see Table 2).

15. Ronald Kline ends his 2015 book *The Cybernetics Moment* with the assessment that the cybernetics movement ended in the mid 1970s, the moment when second-order cybernetics was invented. The great majority of US scientists thought that paying atten-

Author	First-order cybernetics	Second-order cybernetics
von Foerster	The cybernetics of observed systems	The cybernetics of observing systems
Pask	The purpose of a model	The purpose of a modeler
Varela	Controlled systems	Autonomous systems
Umpleby	Interaction among the variables in a system	Interaction between observer and observed
Umpleby	Theories of social systems	Theories of the interaction between ideas and society

Table 2. Varieties of second-order cybernetics in the 1970s and 1980s.

tion to the observer was a return to a subjectivist epistemology, which they regarded as a fundamental error. Von Foerster, Maturana, and other second-order cyberneticians thought that not including the observer was inconsistent with an understanding of neurobiology. The transition from first-order cybernetics to second-order cybernetics is a fundamental transition, as I will discuss in the next part.

Part 2: Von Foerster and a revolution of the general scientific method

16. In order to discuss in more detail the new perspective that was developed by von Foerster from the late 1960s onwards under the name of second-order cybernetics, I will address the following questions:

- What are the characteristics that separate first-order from second-order cybernetics?
- Where do the advantages of second-order cybernetics lie when compared to first-order cybernetics?
- And why should second-order cybernetics be considered as a general model for new scientific operations in other disciplines and for an alternative methodology of science?

17. Von Foerster's distinction between first-order and second-order cybernetics is the difference between excluding the observer from explicit mention and including the observer. Including what had previously been neglected makes for a larger conception of science, a conception that can deal with a wider range of phenomena. In terms of the Law of Requisite Variety (Ashby 1952), second-order cybernetics expands the variety in science and hence the variety in the world that scientists can describe.

18. One answer to these questions would be that von Foerster was adding a new dimension, namely the amount of attention paid to the observer. This change would not be limited only to cybernetics but would potentially affect all of science (Umpleby 2014).

19. However, certainly in his later years, von Foerster did not intend merely to understand cognition and to pay more attention to observers, but to replace the traditional methodology with a new one and with new ways of practicing science. Von Foerster wanted to change the established or traditional scientific methodology as well as the conventional

ways of practicing science, which, in due course, were to be reduced to small niches only. In one lecture, von Foerster described his approach as a demolition.

> Everywhere, in the United States too, the oldest and most beautiful houses are nowadays being demolished and instead steel-and-glass skyscrapers with 36 stories are being constructed. I want to emphasize the reverse process. I start with a 36-story steel-and-glass skyscraper and demolish it. But I am not building a baroque castle instead, but something completely different: maybe a beetle, maybe an ant colony, maybe a family. (Foerster 1988: 20, my translation)

20. The metaphor of the steel-and-glass skyscrapers applied to the accepted scientific method becomes clear in von Foerster's lecture on "Cybernetics of Cybernetics." In these short lecture notes, von Foerster characterizes second-order cybernetics as a fundamental paradigm change, which he did not attribute, as Kuhn suggested, to anomalies and to defects in the older paradigm, but rather to its very "flawlessness" (Foerster 2003a: 284). Von Foerster points to two historical instances in science of elimination of a paradigm by success or perfection. The first case was the Copernican Revolution resulting from "the novel vision of a heliocentric planetary system" (ibid: 284) even though "the Ptolemaeic geocentric system was at its height in the accuracy of its predictions" (ibid: 284). For von Foerster, the second instance was the accepted, hegemonic scientific method with its "flawless, but sterile path that explores the properties seen to reside within objects." The Ptolemaic geocentric system was replaced between the 15th and the 17th centuries. And, according to von Foerster, the hegemonic scientific method we know today is to be reduced significantly in the years and decades ahead and substituted with an alternative, with second-order cybernetics as its prime example.

21. The replacement of scientific methodology was not intended as a nostalgic move towards premodern forms. Several elements of scientific methods, such as the production of hypotheses or theories, experiments, collection of relevant data, the production of new instruments, or empirical testability and falsifiability, need not be changed. But a general replacement of the current method is required for two reasons.

- First, researchers or observers as necessary components in any research process are included in the prevailing research method only in an implicit and hidden way.
- Second, the current research process has the goal of removing this implicit inclusion so that objective accounts can emerge that are strictly independent of researchers or observers. Observer effects, subjective biases, or personal preferences, while recognized and necessary initially, are to be excluded from accounts of the research process as much as possible.

22. Eric Kandel summarizes the traditional scientific method and its emphasis on objectivity in the following way.

> Scientists make models of elementary features of the world that can be tested and reformulated. These tests rely on removing the subjective biases of the observer and relying on objective measurements and evaluations. (Kandel 2012: 449)

23. Thus, von Foerster describes the traditional scientific method as "a particular delusion within our Western tradition," which he characterized by the postulate of objectivity:

> The properties of the observer shall not enter the description of his observations. (Foerster 2003a: 285).

Traditional	New
Appearance	Function
World and I: separated	World and I: one
Schizoid	Homonoid
Monological	Dialogical
Denotative	Connotative
Describing	Creating
You say how it is	It is how you say it
Cogito, ergo sum	*Cogito, ergo sumus*

Table 3. Two paths for interactions with the world/environment.

24. The fear has been that allowing the properties of observers to be included in their descriptions would open the door to subjectivism, biases, and irrationality. Wild pluralisms would be the mildest symptoms in science if researchers and their properties were admitted without further constraints. Nevertheless, human observers are biological organisms. Not to incorporate an understanding of the biology of cognition in our practice of science requires discarding relevant experience and knowledge.

25. At this point, a specification of the foundations for the new scientific method is needed. In the paper "Ethics and Second-Order Cybernetics" von Foerster develops a distinction between two attitudes towards one's environment or world.

> 'Am I *apart from* the universe?' Meaning whenever I *look*, I'm looking as if through a peephole upon an unfolding universe; or 'Am I *part of* the universe?' Meaning whenever I *act*, I'm changing myself and the universe as well. (Foerster 2003c: 293)

26. Von Foerster re-iterated the dualism of apart/a part in the form of the following question – "Am I an observer who stands outside and looks in as God-Heinz or am I part of the world, a fellow player, a fellow being?" (Foerster 2014: 128). From there, he expanded this distinction to an intriguing list of characteristic differences between two fundamentally different ways of interacting with an environment or the world in general.

27. Von Foerster emphasizes one distinction from Table 3 as being very significant.

> For me the most important distinction in the table is between 'Say how it is' versus 'It is how you say it.' These for me are the really fundamental differences between 'standing outside' and 'standing inside' – and here, of course, syntax fits as the set of rules you can see from the outside. Semantics, however, is like a roast that is being prepared and will soon be served. (Foerster 2014: 129)

28. The dichotomy of "standing outside" or "standing inside" can be transferred to the domains of science methods, where the traditional and still hegemonic method follows the practice of science from outside whereas the new alternative corresponds to a science from within.

29. These two epistemic modes of doing science from outside or inside belong to the group of undecidable questions. For von Foerster, analytical questions, such as mathemati-

cal proofs, are already decided, so we cannot decide them. Undecidable questions, such as values or goals, are up to us. Von Foerster characterized undecidable questions with two propositions.

> Only those questions that are in principle undecidable, *we* can decide […] *We can choose who we wish to become when we have decided on an in principle undecidable question.* (Foerster 2003b: 293, emphasis in the original)

30. The new general scientific method as an alternative *modus operandi* and as scientific practices from inside involves activities that are highly interactive and recursive. Louis Kauffman's contribution for the present volume describes the configuration from inside, with researchers as interactive units in reflexive domains. He points out the decisive role of consensus-formation and of the emergence of eigenforms within these reflexive domains. He also provides a fascinating re-invention or re-construction of the scientific method from within. Kauffman's article makes clear that solipsism or subjectivism are not necessary outcomes of operating from within.

31. Finally, von Foerster is quite explicit that the epistemic distinction of doing science from without and practicing science from within refers also to the separation between first-order and second-order cybernetics. The short description of first-order cybernetics as the "cybernetics of observed systems" and of second-order cybernetics as the "cybernetics of observing systems"

> is nothing else but a paraphrase of […] the two fundamentally different epistemological, even ethical, positions, where one considers oneself, on the one hand, as an independent observer who watches the world go by; or on the other hand, as a particular actor in the circularity of human relations. (Foerster 2003c: 303)

Thus, first-order cybernetics becomes the study of cybernetics from without whereas second-order cybernetics becomes cybernetic analysis from within. Any scientific field can be studied in two significantly different epistemic modes, where researchers play highly active roles in the mode "from within" or, as Karl Müller (2016) has described it recently, in an endo-mode as opposed to the still dominant exo-mode of the traditional scientific method. Any scientific observation is addressed to a community of observers.

32. In the decades between von Foerster's introduction of the concept of second-order cybernetics in 1974 and his distinction between two epistemic modes of scientific world-making several years later, the science system was transformed significantly (Gibbons *et al.* 1994; Hollingsworth & Müller 2008). Science became more diversified, complex, and open to new alternatives. Approaches from within, while still not widespread today, emerged especially in the social sciences, in feminist theorizing (Haraway 1988, 1991), in the diffusion of participatory methods (Cooperrider & Whitney 2005; Christakis & Bausch 2006; Umpleby & Oyler 2007), and in the environmental sciences.

33. In the next section, I will focus on a specific institute for social and community research that operated since its beginnings in the 1950s in a mode from within. The next step is important because it shows that the perspective from within is neither utopian nor does it lead to an unrestricted subjectivism, but becomes a feasible way of exploring the social world in a significantly different manner than traditional sociologists are used to.

Part 3: An example of practicing social research from within

34. My goal in this section is to describe a way of doing social research, or second-order socio-cybernetics, that is compatible with an epistemic mode from within. I describe the work of the Institute of Cultural Affairs (ICA) and contrast its work with the usual social science research, which still operates in a conventional mode from outside.

35. In the 1950s, a group of people trained in the ministry had been working with people in the suburbs on "church renewal," encouraging members of the church to take a larger view of the community and to become more involved in community projects. After a few years, they decided to work themselves with people in poor communities. They chose a community on the West Side of Chicago. The financing came from Head Start. One member of a couple would work on the Head Start project and the other person would work on community organizing. When Martin Luther King, Jr. was assassinated in 1968, they were burned out. Rather than give up, they decided to try again. They organized a meeting of a diverse group of people from the community along with a few business people and government officials.

36. They held a week-long conference using facilitation methods. Each day they asked members of the community to answer a question. On the first day: What is your vision for the future of your community? On the second day: What are the obstacles to achieving that vision? Why is it not already present? On the third day: What are some strategies that would remove the obstacles? On the fourth day: What tactics are needed to advance the strategies? On the fifth day: What actions should be taken to implement the tactics? That is, who will do what, when, and how to implement the tactics? At the end of the conference they had a set of plans and committees that worked to implement the plans in the subsequent months. Periodically, at least every 6 or 12 months, they would review what they had planned to do, compare the plans with what had been accomplished, reflect on what they had learned, and then generate a new set of plans.

37. This procedure worked very well in Chicago and a great deal was accomplished. They wondered whether the same methods would work in a developing country. Through contacts in the World Council of Churches, they decided to do a second Human Development Project in the Marshall Islands. Again there was notable success. So, they decided to do 24 projects, approximately one in each time zone around the world. Model villages would serve as examples of what could be achieved. After the first set of projects was done, they decided to do a second set of 24 projects (Umpleby & Oyler 2007).

38. During this time, the people in the Institute of Cultural Affairs (ICA) were also doing other projects. They organized Town Meetings, a one-day event, one in each county in each country having a Human Development Project. They had a program called Global Women's Forum, which brought together women at the local level to share problems and possibilities. Finally, they received a grant from the United Nations Educational, Scientific and Cultural Organization (UNESCO) to bring together people from many Non-Governmental Organizations (NGOs) to share reports on their most successful projects. The final meeting was held in New Delhi, India in 1983. The reports were published in a set of volumes under the title *Approaches that Work* (Burbidge 1988).

39. Many NGOs and church organizations do projects in developing countries. Usually they have a specific focus. They may dig water wells, build churches, build schools, improve housing, or work on specific diseases such as malaria or dysentery. ICA had a different approach. They taught participatory methods, so people would learn how to work

together to define their needs, invent solutions, and cooperate with nearby people who had resources. They created village institutions, usually a farmer's cooperative, a business-man's cooperative, a preschool for children, and a parent-teacher association. Depending on what village businesses already existed, they encouraged the opening of additional small businesses, such as a restaurant, a laundry, a hair-dressing salon, a bakery, and a baby-sitting service. Depending on local needs, they helped to organize sports teams for teenagers and weekly card games for seniors. They organized a weekly market where produce would be sold and people from neighboring communities could see the changes that were happening and sometimes adopt similar initiatives.

40. Each summer, the people in ICA would return to Chicago to discuss the past year and define programs for the coming year. In the fall, they would go to communities around the world to implement the programs they had designed. Local contacts and resources were suggested by members of the World Council of Churches. The next summer, they would return to Chicago for reading and study. They would reflect on what worked and what did not and adopt or invent new approaches.

41. The financial model was that in each project there were two or three couples. One person in a couple would teach in an embassy school; the other would work full-time on community development. In this way they were largely self-supporting, though at a very low income level. Donations to ICA paid for international travel. They recruited agricultural and business advisers from nearby universities and sought donations or loans of labor and equipment to dig wells, install irrigation pipes, provide books for children, etc. Many consultants today use group facilitation methods with clients and continually seek to improve their methods (Cooperrider & Whitney 2005; Bausch & Christakis 2015). To my knowledge ICA was unique in the scale of its work with poor communities. They outgrew their financial resources, and after 1984 the organization devolved into country-based ICA organizations.

42. It is interesting to compare the work that ICA was doing with the way that social science research is done in universities. Currently, the objective in social science research is to test a theory by collecting and analysing data. Experiments should be replicable by others. The researcher conducts the research but otherwise is not mentioned. Research is an effort to find causal relationships among variables at a high level of statistical significance. The goal is reliable theoretical knowledge, not social change or societal improvements directly. Success is measured by number of papers in leading academic journals.

43. ICA did research differently. They read widely, for example Kenneth Boulding, Margaret Mead, Paul Tillich, Ivan Illich, and E. F. Schumacher. They would start with current knowledge and learn by doing. They would change methods as needed and use successful methods with additional communities. Many forms of communication were used – celebrations, a weekly market, a newsletter, signs, and posters. The goal was to improve the quality of life – health, income, and education – as quickly as possible by using available knowledge and expertise (e.g., nutrition advice, irrigation, new crops, fertilizer, and business practices). Success was measured by higher standards of living and the spread of participatory methods to nearby communities. Networks of supportive people were created and maintained.

44. When reflecting on the success of the very practical, grass-roots work of ICA, it is interesting to ask why social science research is so detached from societal problems today. Currently, universities exist around the world and thousands of people are engaged in education and research on social systems. But in the traditional "mode from without," they

work with data and statistical methods rather than with people in communities. And they produce articles with theoretical knowledge without exploring the possibilities of implementing it. I will address these problems again in the final part of this article.

45. The ICA people were deeply involved in communities – living and working side-by-side on a daily basis. They worked to resolve conflicts within the community. They paid attention to emotions, spiritual feelings, cultural beliefs and practices and they worked from within to create – through stories, songs, and symbols – a shared concern for the community and the world, not just individual advancement. In short, they operated as second-order socio-cyberneticians, although the name of the field socio-cybernetics as well as the concept of second-order would be unfamiliar to them.

Part 4: Important consequences of doing science from within

46. For a variety of reasons, second-order cybernetics as cybernetics from within has not been developed to its full potential, especially in social, biological, or cognitive sciences. The early pioneers of second-order cybernetics lost their momentum and did not produce a sufficient number of paradigmatic examples to guide further work. From the 1980s onwards, von Foerster was situated far from laboratories and focused on lecture activities around the world. Glanville provided a transcendental framework for second-order cybernetics (as discussed in Müller 2015), but co-operated mainly with designers and architects. Maturana did not develop his observer-focused autopoietic approach in the direction of scientific methodology, Niklas Luhmann (1997) and Dirk Baecker (2013) concentrated on an abstract framework of second-order observations and societal differentiations in an exo-mode, and other leading figures such as Pask, Varela, and Beer died between 1996 and 2002. In my view, second-order cybernetics as a mode of research from within still has a significant future, especially in the social sciences, the life sciences, and cognitive science.

47. In the fourth part of my article, I show that the new science configuration from within, which was initiated by von Foerster under the name of second-order cybernetics, is part of a continuing evolution of science and produces significant advantages in relation to the traditional approaches from without. In this part, I focus on four consequences of this on-going transformation from exo- to endo-science that are not visible at first sight, but that are important for further research in second-order cybernetics.

48. For the science system in general, the reflexive turn to a mode from within, or an endo-mode, can yield at least four groups of new opportunities, namely

- the expansion of types of scientific problems and problem solutions
- a stronger emphasis on scientific knowledge production in applications and implementations and in translational or extension activities
- more diversified and more societally-oriented career paths for individual researchers
- new possibilities for linking scientific research with ethical considerations.

49. In all four areas, second-order cybernetics can play significant new roles in promoting these various shifts and can build a new cybernetic research agenda across different disciplines in the natural and the social sciences.

Diversification of scientific problems

50. The example of the ICA in the previous section and its operating mode from within makes it clear that the domain of scientific problems and scientific problem solutions is becoming more diversified and complex. Operating within this new general science method from within will be accompanied by a widening of scientific problems and problem solutions from a predominantly theoretical focus to a mixed form of theoretical and practical problems, where problem solutions are aimed at eliminating or significantly reducing societal or environmental problems. The theoretical background for this type of practical problem-solving goes back to the 1950s and 1960s, where a new approach under the name of "action research" was propagated, especially within Latin America (Fals Borda 1978). Since then, action research has institutionalized itself, albeit in rather marginal proportions (Greenwood & Levin 2007; Reason & Bradbury 2001).

51. Under the old regime of the traditional scientific method, a societal problem was solved once this problem was successfully modelled or explained. The theoretical problem solution could allow for additional features such as forecasts or scenarios. Practical problem solutions require the reduction or elimination of a societal or environmental problem, and this practical problem solution must have observable and positive consequences for the well-being of affected groups or communities. Practical problem solutions come about through co-operation and interaction between target groups and scientists until the goal of problem elimination or reduction is reached. Obviously, practical problem solutions will not replace theoretical ones in the years ahead, but practical problem solutions will gain in importance and ideally will be seen as a viable alternative to purely theoretical accounts of system behavior.

52. Table 4 presents an overview of the main differences between theoretical and practical types of societal problems and problem solutions. Under the traditional regime from without, only the left column was admitted for (first-order) socio-cyberneticians. In the alternative mode from within, both types of problems and problem solutions can be pursued by endo-social scientists.

53. Second-order social cyberneticians can focus their work on new methods, instruments and heuristics for the practical solution of societal, environmental, medical, or technological problems and can create new methods for more participatory and more active types of science.

Advances in the knowledge of implementation

54. A second advantage of doing science from within is to improve our knowledge of implementation. Von Glasersfeld (2005) described knowledge as a combination of theoretical elements together with the know-how of their applications. In recent decades, many research facilities have been established worldwide as applied universities, as faculties or institutes for social work, for education, for management science, and for other applied fields. Under the traditional scientific regime, these institutes, faculties, or universities were considered as second-rate compared to the research universities, with their focus on basic research. The traditional scientific method emphasizes theoretical advancements and views applied work as lying outside science. If implementations were not progressing well, that was not a scientific problem. In addition, the interactions between basic and applied science were not strongly developed, due to the complex multi-disciplinary nature

Theoretical: Research problems (RP)	Practical: Societal or environmental problems (SEP)
Mode from without (objectivity, observer-free)	Mode from within (observer participation)
Building a theoretical model/ framework for RP	Assessing SEP (history, problem solutions in the past, target groups)
Collecting relevant data	Establishing and carrying out a work plan
Theory or model-testing	Building a workforce and procedures for solutions of SEP
Explanatory account for RP	Effective reduction or elimination of SEP
Success through publications and scientific impact	Success by client groups acquiring skills for self-improvement of quality of life
Reliable theoretical knowledge	Robust practical knowledge

Table 4. Two types of scientific problems and problem solutions.

of applied problems relative to theoretical research. In the old regime, the applied work remained under-developed and practical problem solutions seemed outside the realm of science.

55. Many societal or environmental problems are solved, in principle, in their scientific dimensions, but lack corresponding implementation stages. A developmental study showed, for example (Suri 2008), that the problem of global poverty and hunger does not need enormous financial transfers from the North to the South. The crucial missing element is local knowledge and adequate knowledge distribution with respect to seeds, farming, and marketing. In the new regime of a science from within, agricultural universities, faculties, and research institutes worldwide could develop a global program of poverty reduction through knowledge diffusion and concerted translation efforts worldwide.

56. The present time is characterized by an abundance of societal and environmental problems locally, nationally, and globally, where a high accumulation of theoretical scientific knowledge is accompanied by a deep deficiency in extension, implementation, or translational knowledge. To a significant degree, the current legitimation crisis in science (Nowotny, Scott & Gibbons 2001) is connected to the problem of under-developed implementation of new and avant-garde knowledge and technologies. This asymmetry can be attributed partly to the once-dominant mode from without and to the resulting lack of practical solutions.

57. Second-order cybernetics in particular can become a vital element in pursuing research to close these asymmetries and deficiencies and to create an abundance of new bridges for transfers between basic and applied research and for the rapid diffusion of new technologies in areas such as health, information, or industry.

Strengthening and diversifying science curricula

58. A third advantage of pursuing research from within is to transform the curricula, especially in the social or the environmental sciences, and to build these curricula on two

	Economics	Sociology	Political science	Social psychology	Public opinion
1949/50	5.7	24.1	2.6	22.0	43.0
1964/65	32.9	54.8	19.4	14.6	55.7
1979/80	28.7	55.8	35.4	21.0	90.6
1994/95	42.3	69.7	41.9	49.9	90.3

Table 5. The rise of survey research from 1950 to 1995, in percent (from Saris & Gallhofer 2007: 2).

different pillars. Shifting to a mode from within will add to the cognitive diversity in these fields in significant ways.

a. Researchers in the social or the environmental sciences are currently losing their direct contact with their fields of investigation. The predominant method of research used in journal articles in areas such as sociology, psychology, or political science is now based on survey data. Table 5 shows that during the period from 1950 to 1995, articles in high-quality journals for economics, sociology, political science, social psychology, or public-opinion research are based more and more on the analysis of survey data. A mixed approach of dealing with theoretical and practical problems provides much richer perspectives on societies and their environments than an almost exclusive focus on survey data.
b. Students and future scientists, as knowledge producers, will gain in their understanding of societies if they can shift between theoretical and practical work and can avoid, to quote Ludwig Wittgenstein, “a one-sided diet” (Wittgenstein 1967: §593). Science from within and its variety of problems and problem solutions as well as its strong emphasis on implementation knowledge should allow more cognitively diversified career paths in research work.
c. A more diversified and complex perspective on society and the environment should have a positive effect on the theoretical or the interpretive work of social or environmental scientists. Due to the more complex nature of practical problems and solutions, it can be expected that in the long run, theories or models will become more practical and the implementation or extension work will become more theoretical.

59. Again, second-order cybernetics can play a significant new role in the mediation of theoretical and practical problem solutions in the social and the environmental sciences, and in finding intelligent new mixes in the construction of curricula and PhD programs in these fields.

Recombining science and ethics

60. The fourth advantage of the science mode from within is that it makes scientific operations more dependent, aside from the goals of researchers, on urgent societal prob-

lems, on the needs of different groups and populations, or on local or national values. Here, science from within becomes significantly more negotiable and open to different normative contexts and more sensitive and receptive to ethical considerations.

61. Without losing its scientific credentials, second-order cybernetics could become a new instrument for fine-tuning successful theoretical and practical problem solutions and the adaptation to local or national needs. In this way, second-order cybernetics could operationalize von Foerster's ethical imperative:

> Act always so as to increase the number of choices. (Foerster 2003d: 227)

62. Following this general direction of an expansion in the number of choices, second-order cybernetics as cybernetics from within could develop a new function as a lever for humanizing science in a significant and sustainable manner.

Part 5: New horizons for second-order cybernetics

63. The research agenda for second-order cybernetics has been restricted so far to the transformation from exo-science to endo-science and to special niches within these transformation processes.

64. But second-order cybernetics as a trans-disciplinary field for the study of communication and control from within can be advanced in a more general and systematic way. Here I will mention four broad domains with new objectives for second-order cybernetics that can be reached independently and outside the four areas in the previous section. These four general targets constitute new long-term research paths that could be undertaken in the years and decades ahead, namely:

a. shifting from the traditional epistemology of cybernetics and radical constructivism to the building of more advanced epistemic frameworks
b. providing the foundations for actor-based disciplines such as psychology, management, and the social sciences
c. widening of methodologies, especially for the cognitive, the social, and the biological sciences
d. recognizing a multiplicity of scientific contexts, their differences, their implicit status, and the need for mediations.

These four general and systematic domains for contemporary second-order cybernetics research and development can be described in the following manner.

Building viable epistemologies for radical constructivism

65. The epistemology of radical constructivism as a research tradition that includes second-order cybernetics (Müller 2010) is mostly understood as being another variant of social constructionism or postmodernism. Among the group of radical constructivists, von Glasersfeld probably moved closest to philosophical domains. He established a radical constructivist epistemology directed against scientific realism, against an objective and observer-free reality, with an emphasis on the active role of observers in constructing their realities, and, finally, with viability as an alternative criterion for the contingent acceptance of statements, hypotheses, or theories (see Glasersfeld 2007; Müller 2011; Steffe 2007).

In sum, second-order cybernetics and radical constructivism have so far relied on an epistemology that appears less refined when compared to the highly sophisticated approaches within contemporary epistemologies.

66. As a research program on cognition, second-order cybernetics, as well as radical constructivism, offered various empirical explanations of the stability of reality constructions, especially through von Foerster's postulate of epistemic homeostasis:

> The nervous system as a whole is organized in such a way (organizes itself in such a way) that it computes a stable reality. (Foerster 2003b: 244)

67. But the relations between radical constructivism as an empirical research program and scientific realism as a hegemonic epistemological paradigm have been restricted to a non-dialogue that can be summarized in a slightly paradoxical manner: radical constructivism provides an explanatory account of why realism seems so natural or obvious, while scientific realism continues to argue for the fallibility of radical constructivist arguments.

68. But a future epistemology for second-order cybernetics or radical constructivism should be linked to at least two recent epistemological approaches, going by the names of social epistemology (Goldman & Blanchard 2015) on the one hand and epistemological contextualism (Rysiew 2016) on the other hand. These were advanced without explicit references to second-order cybernetics or to radical constructivism and deal with a variety of problems also relevant for the traditional approach by von Glasersfeld.

69. As a challenging future objective, second-order cybernetics as cybernetics from within should produce new state of the art epistemologies that include elements from social and contextualist epistemologies and that produce more refined new conceptual frameworks for dealing with epistemological issues within the tradition of radical constructivism.

Providing foundations for actor-based disciplines

70. A very large task for second-order cybernetics lies in its role as a founder of last resort for academic disciplines with strong involvements of observers, namely for the social sciences, for the life sciences, or for the cognitive sciences. This task may seem far-fetched and beyond the scope of second-order cybernetics, but Bernard Scott (this volume) has provided an intriguing example for the foundations of psychology.

71. Scott characterizes second-order cybernetics as cybernetics of the observer and presents a study in second-order cybernetics on the foundations of psychology. He uses the framework of Pask's Conversation Theory (Pask 1975a, 1975b), with its differentiation of psychosocial P-individuals and biomechanical M-individuals and a variety of interaction patterns between these two units. This type of study could be undertaken by second-order cyberneticians for a variety of academic disciplines, including my own discipline of management.

72. Another example for fruitful future work for second-order cybernetics lies in the foundations of the cognitive sciences, where second-order cybernetics could emphasize three necessary requirements for cognitive science investigations that should be fulfilled simultaneously.

a. Studies on cognition should be based on von Foerster's unity of cognition postulate, where cognition is conceptualized with a multiplicity of faculties that cannot be isolated functionally or locally (Foerster 2003e: 105). These faculties go well beyond the domains of perception, memory, or inferring (Foerster & Müller 2003) and also include learning, evaluating, and movements.

b. In Varela's afterword to *The Tree of Knowledge* (Maturana & Varela 1992), enaction was promoted as the middle ground between solipsism and representationalism, and a variety of e-properties such as embedded, embodied, enacted, environment, etc. have become essential building blocks for cognitive science studies (see also Vörös, Froese & Riegler 2016).
c. First person approaches should not be excluded but must become a significant and necessary element in the study of cognitive processes. This line of research was already pursued by Varela, who argued vigorously for the necessity of these first-order approaches, especially in studies of consciousness (Varela & Shear 1999).

73. Second-order cybernetics can provide new and more complex frameworks for cognitive science and can become a constant reminder that the foundations of the cognitive sciences are best developed in a way that remains consistent with these three requirements, which seem to get lost more and more in the wider stream of contemporary cognitive science research and publications.

Creating methods for endo-research across disciplines

74. A particularly fruitful task for second-order cybernetics in the future lies in the reconfiguration of available methodologies, especially in the social sciences, the life sciences, and the cognitive sciences, from their traditional exo-designs to new endo-designs.

75. Over the last few decades, I have worked on a number of methods for quality improvement, for process improvement, and for service learning that were organized largely in an exo-mode and that belong to the methodology of management science. Second-order cybernetics can interpret these and other methods in an endo-format across a broad range of academic fields and can focus on the comparative advantages of these changes. This work would be consistent with the current interest in teams.

76. The tasks ahead in the fields of methods and methodologies are challenging. The primary objective lies in a stream of studies that use the revised endo-methods and endo-methodologies. In this way, the potential strengths of these reinterpreted methods can be specified concretely and multiple comparisons between the new endo-methods and the old approaches can be undertaken.

Mediating between different contexts in science

77. An important research field for second-order cybernetics as cybernetics from within lies in an enrichment of contextual differences, due to a new transparency of explicit differences in the goals of researchers, their academic specializations, and their cultural backgrounds, especially because differences in contexts have been rather ignored so far (Lissack & Graber 2014).

78. I have conducted this type of research since the 1980s, when I worked to facilitate communication between cyberneticians and systems scientists in the United States and the Soviet Union (Umpleby 1987). During this period, a very important theoretical framework was established by Vladimir Lefebvre (1982, 2001) under the name of reflexive control. It described two different ethical systems that characterized the United States and the Soviet Union.

79. The variety of issues in contextualizing science will produce challenging new research tasks for second-order cybernetics, which can become a mediator across different contexts and across different fields of scientific research. Currently, the ASC has moved the problems of contexts and contextualizations to its primary agenda, so this work is being conducted already.

80. In my view, these four roads ahead for second-order cybernetics offer a rich and diverse research program that can be undertaken beyond the frameworks and approaches inherited from the pioneers of second-order cybernetics.

Conclusion

81. Early work in cybernetics provided a theory of circular causal, regulatory phenomena that occur in biological and social systems and in some machines. It offered a way of explaining goal-seeking and goal-formulation. A general theory of perception, learning, cognition, and adaptation was created that influenced many fields and that helped to create the information age. This first wave of cybernetics ended around 1975 (Kline 2015).

82. Second-order cybernetics pursued a more ambitious goal that has not yet been taken up by the broader scientific community or, more paradoxically, by second-order cyberneticians either. As I have described, second-order cybernetics has attempted to establish a new way of operating scientifically from within, by noting that observing systems observe systems from within, as opposed to the traditional scientific approach from without.

83. By proposing the idea of second-order cybernetics, von Foerster challenged a key assumption in the methodology of science, namely the goal of objectivity to be achieved by eliminating observer-effects. He showed that scientific disciplines or fields in general can be organized in two ways. Many scientific fields still use the traditional approach of observing from outside. Second-order cybernetics questioned this orthodoxy vigorously. In doing so, von Foerster initiated a still-unfinished revolution in science. The new general methodology of science from within changes the status of the researcher from a hidden factor to an active participant within a highly interactive system. Some fields, such as physics, will retain their traditional methodology due to the focus on inanimate objects in the case of physics. But many scientific fields can gain significantly by shifting to a mode of observing from within. In this sense, second-order cybernetics provides a role model for operating scientifically in a new way that offers advantages in terms of problem solutions, knowledge production, and robust scientific outcomes.

84. Finally, second-order cybernetics can expect a bright future by moving along the four new roads outlined above towards an advanced epistemology, to foundations and endo-methodologies for the social, the biological, or the cognitive sciences as well as towards contextualizing science. In this way, second-order cybernetics can still act as an avant-garde model for humanizing science and for making science more receptive to societal needs at the local, national, and global levels.

Acknowledgement

This article benefited from many conversations with Karl H. Müller. Whereas I worked with von Foerster during the early years of second-order cybernetics, the 1970s and 1980s, Karl worked more closely with von Foerster during the 1990s and 2000s.

Open Peer Commentary: **Obstacles and Opportunities in the Future of Second-Order Cybernetics and Other Compatible Methods**

Allenna Leonard

1. In this commentary, I would like to broaden the context of Stuart Umpleby's target article to look at some of the other political, economic and social factors that have had an impact on the take-up of both first- and second-order cybernetics and on parallel developments in contiguous fields. The promise of second-order science in the social and behavioural sciences and the potential for it to address issues of implementation or enaction is constrained by the general lack of comprehensive social science research and by the lower sense of common values that, to a greater or lesser extent, now characterizes our communities and nations.

2. The challenges of post-war reindustrialization and the introduction of computers and other machines drew heavily on the contributions of cybernetics, operations research, psychology and human factors analysis that had been effectively utilized during World War II, particularly in the US and the UK. It received funding from governments and from industry that also included action research, research on the human/computer interface and communications as well as concerns about the impact of technology and the quality of working life. In addition to those associated primarily with cybernetics (Norbert Wiener, Ross Ashby, Warren McCulloch, Gregory Bateson, Heinz von Foerster, Stafford Beer, Gordon Pask, Jay Forrester, Humberto Maturana and Francisco Varela), there were others who were more associated with the general systems community, who included, among others, Kenneth Boulding, C. West Churchman, Russell Ackoff, Chris Argyris and Donald Schön. Almost all of them sounded alarms about the consequences of the social and economic trends that were perceived from Wiener's *Human Use of Human Beings* (1950) onwards.

3. Von Foerster's description of second order cybernetics was, perhaps, the most elegant and rigorous, as it was firmly based in neurophysiology, but the appreciation of perspective was evident in other fields as well. George Kelly's personal construct theory was published in 1955; Argyris and Schön's *Theory in Practice* was published in 1974. The maxim "where you stand depends on where you sit," otherwise known as "Miles' Law" appeared in *Public Administration Review* in 1978. In addition to the community development work of the Institute for Cultural Affairs, the work on socio-technical systems theory of Eric Trist, Fred Emery, Enid Mumford and others associated with the Tavistock Institute worked with group process and multiple perspectives. Since then, a number of group processes have been developed and used that have been very effective in improving equity, reducing risk and generally increasing inclusion.

4. All of these approaches have one thing in common: they allow for multiple constructs of stakeholder positions, cultural and social contexts and time sensitive frameworks. And they challenge the status quo of existing power relationships. To relate one anecdote, colleagues of mine who were offering to do a syntegration (Beer 1994) had convinced middle management that the challenges their company faced could best be addressed by a planning process that would bring the multiple perspectives and collective knowledge of the group together. The president of the company rejected the proposal on the grounds that he did not want to establish the precedent of listening to his employees. Beer himself had many comparable stories from his years in the steel industry and consulting. His experience was that many people in charge of companies would rather go out of business than relinquish management prerogatives or even instigate improvements if the suggestions had come from the shop floor.

5. If we look to the present, the challenges faced by second-order (and first-order) cybernetics are part of a larger problem with science as a whole. These are times when even Karl Popper's "Science One" is under attack if the conclusions reached challenge economic or political interests. We see this most publically in the denial of climate change and environmental damage from pollution, but there have been many others. Even the evidence that smoking is bad for one's health was disputed for many years by tobacco companies. We have also seen the United States Congress pass legislation forbidding the Center for Disease Control from engaging in research on gun violence in the United States – a clear example of unwarranted and damaging interference.

6. Answers to Umpleby's questions about the characteristics separating first- and second-order cybernetics and the advantages of second- over first-order cybernetics depend on the context. After the perspective of the observer/researcher is made explicit, there will be differences depending on whether or not the researcher is a participant observer. When the subject of the research is not one where the observer has any direct contact or influence or where the researcher is fulfilling a specific contract, the process may be closer to first- than second-order cybernetics. However, the extension into new areas of research on implementation will necessarily incorporate more second-order characteristics. In both cases, hypotheses, data collection and interpretation and falsifiability come into play, but in somewhat different ways. The one aspect of the traditional approach to science that is very difficult if not impossible to achieve when researching implementation projects is reproducibility. There is too much variety in human social activities.

7. The backlash against the universities that led to the closing of von Foerster's Biological Computer Laboratory was part of a process that saw, over time, reduced government funding for universities and university research across the board and a corresponding increase in the percentage of costs paid for by tuition and fees. A majority of undergraduate courses are now taught by adjunct faculty who are no less qualified than their predecessors but who must make do with short-term contracts – sometimes assembling a portfolio of courses in different universities. Such faculty members have little or no job security and are seldom in a position to commit the time to do or assume the costs of doing field work. Surveys and correlations based on publically available data are essentially the only affordable means of doing publishable research. Implementation falters as well: it is not the knowledge that is missing; it is the resources and the political will to use that knowledge that are lacking.

8. It must be noted that without sponsorship, the research conducted by individuals is necessarily limited in its scope. Sponsored research, whether sponsored by governments or

foundations, usually has very specific goals and criteria. My own experience of working in a university development office and later as a volunteer in an environmental organization has been that a proposal's likelihood of success is directly linked to how closely it mirrors the funder's criteria. In turn, funding agencies, especially in government, favour projects with straightforward measures that are easy to defend against charges of wasting taxpayer money. Needless to say, the higher the variety of an issue or problem, the more difficult it is to achieve unambiguous or apolitical results.

9. As the 21st century finds its feet, the need for more use of cybernetic and systems-based approaches will become clear as the unsustainability of current practices becomes more and more evident. Indicators of ecological, social and economic instability are already apparent. It is to be hoped that it will be possible to mobilize to meet those threats to human life as effectively as it was possible to mobilize for war. Appreciation of that risk could become enough of a common value to move forward.

10. If and when that happens, a continuum between first- and second-order cybernetic approaches will be needed; although looked at broadly, almost all applications that are not mechanical will fall somewhere along the line toward a second-order approach. Any construction of a cybernetic model is going to select some variables and ignore others, based on the purposes of those constructing or commissioning the model and the goal of the research. In most applications, whether designated as first- or second-order approaches, there are preliminary discussions regarding the scope of the model and the assumptions made about the situation and the variables that are relevant. Constraints may be imposed by time, resources or the boundaries of the client's concerns. If the context is complex or in flux, periodic reexamination of the project goals and model fitness is or should be done and further discussions held. Whether the modeler stands "outside" or "inside" may also depend on perspective. If one is using a model to predict possible consequences of a government program that will have an effect on the modeler or the modeler's client, that person may be "endo" with respect to their own interest in the outcome but "exo" with respect to the actors who designed and will carry out the program.

11. As Umpleby notes, there is a great deal of research to be done regarding the further development of second-order science. Some is based in cognitive science and could explore the distinctions that we share as human beings and the dimensions of their individual perceptions. As we accept the neurophysiological components of individuals, we need to know more about how these individual perceptions and constructs become shared. The case of the Institute for Cultural Affairs that Umpleby describes is interesting because it began with a primary emphasis on implementation. Theory was explored to inform the doing. This approach has the advantage of incurring very modest additional costs and being relatively insulated from other agenda. That would be a particularly valuable change as it would not only extend the reach of issues addressed by science but would also be much more inclusive with respect to the type of researchers who would become involved.

Open Peer Commentary: **Connecting Second-Order Cybernetics' Revolution with Genetic Epistemology**

Gastón Becerra

1. In this commentary, I will discuss Stuart Umpleby's target article about the "scientific revolution" he attibutes to Heinz von Foerster's work. I will draw some parallels with another constructivist stream that the author does not mention, at least not in the references listed in §46. I am referring to Jean Piaget's genetic epistemology, both in its original version (preceding the radical and cybernetic constructivism) as well as the most recent reformulation, proposed by Rolando García.

2. A few words are necessary to introduce this last author. García (1919–2012) was an epistemologist who collaborated with Piaget at the end of his work (e.g., Piaget & García 1982, 1988). After Piaget's death, García sought to create a new synthesis that could organize the different aspects considered by genetic epistemology (García 1987, 1992, 1999, 2000) as well to expand its scope by presenting it as a tool to interpret current problems and challenges of science (García 1997, 2006). In my understanding, Umpleby has similar intentions towards von Foerster's work.

3. Here, I seek to broaden both the context and the image of the revolution that Umpleby offers. Specifically, I will show that the movement propelling it was a more comprehensive constructivism than the cybernetic-radical tradition. I also suggest that the different philosophical assumptions from these variants of constructivism continue to condition its achievements and challenges when it comes to reflecting on science. All this will allow me to introduce García's work to the readers of *Constructivist Foundations*.

A broader approach towards epistemology

4. In Table 1 and §5, Umpleby situates von Foerster's view on science versus the more classical view in its normative (Karl Popper) or sociological (Thomas Kuhn) aspects. The main element of this "revolution," in the author's opinion, is the inclusion of the question of "how an individual constructs a reality," which von Foerster addresses by reflecting on neurophysiological research from the 1960s. If we go back to the program posed by Piaget, this is where we can draw the first parallel.

5. As is well-known, the main focus of Piaget's work is epistemological:

> Genetic epistemology attempts to explain knowledge, and in particular scientific knowledge, on the basis of its history, its sociogenesis, and especially the psychological origins of the notions and operations upon which it is based. (Piaget 1970: 1)

For Piaget, such a project entails integrating several disciplines and methods: on the one hand, it resorts to formalizing analysis to deal with matters of knowledge validity, and on the other, to historical-critical and genetic analysis in the fields of history of science and developmental psychology to deal with the issue of knowledge constitution. Genetic epistemology holds that the explanations constructed around the development of individual knowledge can shed light on the development of scientific knowledge. The core hypothesis is that there is a "functional continuity" between both domains that would enable their transformation to be explained by resorting to the same constructivist processes and mechanisms (Piaget & García 1982: 31).

6. The underlying question furthering Piaget's project can be grasped via a terminological clarification. As García (2000: 25f) recalls, Piaget used the term "épistémologie" in its French sense, which designates the critical study of scientific knowledge, i.e., the analysis of its logical foundations and fields of action. The definition of genetic epistemology maintains this original sense of the term although it traces its genesis back all the way to the most elementary individual forms, thus broadening the term's sense towards a general theory of knowledge. In García's opinion, this move deserves to be described as "revolutionary." Such clarification is important for us because it shows how different objectives that are usually grouped under the same "epistemology" tag are connected. Jeremy Burman (2007: 722) provides a clear summary when, turning away from the ambiguity of such a tag, he suggests understanding Piaget as a "meta-epistemologist" or a "philosopher of knowledge" who, based on the empirical consideration of cognition, is interested in making philosophy of science a scientifically informed analysis.[1]

7. The parallels between Piaget's and von Foerster's project can be found not only in their respective views on science but also in their concurrence on some questions regarding the theory of knowledge, something that both authors acknowledged (see von Foerster's 1981 contribution to Piaget's *festschrift* and the honoree's answer in Inhelder, García & Voneche 1981). This basic concurrence is found in the common denominator of "constructivism of the epistemic and cognitive theory," which could also be applied to Ernst von Glasersfeld, Humberto Maturana, Francisco Varela and Siegfried Schmidt, to name the authors pointed out by Karin Knorr-Cetina (1989) and Niklas Luhmann (1990), from whom I am adopting the denomination. All of them made their contribution in a wider movement in favor of the "naturalization" of epistemology and science analysis.

8. In considering this wider constructivist tradition, I do not want to ignore the significant differences that can be found in terms of philosophical assumptions, such as the characterization of the cognizant subject or they way they understand society. Here, it is my interest to suggest these assumptions should be considered when evaluating the constructivist philosophies of sciences' accomplishments and challenges. Specifically, by including two professed realists such as Piaget and García, I seek to introduce a difference in Umpleby's considerations, whose analysis seems to focus only on the less conflictive dialogue between von Foerster and radical constructivism authors.

Challenges and achievements

9. Among the "significant advantages in relation to the traditional approaches" to analyzing science, Umpleby points out that there is a preeminence of the social aspect: §50 deals with shifts from theoretical problems to "practical" ones – e.g., social and environmental ones – as well as with "action research"; §§56f address the social contexts

1. Coincidentally, there are reasons to believe Thomas Kuhn also used "epistemology" in an analogous sense, associated to the theory of knowledge (see, e.g., Kuhn 1970: 96, 126), although, instead of a constructivist standpoint, a certain empiricism and psychological behaviorism can be observed (Becerra & Castorina 2015). In any case, what distinguishes epistemologists such as Piaget, Ernst von Glasersfeld or von Foerster from philosophers of science such as Kuhn and Popper is that the former thematize cognitive mechanisms, which would eventually influence how the "social" dimension is understood.

of applied knowledge; §59 calls for giving academic value to these operations; §61 calls for contemplating the ethical dimension of research. Such preeminence is sustained in the "new horizons" the author sets out for second-order cybernetics: in §76, the step towards endo-research that is referred to can be understood as considering the context of knowledge production; and §77 points out the importance of comparing objectives and assumptions to evaluate prospective dialogues and risks when integrating "contexts and disciplines." This is where we can draw a second parallel with genetic epistemology.

10. In *Psychogenesis and the History of Science*, Piaget & García (1982) sought to reevaluate the role that social context plays in the field of knowledge. In their conclusions, they propose the existence of invariant mechanisms explaining the emergence of knowledge, along with social, cultural and historical contexts of meaning conditioning the directionality these mechanisms take. Piaget & García refers to this conditioning as an "epistemic framework" (1982: 228). In his later works, García re-defined the epistemic framework as the "boundary conditions" modulating the intrinsic activity of a cognitive system (García 1992: 31). According to this view, knowledge evolves by reorganizations fed by the exchanges between the (cognitive) system and its (social) environment (García 1999: 179).

11. Eventually, with García's revision (2000), the scope of his analysis of constructivist theory was widened, which makes it possible to state different ways of analyzing the "epistemic framework" (Becerra & Castorina 2015).

12. A first type of analysis is the one Piaget & García (1982) called "sociogenetic," which is intended for the history of science field. Here, the epistemic framework refers to a worldview (*Weltanschauung*) resulting from philosophical, religious and ideological factors that influence the contents of theorization by enabling or inhibiting our questions (Piaget & García 1982: 228–234). García provided an example:

> Ohm, in Germany, discovered the first quantitative law in electricity … The reaction of the *Naturphilosophers* was very consistent: What was the point of *measuring* such phenomena? Electricity is something very immaterial – how can one measure it? Experiments and mathematics were considered entirely irrelevant to obtaining a true understanding of Nature. … This was the *Weltanschauung* involved in the Romantic movement … In this case, a particular cultural pattern enters in a very concrete way into the shaping of science in a particular society at a particular time. In the case we are considering, it acts as an epistemological obstacle, to use Gaston Bacherlard's expression. Here, the social component is not merely providing directionality to scientific research; it enters deeply into the conceptualization of science. (García 1987: 136)

This raises the question of whether a similar conceptualization has been developed by second-order cybernetics.

13. A second analysis relates to the "psychogenetic" field covering the development of knowledge from childhood to adult thinking. The epistemic framework here refers to the social meaningfulness and cultural practices that make certain phenomena or objects visible or invisible in a shared social world (for a similar idea, see Overton 1994). Surprisingly, Umpleby has not given further consideration to social context and social meaningfulness when arguing about the cognitive sciences (§§72f).

14. Two further analyses may be more in line with the Umpleby's intentions. One of them is the metatheoretical analysis, i.e., the analysis of assumptions underlying a theory, which Piaget (1979) used to include in the "internal epistemology" of the sciences. The epistemic framework lies here in the history of a specific theoretical or disciplinary field,

which conditions future developments through the dialectic relations among the different levels of theorization, data selection and interpretation, explanatory models, etc. This type of analysis can also be applied to contemporary scientific problems, thus becoming a fundamental tool for enabling interdisciplinary research of "complex issues" such as the societal and environmental ones (for a brief review of these research programs in contact with the sociocybernetics approach, see Becerra & Amozurrutia 2015). The epistemic framework here is expressed through the researchers' set of social and political values making up the multidisciplinary team, say, in the way the social and political need emerges, marking off the intervention direction (García 2006). Only recently, the interdisciplinary literature has started to consider the sociopolitical element as a key factor in succeeding in this type of endeavor (Boix-mansilla 2006). I think these projects bear a spirit in terms of challenges and achievements that is similar to the ones Umpleby highlights in §§50, 61, and 76f.[2]

15. García suggests that if the approaches used to address complex issues are to be improved, then actions must be taken to integrate knowledge and to co-construct the study object among the multidisciplinary team members. A mere call to interdisciplinarity is not enough. What is needed is a new methodology, explicit lines of work, and new tools and techniques easing such integration. García proposes this methodology by reflecting on his research experience on climate change, drought and famine through a constructivist lens (García 2006). As far as I know, cybernetics and radical constructivism have not made much progress in designing this kind of proposal (although Hugo Alrøe and Egon Noe 2014 provide a good discussion that, in many aspects, is in line with García's considerations). Perhaps the lack of greater integration with empirical social research presents an obstacle. I think Umpleby's remarks in §§65–69 suggest a similar diagnosis. I can only hypothesize as to what extent such lack of integration is due to an unclear stance on society's "reality status," and subsequently, its effective conditioning on knowledge (Glasersfeld 2008; Müller 2008). In any case, García's work could be a fine case for observing how a constructivist perspective that acknowledges social forces and social structures can indeed make a contribution on this matter.

Open Peer Commentary:
Shed the Name to Find Second-Order Success: Renaming Second-Order Cybernetics to Rescue its Essence

Michael R. Lissack

1. Stuart Umpleby's target article highlights the intellectual progress of second-order cybernetics and its related branch of constructivism while at the same time making note of (and slightly bemoaning) its lack of implementation in both contemporary academic

2. It is worthwhile mentioning that Umpleby provides a deepened elaboration on the organizational dimension of science, which has been poorly elaborated in García's work. A special note must be given to the ethical considerations found in von Foerster and Piaget, which García seems to have replaced with political and strategic considerations.

thought and practitioner practice. His hopeful expressions of the paths second-order cybernetics might take contrast with the field's lack of progress for the past two decades or more. In this commentary, I will rephrase Umpleby's proposed pathways by making explicit the main obstacle to their implementation: the very words, labels, history, and jargon that cyberneticians use to define their field and to encourage the uptake of its perspective by others.

2. The reader should take careful note of an important irony here. In my role as the President of the American Society for Cybernetics I am tasked with preserving, evangelizing, and promulgating the essences of the field. To do this successfully, I believe that we need to recognize the context in which the very label "cybernetics" functions. The word has shifted in its meaning. The two-syllable conjunction "cyber" is now associated with computers and computation. While the old meaning of "steering" remains in the dictionary, it is lost on those whom practitioners in the field need to reach. If we are to further cybernetics, and especially second-order cybernetics, as a field of intellectual inquiry, I believe we as a community need to accept that we have lost the battle of "the word." What matters in successful communication is how the listener receives the signals being transmitted and then converts those signals into personal meaning. Our insistence on making use of the 1950s and 1960s meaning of words such as "cybernetics" is getting in our way. Our desired listeners struggle to grasp our intended meaning. To "save" cybernetics so that it may live and prosper, I believe that its very name needs to be relegated to "historic label" and that we, as a community, need to find new ways to express our essential thoughts.

3. It is on one of those essences that I will focus herein – the rôle of always asking about how context matters. Context here must be viewed in its broadest sense. Not just the material, social, and physiological opportunities, boundaries, and constraints that may serve to describe a given situation, but also the intellectual, semiotic, and lexical triggers that affect how any given participant or observer mentally processes that situation. As Umpleby quotes Thomas Kuhn: different participants/observers

> see different things when they look from the same point in the same direction. [...] Both are looking at the world, and what they look at has not changed. But in some areas they see different things, and they see them in different relations one to the other. (§4)

4. In response to the Kuhn quote, second-order cybernetics and constructivism would also add that indeed what they look at has changed. Each participant and observer has their own set of mental constructs, and they can only "see" what they possess the constructs for. In §7, Umpleby notes:

> Since people have different experiences – language, home life, culture, religion, academic training, and job experiences – each person's 'reality' is in some respects unique, though our knowledge of the physical and social world has many common features.

As Kuhn (1970: 48) put it: "You don't see something until you have the right metaphor [model] to let you perceive it." And as Daniel Kahneman (2011: 87) elaborates: "We often fail to allow for the possibility that evidence that should be critical to our judgment is missing. What we see is all there is."

5. Each participant and observer is thus perceiving, dealing with, processing, reacting to, and enacting their own "private" world. As Karl Weick claims, when people "enact" the environment,

> they construct, rearrange, single out, and demolish many 'objective' features of their surroundings. They unrandomize variables, insert vestiges of orderliness, and literally create their own constraints. (Weick 1995: 30f).

In other words, they attempt to reduce the "world" to their "model" and labels. In so doing, they are making personal choices.

6. Cyberneticians and constructivists have insight into the boundaries and constraints imposed by each individual's frame of thought. But, all too often, both cyberneticians and constructivists fail to recognize the imprisonment of their own personal frames of thought. The centrality of those personal frames was, of course, underscored by Ernst von Glasersfeld's "substitution of 'viability' or 'functional fit' for the notions of Truth and objective representation of an experiencer-independent reality" (Glasersfeld 2001: 31). Somehow, members of our community all too often fail to reflect on the idea that it is they who are drawing the frames and thus determining viability. "Fit" is recognized as a function of context. But, "fit" is also a function of personal choices. All of the factors that contribute to personal choice must be included in any definition of context. What we easily attribute to others is quite often hard to see in oneself – including, in this case, the role of personal choices.

7. Both cyberneticians and constructivists forget the lessons of Hans Vaihinger's (1924) *Philosophy of "As If."* They seldom discuss the power of "enabling constraints" (Juarrero 1999). They overlook the criticality of Robert Rosen's (1985) models. They confuse how to apply the dictum "make everything as simple as possible, but not simpler" (which is almost always incorrectly attributed to Einstein) with von Foerster's imperative "act always so as to increase the number of choices" (Foerster 2003d: 227). And when discussing differences in mindsets with realists, they all too often forget Richard Rorty's dictum: "Knowledge is not a matter of getting reality right … but rather a matter of acquiring habits of action for coping with reality" (Rorty 1991: 1). Coping means finding meaning where one can by making choices.

8. One of the tenets constructivism shares with several other approaches such as semiotics is that meaning is not embedded in language as if that language was merely a look-up table. Collectively, in everyday life, words and phrases often emerge from concrete situations in which participants jointly work out ways of describing what is going on. New terms, symbols, or images are situated; they acquire meaning through collective use in real situations. They are the product of a never-ending web and network of intersecting personal choices. Those choices get simplified so that the situation can be indeed reacted to and moved on from. All too often the choices made will fail to reflect their own nuanced environment and instead demand coherence with a simpler exogenous model. "We need models to explain what we see and to predict what will occur. We use models for envisioning the future and influencing it" (Derman 2011: 43). Sometimes this approach works. Oft times it fails. But note – both success and failure are rather clear cut when they occur.

9. Following the observations of such success and failure come the attempts made to explain the results. It is here where second-order cybernetics and constructivism can have their greatest impact. Umpleby notes: "Under the old regime of the traditional scientific method, a societal problem was solved once this problem was successfully modelled or explained" (§51). But in today's society, a solution is not accepted without practical application. The "authority" of science conducted per the "scientific method" has eroded. Practical

applications demand context. They can only be explained in light of the full context as I defined it above.

10. Public acceptance of solutions is now dependent upon the public's willingness to accept restricted definitions of context such that the problem appears to be solved in that restricted context. Mere modelling or theoretical explanation is insufficient. In the absence of a defined context where a practical solution is demonstrable, problems are not "solved" – regardless of the elegance of a model or an "explanation." The public has, in effect, chosen to disregard "science" as the source of authoritative solutions.

11. This choice stems naturally from our public tendency to accept a narrowing of the information we consider. "We take up only those actions and solutions that have an immediate effect on the situation, and always as they have been framed for us" (Piattelli-Palmarini 1996: 58). The frames used to view the problem are usually provided by others (be they politicians, the media, or "opinion") and they are usually missing information. "We, therefore, fail to note important items in plain sight, while we misread other facts by forcing them into preset mental channels, even when we retain a buried memory of actual events" (Gould 2010: 223). More critically, we all too often fail to realize that only "true models" in the Rosen (1985) sense allow for interventions to be "rehearsed." We instead allow others to frame mere descriptions for us as if they were models – and then are surprised when the anticipated affects of interventions go awry.

12. What we choose to see will affect what we then pay attention to, which then affects the processes we call upon to make sense out of those attended to items. As Deborah Lupton points out, we then need to "clean up" the attended to data and its resulting story – removing the anomalies and ambiguities, and leaving behind a "simple story."

> All cultures have ways of dealing with these anomalies and ambiguities. One way to deal with ambiguity is to classify a phenomenon into one category only and maintain it within the category, thus reducing the potential for uncertainty. Another method of dealing with anomaly is to physically control it, removing it. A third way is to avoid anomalous things by strengthening and affirming the classification system that renders them anomalous. Alternatively, anomalous events or things may be labeled dangerous. (Lupton 2013: 62)

Simple stories are usually not "true models" (a la Rosen) but are mere descriptions. But we use them as if they were models – models devoid of nuance, ambiguity, and context.

13. When this succeeds, it is as Stephen Hawking and Leonard Mlodinow describe it, "as-if" our simple story was the very reality we need to deal with.

> The only meaningful thing is the usefulness of the model [...] When such a model is successful at explaining events, we tend to attribute to it, and to the elements and concepts that constitute it, the quality of reality or absolute truth. (Hawking & Mlodinow 2010: 7)

Empowered by such success, we overlook its context-dependence. But our very definition of success has been intimately tied to the boundaries and constraints (cf. Juarrero 1999) that we imposed so as to frame the situation. In other words, our choices about context help to determine the success or failures of our models.

14. We must remember the lessons from Vaihinger: "The object of the world of ideas as a whole is not the portrayal of reality – this would be an utterly impossible task – but rather to provide us with an instrument for finding our way about more easily in this world" (Vaihinger 1924: 15). and of Rorty. Our simple story is the result of choices we each make and have made. We make those choices to "make sense" out of a situation, to help us in

our way-finding and in our coping. We always have the option of making different choices. But, the reality with which we deal will be the one we choose to deal with.

15. This is the realm of "pragmatic constructivism" (Lissack & Graber 2014). The "pragmatic" here refers to the process of how we go about explaining a situation to someone else and the process of how we reach an understanding of that situation ourselves. When simple perceptions are inadequate, the need for tools that enable better access to the "what, who, and how much" that one needs to know in order to act becomes painfully obvious. The pragmatic constructivist is happy to accept the scientific realists' models as a base that must then be modified to account for boundaries, constraints, and the manifold possibilities inherent in the interactions of large numbers of autonomous and semi-autonomous agents. Such modifications are rooted in the observer/actor's understanding of the situation at hand – an understanding that itself can be molded by the interactions it observes and participates in. Explaining is the ability to relate a narrative to the questioner, which, at a minimum, allows a "fit" between the question asked and the "attended to" context and, in depth in the form of acquired understanding, allows the explainee to apply such narrative to new contexts and new questions.

16. When we are explicit that we are choosing the realities we deal with, the problems we recognize as problems, and the boundaries and constraints that enable solutions, we are not only accepting some form of constructivism, but we are also accepting that we each have a sense of responsibility regarding such choices. Cybernetics has served to produce great insight into how we might manage these responsibilities, including:

- the role of the observer (von Foerster 2003d),
- the law of requisite variety (Ashby 1958),
- the importance of the observer in cognition (Maturana & Varela 1980),
- the use of Black Boxes (Glanville 1982),
- the idea that all action is in some ways a conversation (Pask 1975a),
- the importance of recognizing that "true models" (in the Robert Rosen sense, cf. Lissack 2016) differ from descriptive representations,
- the importance of narratives (Clarke 2014).

These insights can be reduced to a fundamental essence: it is critical to ask and explore how context in its fullest meaning matters. Second-order cybernetics is at essence the science of exploring how context matters.

17. Exploring how context matters is a second-order concept. It is the essence of everything written above. Indeed, it may be the essence of everything in this volume. What is important is that "exploring how context matters" is not jargon, is not domain or intellectual foundation restricted language, is not hard to grasp. Exploring how context matters is a question that can be applied to every scientific exploration, every strategic business decision, every social issue, and nearly every personal choice. If we can focus our second-order cybernetics efforts on getting others to "explore how context matters," we can reintroduce a ubiquity to the cybernetic/constructivist endeavour.

18. Umpleby (§82 and §84) poses a challenge to the second-order cybernetics community: find relevance or risk death. His list of issues where a second-order cybernetics approach may yield valued results is both lengthy and practicable. But Umpleby (like the other authors in this volume) has minimized our actual dilemma: if we continue to use jargon to which others cannot relate, we fail.

Dimension	First-order	Second-order
	Commonly cited distinctions	
Approach	Reductionism	Holism
Method	Analysis	Synthesis
Primitives	Entities	Relations
Processes	Deterministic	Probabilistic
Relation to designer	Controlled	Autonomous
Relation to designer	Designed	Self-organized
Embeddedness	Context free	Context dependent
Role of observer	Observer-free	Observer-dependent
Position of observer	Outside observed system	Embedded in observed system
Theory dependence	Theory-free phenomena	Theory-determined phenomena
Metaphysics	Ontology	Epistemology
Working framework	Naïve or critical realism	Pragmatism, radical constructivism
Aim	Understanding (prediction)	Action (intervention)
Definition	Clarity, operational definitions	Ambiguity as opportunity
Subjects	Inanimate objects (nonhuman)	Thinking participants (human)
Reflexivity	Unreflexive	Reflexive
Medium of interaction	Physical	Language
Mode of interaction	Communication	Conversation
	Common conceptual rubrics	
Historical period	Early cybernetics (1930–1975)	Late cybernetics (1975–present)
Target systems	Artificial devices	Human systems
Disciplines	Engineering & natural sciences, medicine (bionics)	Social, psychological, therapeutic arts & sciences, management, arts, organizational & social change
	My perspective	
Purposive systems (systems with embedded goals)	Study, design, construction and use of purposive systems (natural and/or artificial)	Study, design, construction and use of interactions between purposive systems (natural and/or artificial)

Table 1. Some dimensions of the proposed first-order/second-order cybernetics transition.

19. Many other intellectual communities are doing work that falls within the domains of second-order cybernetics and what I prefer to call "pragmatic constructivism." We can bring members of these communities "into the fold" if we begin to use language that gives them meaning. Together we can co-construct a new science of context. Exploring how context matters is just a beginning.

Open Peer Commentary: **Beware False Dichotomies**

Peter Cariani

1. I am in general agreement with Stuart Umpleby's thoughtful historical and conceptual review of second-order cybernetics, and do agree with him that widespread adoption of the epistemological stances of second-order cybernetics would constitute a fundamental revolution in how most scientists think about and do science. The proposed revolution involves a shift from realist ontology to constructivist epistemology, from realism to pragmatism, and from observer-free to observer-aware descriptions. That is all fine and good.

2. Missing, however, from Umpleby's account (e.g., the Popper-von Foerster-Kuhn trichotomy in Table 1) is an acknowledgment of parallel paradigmatic realist-pragmatist debates that have played out in science and in the philosophy of science. The cyberneticists have not been the only ones to formulate alternatives to realism. And not all scientists are either naïve-realists or Popperian falsificationists. There have always been pragmatists and operationalists (e.g., Heinrich Hertz, Ernst Mach, Percy Bridgman, and Nils Bohr) who have regarded scientific models as provisional, observable-dependent, predictive models rather than as correspondences with some underlying absolute, knowable reality. In the debates over quantum mechanics and the foundations of physics debate, Bohr and others sought to explicitly include the observer in the predictive modeling process (Murdoch 1987). Similarly, in the philosophy of science, there are well-developed, full-blown pragmatist and radical constructivist alternatives to realism (Van Fraassen 1980; Munévar 1981; Glasersfeld 1995). A tactical question is whether second-order cybernetics is the best banner under which to advance this alternative understanding of science, or whether a broader front united with like-minded scientists and philosophers of science would be more effective.

3. A second, related point is that normal practice in empirical science *does* to some degree take into account the methods and purposes of the observer. Although most individual scientists are not very reflective about their ontological and epistemological assumptions (in many fields, especially in biomedicine, discussion of "larger philosophical questions" is palpably maladaptive for securing funding), in practice empirical science does not automatically make the naïve realist assumption that what is being measured is completely well-characterized (i.e., the assumption "it is what we think it is"). With the exception of theoretical discussions and commentaries, every scientific paper has an introduction section where the observer's reasons for considering a phenomenon are presented and a methods section where the experimental procedures of the observer are spelled out to the extent that the observations made can be reliably replicated by others (i.e., intersubjectively verifiable). The removal of subjective biases and the striving for "objective" measurements and evaluations (as in Kandel quote, §22) is not to assert realist objectivity but to assure conditions for pragmatist intersubjective verifiability. By "objective," Kandel means not that the measurement is objectively true, but that, to the greatest extent possible, it is not being intentionally biased by the observer.

4. The observer determines which measurements will be made and how they will be made, but the specific outcome of each measurement cannot be under the control of the observer if measurements are to be replicated by others (who have potentially different intentions). A realist says "say how it is," and Heinz von Foerster replies "it is how you say

it" (§27 and Table 3), but a pragmatist-operationalist would say "you will see it this way, if you construct these lenses for observing it."

5. The observer alone does not determine what he/she/it observes. Exactly what is observed is co-determined by the experimental preparations and measuring devices chosen by the observer (the observational frame) and how the measuring devices interact with their surrounds. We construct our own epistemic realities to the extent that we choose how to view the world (the nature of the measurements made), but beyond this, we do not control specifically how the world will appear to us (the outcomes of those measurements) once the choice is made. If we did control specific outcomes, there would be no reduction of uncertainty for the observer (in Ashby's sense). It would also mean that observers with different intentions would not be able to use each other's data in building models and theories.

6. Even the most reductionistic molecular biologists, who tend to eschew philosophical considerations entirely, are keenly aware that interpretation of a given body of data depends critically on exactly how the system was prepared and the measurements made. They understand that they do not know exactly what is being measured in their assays. A great deal of time and mental effort is spent *not* taking data at face value, and acknowledging that different observational frames, even if they are supposedly measuring similar or related things, are not necessarily commensurable. Rather than caricaturing most normal scientists as unsophisticated naïve realists, it may be more persuasive to emphasize those parts of scientific practice that do explicitly take the observer into account, thereby showing that many aspects of the new perspective are already valuable parts of current practice.

7. "Second-order cybernetics" is often contrasted with "first-order cybernetics" along various dimensions of difference. Many can be seen in my Table 1, which includes distinctions from Umpleby's Tables 1–4 and additional ones from recent related discussions with Umpleby and other cyberneticists. These are not repeated verbatim, so they reflect my understanding of the distinctions. They also include some ideas from Eric Dent, Umpleby's former graduate student and collaborator.

8. The commonly cited distinctions in the table relate to the dominant subject matter, philosophical stances, aims, and methods of the two proposed forms or modes of cybernetics.

9. Although this dichotomy of distinctions may fairly characterize second-order cybernetics, I think it seriously misrepresents the concepts and practices that most people would tend to label as first-order cybernetics. For example, three major artefacts of the early period of cybernetics – Ross Ashby's adaptive homeostat, Grey Walter's autonomous robotic tortoise, and Gordon Pask's self-organizing electrochemical device – do not fit well into the caricature of first-order cybernetic devices as deterministic, controlled, predictable, or even well-defined (Cariani 1993, 2009; de Latil 1956; Pickering 2010). The major proponents in the early phase (e.g., Norbert Wiener, McCulloch, Ashby, Walter, Pask, Stafford Beer, von Foerster) held pragmatist, not realist, epistemologies heavily informed by neuroscientific and psychological perspectives.

10. The dichotomy of first- vs. second-order cybernetics too easily lapses into a historical break (early vs. late cybernetics), a difference between nonhuman (natural sciences and engineering of artefacts) and human systems (socio-psychological interventions), and a disciplinary divide (hard scientists and engineers vs. soft social and psychological scientists and human systems people). Do we want to divide cybernetics along these lines? Perhaps the distinction can be coherently construed as a historical transition from one kind of research to another. But, if so, we should avoid the error of thinking that the first wave

of cybernetics was epistemologically less sophisticated, either because they came earlier or because many of them studied and designed feedback control systems or because many of them were biologists and neuroscientists. If a revolutionary flag is to be raised, the distinction should instead be made between an observer-aware second-order cybernetics and traditional, realist conceptions of science and engineering. There is no need to use the distinction to divide cybernetics itself.

11. I do agree that a historical and sociological transition can be identified in the cybernetics movement, but this is not so much a revolution of ideas but a shift in who continued to work explicitly under the banner of cybernetics and what kinds of systems they studied ("first-order" and "second-order" are misleading labels for this kind of transition because they imply an essential, conceptual distinction rather than a sociological one). During the 1950s and 1960s, there had been a series of scientific conferences and technical publications related to cybernetics, self-organizing systems, and bionics, but by the late 1960s, these had ceased entirely. It appears that government funding for the cybernetics of natural and artificial systems dried up, causing scientists and engineers to leave the field. The center of gravity of the field then shifted to the cybernetics of human systems.

12. In part, the loss of funding can be attributed to the Mansfield Amendment (Umpleby 2003). But perhaps more importantly, by the late 1960s, proponents of computational approaches to artificial intelligence had also achieved control over the major sources of American defense funding (ARPA, ONR) and did not hesitate to defund competing research programs. Funding for research in cybernetics, neural networks, trainable machines, bionics, and self-organizing systems all abruptly came to a halt. Many people attribute the cutoff of funding for research on neural networks directly to the influence of Marvin Minsky on funding decisions (Dreyfus & Dreyfus 1988; Boden 2006). In conversations with his colleagues and students, I have learned that von Foerster expressed similar beliefs about why funding for his Biological Computation Laboratory had dried up. More light needs to be shed on this history.

13. There is also the question of whether the dichotomy divides cybernetics in such a way that undermines the maintenance of diversity (perspectival variety) within the cybernetics movement. Projection of realist and reductionist (purportedly first-order) beliefs onto engineers working on artificial systems or scientists studying natural systems has the effect of discouraging their participation in the cybernetics movement. It is probably unhealthy for the movement to nourish an identity politics of who is second-order or whether a given person's perspective is sufficiently second-order (i.e., "politically correct"). As a neuroscientist and theoretician, I constantly wonder about myself and exactly where I stand vis-à-vis the tenets of second-order cybernetics.

14. We need an inclusive big tent rather than a divisive faction fight. I think the cybernetics movement will be enriched if it brings in participants from *all* fields that deal in some significant way with purposive systems (Table 1, bottom row), i.e., those systems that have internal goals that they pursue (Ackoff & Emery 1972). These include fields of endeavor that deal with the broad range of purposive artificial, natural, and social systems: engineering, natural sciences, neural & psychological sciences, social sciences, therapy, management & policy sciences, the arts, and movements for social change.

15. There will be those who deal with how such systems are organized so as to effectively pursue their goals (first-order), and others who deal with how such self-directed systems interact with other such systems (second-order) to cooperate, compete, and converse. Somehow we will all get along and learn from each other.

Open Peer Commentary: **Second-Order Cybernetics Needs a Unifying Methodology**

Thomas R. Flanagan

Introduction

1. In his target article, Stuart Umpleby's current review of the evolution of the epistemology of second-order cybernetics draws our attention first to a neurophysiological (neuropsychological; cognitive neuroscience) consideration of the sensory system. Sensory-input, the first step in sense-making, is both organically selective and autonomously filtered. Meaning, which is the substrate of cognition, is individually derived from direct and vicarious experiences via the senses. Our shared understandings are meanings that have become socialized through communication. And communication again involves selective and filtered sensory channels. Sensing and sense-making is, Umpleby asserts, the foundation of second order science … and for all other sciences too. The specific relevance of sense-making to the practice of second-order cybernetics is asserted to be based upon a (r)evolution in the construction of social meaning and a concurrent enhancement of action taken within living systems.

2. The revolution is stated as an altered perspective. The cybernetic scientist is immersed within the system with the inference that being positioned within a system will alter the experience of the experimental observations that the scientist gathers. Implicit in this expectation is the belief that the scientist's way of seeing and interpreting observations into meaning changes as a function of the altered observational perspective. This theme warrants considerable exploration given that sense-making (and higher meaning-making) have autonomous features forged by formative experiences beyond voluntary control as well as reflective features that are more obviously and directly under voluntary control. The point that I am exploring here relates to autonomous features of a researcher's sense-making capacity that may be resistant to un-learning and re-learning without specific methodological support.

Expanding the research arena

3. A fundamental feature of an observer's immersion within a sapient living system relates to the exchange of emerging understandings with and among actors in that system. When sense-making is a collective, subjective process (rather than solitary, objective work), the product of the sense-making is co-constructed. The second-order cybernetic researcher cannot be viewed simply as a cultural anthropologist artfully hidden behind a one-way mirror positioned in the center of an interactive sapient system; the cybernetic researcher is interacting directly and is being directly influenced by interactions. The researcher also cannot afford simply to be an arbitrator for collisions among differing theories of how things work in the system under study. Interdisciplinary perspective is not a matter of theoretical reconciliations, but rather it is a co-construction of new theory through the reconfiguration of meaning drawn from joint consideration of primary observations. This recombination can be considered as homologous to the synthesis of new chemical enti-

ties through reactions that facilitate recombination of elements into new coherent wholes. This letting go and rebuilding cycle can be particularly painful for researchers who cling strongly to favored theories.

4. Input that is intentionally drawn from interactive recombination of ideas from a community of actors changes expectations of what constitutes a scientific finding. The finding is a complex function of:

a. the observational and communication dynamics within the system under study,
b. the conjoint sense-making methodology selected for use by the researcher and fellow actors,
c. the focus and boundary conditions of the inquiry specified by the researcher,
d. the adequacy of the reporting narrative (see next section).

5. Umpleby reports that second-order cybernetics research was used in the 1950s to engage tactical response from within a community (i.e., the Institute of Cultural Affairs, ICA; §34). The wisdom behind this use was based in the belief that those closest to problems would have the best insights into how problems could be addressed.

6. Indeed, input from the public could be gathered to contribute to the co-construction of a model of how things could work. A fine distinction should be drawn about the role of the researchers who are practicing second-order cybernetics in the community: are they contributing to the blended emergence of new scientific meanings related to how to understand and then solve tactical problems (in the sense of second-order cybernetics), or are they more modestly manipulating and observing the natural evolution of thinking that occurs in a sapient system once that system has engaged in collective action (in the sense of expert analysis). Without reference to specific methodological interventions, it is difficult to extract the extent to which researcher exchanges have been critically catalytic for new ways of thinking or to which researcher exchanges have been sampling ongoing innovative action. Methods as modest as hosting discussions represent an intervention within which exchanges among citizens (to the exclusion of exchanges with researchers) might catalyse inclusive citizen sense-making and design of new response tactics without impacting the science. Had citizens been involved in co-designing an intervention program that included when, where, and how to host specific types of exchanges, researchers and citizens might more convincingly demonstrate their inclusion in second-order cybernetic work. This level of inclusion would represent co-design of "collective choice rules" (after Ostrom 1990). If the citizens were to be involved in specifying how the inquiry into the system had been designed (e.g., the questions that were being asked and the venues that were to be used in citizen-researcher co-engagement), they would have been participating at the "constitutional level" (again after Ostrom 1990).

7. It can be argued that only when citizens do participate at the constitutional level will they provide input into the "theories" upon which the second-order cybernetic interventions would operate. This level of participation is not an easy starting point – which will be obvious to any practitioners with experience in the arena. Work at this level is critical because activity at this level is the most direct way that experiences from the arena feed back into the corpus of theory (see the domain of science model, below). It is my belief that citizens will only contribute to the corpus of the theory of practice when they are supported with methodologies that provide compelling demonstrations of the coherence of their collective sense-making activity.

Developing a sense-making methodology for second-order cybernetics

8. Research method matters. To amass a coherent body of findings for comparisons among and across sapient systems, the scientific community supporting the research (as well as supporting the evolution of the science behind the research) will need to minimize the ambiguity of the findings. The dynamics within the systems themselves will be uncontrolled variables, while the word-use and the reporting style of the resulting narratives must be matched to the needs of specific audiences. Improvement then depends upon refinements to the precision and boundary conditions of the inquiry and to the sense-making methodology. A focus on codification of the topic and boundary of inquiry along with a concurrent focus on a codified sense-making methodology are needed to reduce ambiguity that will impede the evolution of the field.

9. Much as Umpleby in his article focused on fundamental limitations of neurophysiological processes to identify the foundation of second-order cybernetics, establishing the foundation for a dialogic sense-making methodology benefits from first considering fundamental limitations of the sense-making process.

10. Collective sense-making is a cognitive communicative task with expected and unexpected costs and expected and unexpected benefits. As such, the sense-making process can be modeled as an unspecified economic system (Charter & Loewenstein 2016). In Nicholas Charter and George Loewenstein's model, natural language is the fundamental means of negotiated exchanges. Natural language as a medium of exchange is problematic due to its idiosyncratic evolution in all individuals (as discussed above). The embodied cognition theory posits that each lexicon of natural language results from an accumulation of combinations of early experiences captured into locally used words (Lakoff 2012). In this view, words are compounded metaphors for experienced meaning, and each coding of a complex, multidimensional cognition into a linear flow of words constitutes an idiosyncratic abstraction that unavoidably leaves some meaning behind. Languaging addresses the matching of meaning with word use. Iterative explanation allows initially succinct verbal expressions to be augmented with some of the meaning that was stranded during a prior expression. Iterative explanation is only possible, however, when the sense-making methodology preserves both the authenticity of an original author's statement and the real-time linkage of that specific author to that specific statement. This linkage assures that the augmented meaning will remain coherent with the original cognitive thought. Second-order cybernetic methods will need to evolve to support and to improve the efficiency (reduce the cost) of group-level sense-making.

11. A collective sense-making methodology for second-order cybernetics must include provisions for languaging because people use language that is uniquely coded for expressing only certain parts of their immediate needs (see Christakis & Bausch 2006). The meanings behind statements need to be decoded and clarified within a consensual linguistic domain so that parties engaged in collective sense-making can accurately share understandings. The language used within the sense-making process evolves when it is discovered to carry additional meaning. The meaning behind linguistic labels for complex thoughts need to be captured in the language of the group and presented as an evolving reference for real-time support of the parties engaged in collective sense-making. Languaging among groups is difficult, and is most difficult when applied within highly diversified groups, as prescribed through Ashby's law of requisite diversity. A methodology for languaging will be critical in enabling the expansion of second-order cybernetics.

12. The essence of sense-making is the process of pulling disparate ideas into a coherent assembly. Ideas are connected by virtue of the paths through which they exchange materials in classic system dynamics modeling and by linkages through which they exert influence in one of the multiple forms of interpretive structural modeling (see Christakis & Bausch 2006; Warfield & Cardenas 1994). These two methodologies are complementary examples of codified means of making connections between ideas. Connections could be made using any link between any pair of nodes within a systems map. The coherence of the overall structure will depend upon the ease with which the structure is read.

13. Different audiences can be expected to have different skills or preferences for interpreting system models. Participants in the community under study need to share mental models to make sense of their situation so that they can design interventions with a cybernetic perspective. Interpreting and internalizing a systems model must be balanced with benefits from reduced executive and management costs in

a. mobilizing action, and
b. monitoring and coordinating goal-directed activity.

A model building or problem structuring methodology for second-order cybernetics should include an agile method for building and updating models at levels of detail appropriate for community use.

14. Whether we are based in a scientific community or the civic community at large, we socialize ourselves through the narratives that we share. When working with complex sapient systems, narratives will spontaneously emerge, and carefully crafted narratives are subject to the intended or unintended influences of those who retell the narrative. When a narrative is based upon a systems model, the narrative can be anchored to a learning artifact that was constructed through a conjoint sense-making process from within the community (Flanagan 2008). The artifact has legitimacy to the extent that its designers are recognized as authentic voices in the community, and it has durability to the extent that it is current (i.e., can be readily updated as new information is discovered or as prior ideas become irrelevant). The methodology for crafting narrative with explicit reference to learning artifacts from sense-making work will enhance and sustain social impact mediated through second-order cybernetic interventions.

Advancing the science of second-order cybernetics

15. Learning within a complex arena is frequently reported as a case study narrative. This mode for contributing data into a reservoir of comparable scientific information could be improved if case study methodology was codified. Codified case studies would contain a consistent set of elements, which would facilitate inter-case comparisons. Codified cases could evolve to include self-assessments by authors on the extent to which specific elements of the case were observable, and suggestions for improving observability in future cases. Without agreement on case study design within the scientific community, each case study is a narrative (i.e., an incomplete collection of all potential observations of a system coherently expressed as a story). Compelling narratives are artfully crafted with features that appear immutable because they are embedded in networks of dependent findings. An easily read and easily compared graphic alternative to (or addition to) a thick narrative would render relations among salient findings more explicit. This will benefit both the scientific community and the sponsors of civic works. Second-order cybernetics reports

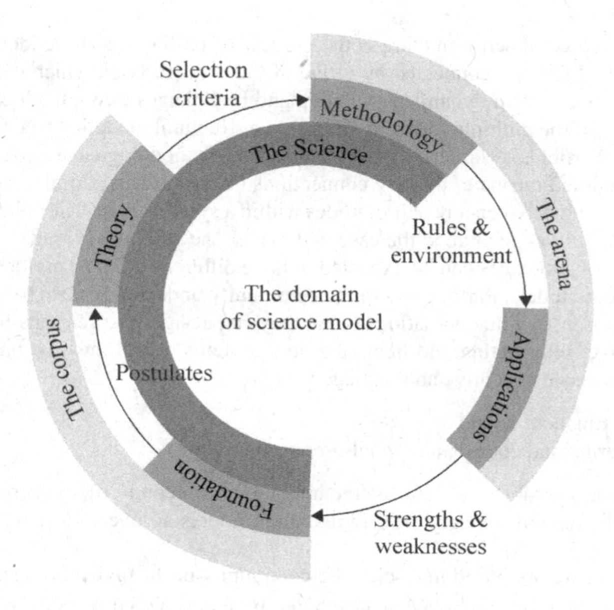

Figure 1. The Warfield domain of science model.

should use network analysis approaches to model the systems that are analyzed. If the inclusion of a systems model were part of a codified reporting scheme, then, over time, that model-construction process could powerfully contribute to learning by facilitating meta-analysis across accumulated cases.

16. John Warfield (1986, 1987) advanced a model wherein philosophy, principles, axioms, tenants, or laws constituted a foundation from which theories of action were devised (see Figure 1). To test the theory, specific methods were adopted or developed with attention to specific conditions of the arena within which the methods would be used. Foundational principles, causal theory, and applied methodology were framed as components of science. In this model, the methodology is the means through which the science connects to the arena of application (Bausch & Flanagan 2013). Within the arena, methods specify agents, venue, and schedule, resulting in applications. Lessons from the impact of specific applications in the arena are envisioned to connect back to parties who maintain an active surveillance of the relevance of the principles that are the foundation of the science. Learning that occurs in the arena thus informs the science. The specific challenge is the imperfect understanding of lessons from the isolated perspective of the arena, and the imperfect link (through the scientist) from the arena back into the corpus of the science.

17. Second-order cybernetics is an approach that seeks to fuse the knowledge of the science with the wisdom of the arena. The second-order cybernetic scientist plays a critical role as a sapient agent of the science in a sapient system of the arena. At this juncture, sense must be made of a situation or opportunity that couples the arena to the science so that les-

sons from the use of the application of prescribed methods in the arena will efficiently and authentically be communicated back into the corpus of the science.

Conclusion

18. Second-order cybernetics is part of a slow cultural (r)evolution toward greater civic inclusion. The transition will require expansion of civic capacity for participating in pluralistic investigations of complex situations on one hand and governance confidence in civic participation on the other hand. Without delegating the authority to design parts of the future to local authorities, the political will that is needed to engage in such a process inclusively will not peacefully emerge (Ostrom 1990; Flanagan 2014). Second-order cybernetics makes a critical contribution to learning experiments that advance this transformation. I strongly share Umpleby's view that "second-order cybernetics as a mode of research from within still has a significant future" (§46); however, realizing this potential requires a robust sense-making methodology (see also Umpleby 2002). The methodology is not only needed to enhance impact within the arena but also to bridge the arena back to the corpus of the science (Figure 1). Structured dialogic design (Christakis & Bausch 2006) is one sense-making methodology that could be used as such a bridge.

Open Peer Commentary: **Viva the Fundamental Revolution! Confessions of a Case Writer**

T. Grandon Gill

Introduction

1. As a teacher and researcher in information systems, I have always been perplexed by a paradox. The authentic discussion case study, a meticulously prepared in-depth description of an actual decision situation faced by a protagonist (key decision-maker), has long been recognized as the premier means of incorporating constructivist learning into the business classroom. The process of writing these cases often involves lengthy visits to an organization, extensive interviews of the protagonist and other stakeholders in the decision, the gathering of archival data from diverse sources – both paper and electronic – and, finally, synthesizing these into a document that is accessible to a broad range of students, who can be expected to come at the decision from diverse perspectives and very different levels of expertise. The paradox is as follows: despite the nature of the activities involved in developing a quality case, few academic researchers in business today are willing to characterize the case writing process as "research." They would be absolutely horrified by the notion of anyone proposing case writing to be a form of "science."

2. Stuart Umpleby, in his target article "Second-Order Cybernetics as a Fundamental Revolution in Science," nicely captures the notion that there need to be alternative approaches to science. For different domains, different approaches will necessarily dominate.

In my commentary, my objective is to make two key points. The first is that the process of developing authentic discussion cases maps nearly perfectly to the underlying philosophy of second-order cybernetics. The second is that most of the environments that are studied in business and management environments could benefit greatly from more – not less – of this type of science. I begin, however, by providing some background on case studies.

Background

3. To understand the current state of business research and the role played by case studies, particularly in the US, it is useful to go back to the 1950s. Up to that point in time, graduate schools of business were generally treated as quasi-disreputable institutions whose main purpose was to provide an academic pathway for wealthy individuals whose intellectual stature or morals were insufficient for serious intellectual endeavors. At Harvard, for example, legend long has it that bloody pitched battles between the business school faculty and their more serious colleagues at the Graduate School of Arts and Sciences were avoided only through the moderating presence of the Charles River between. While the actual situation doubtless differed from this caricature, it is nevertheless fair to assert that the typical products of business research in those days differed considerably from what is considered respectable research today. Specifically, business researchers were much more likely to produce practice focused artifacts – such as case studies and industry reports – than the theory-driven research of today.

4. In the late 1950s, this situation culminated in a couple of foundation reports on business education and research that were highly critical of business schools. Two recommendations of these reports were particularly noteworthy. On the education side, institutions were encouraged to make greater use of the case method, as practiced at Harvard Business School (HBS). That produced an extensive program of funded case method training in the 1960s. On the research side, the reports advocated adopting approaches of social science modeling, particularly applauding the practices of Carnegie Mellon University.

5. One (possibly) unintended consequence of this bifurcation of paths was that case studies became associated with "education," whereas what was considered "research" moved towards the philosophical positions of Popper and, to a lesser extent, Kuhn (as succinctly set forth in Umpleby's Table 1). Moreover, since the 1960s, the research activity has increased in relative importance at many business schools, particularly those that are part of research universities. For this reason, the perceived value of case writing as a vehicle for career advancement continues to decline.

Second-order cybernetics and the case writing process

6. In order to understand the relationship between the case writing process and Umpleby's article, it is useful to describe the case artifact and case writing process briefly. As noted previously, a discussion case study is an in-depth description of a decision situation facing a protagonist. The ideal case study combines a number of characteristics:

- It is *authentic*, meaning the process through which it has been developed involves a systematic assessment of the situation, triangulated through multiple sources. While some disguising, such as changing individual or organization names, is often employed, the best cases do not materially change the setting or decision context.

- It is *open-ended.* Discussion cases present decisions that need to be made. It is also very unusual for a discussion case to have a single "right" answer. In fact, many facilitators – including myself – are indifferent to what decision was actually made by the organization. Just as in the "real world," we believe that there will always be environmental factors or elements of the context that could make a well-considered decision turn out badly or an impulsive decision turn out well. The focus, therefore, is on the rigor of the process used to reach a recommendation. It is this aspect of discussion cases that makes them such beneficial artifacts for constructivist learning.
- It is driven by the *protagonist's perspective*. Because we ask participants in a case discussion to view a situation through the eyes of the decision-maker, the case must present the world as the protagonist sees it. This can have significant implications for the case writing process.

7. In order to develop a case study, the case writer must proceed through a series of steps, described in detail in my book *Informing with the Case Method* (Gill 2011). The skeletal version of a typical discussion case development project might proceed as follows:

- *Site identification.* The case writer perceives a need for a case study on a particular topic or involving a particular setting (e.g., industry, function) and reaches out to organizations that might be an appropriate fit. This might involve either using the author's own contact network or "cold calling."
- *Protagonist and decision identification.* A potential protagonist is identified within the organization and the case writer interviews that individual to determine what decisions are currently being made. The goal here is to identify a decision that relates to the perceived need. Eventually, both the protagonist and the case writer need to agree on the decision that will become the focus of the case.
- *Data gathering.* Through a series of interviews with the protagonist and other individuals influencing or impacted by the decision, the case writer develops a rich view of the context of the decision. In parallel, the case writer acquires a collection of archival information (e.g., memos, white papers, brochures, images, emails) and may also request reports generated from the organization's information systems. In addition, the case writer often gathers information from publically available sources relating to the industry, region, product or decision. These often become supporting exhibits for the case.
- *Writing.* The information that has been gathered is synthesized into a 5–15 page document that is supplemented with exhibits. The case then goes through a series of drafts that must be approved by the protagonist and other individuals within the organization.
- *Publication.* At the end of the case writing process, the pre-release version is often tested in a classroom setting prior to publication. Once all parties are satisfied, the case is released for publication.

8. Arguably, many elements of the process – particularly the use of triangulation in data gathering – follow the same practices that would be employed in developing a research case, as presented in Yin (2009). There are, however, a number of key differences. First, most research cases are built around a body of theory, and seek to test hypotheses. Most discussion cases are driven by decisions that need to be made. Second, research cases

seek to adopt an objective perspective. Discussion cases must necessarily adopt the perspective of the decision maker. Although adopting an objective view of a decision setting in order to contrast that with the protagonist's view can be a useful element of the discussion of the case, if the protagonist does not believe a fact to be true or relevant, that fact is unlikely to influence the decision.

9. The third and most important way in which the development of discussion cases diverges from more traditional case research is in the researcher's engagement with the decision. Best practice for traditional case research involves as little observer interaction as possible. In discussion case development, the opposite more often applies. Motivation for such engagement comes from both sides. One of the principal benefits that a protagonist perceives from participating in the development of a case is access to "free" observations from the case writer. Typically, the case writer is more than happy to provide these. In the process of doing so, the decision can be impacted. From the case writer's side, the goal is to prepare a logically consistent document. In most instances, this will involve questioning the protagonist with respect to the rationale of the possible decision options being considered. In the course of this process, the case writer may introduce new options or may even make the protagonist rethink and reject previously considered options.

10. These differences between research and discussion case development are major contributors to the academic unwillingness to view discussion case development as research. But consider them in terms of Umpleby's Table 1 distinctions between research philosophies. Discussion cases are built around the protagonist's construction of a decision, as opposed to any objective reality. The observer – in this context, the case writer – plays an active role in the process being observed. Any notion that there is a "right" or "true" solution to a problem is discounted; instead, the discussion case writer accepts that many forms of uncertainly will not be reducible and that many facts provided by the protagonist and others are likely to include a considerable (but unknowable) amount of opinion. Indeed, the whole concept of objective knowledge is of questionable relevance in the decision-making process; what drives the protagonist's decision is what she thinks is known. Which, of course, is often influenced by the case writer during the process.

11. The analysis just presented proposes that if second-order cybernetics became broadly accepted as a research philosophy, discussion case writing *would* be considered research. Unfortunately, that does not address the more value-laden question: *should* second-order cybernetics be considered a valid philosophy for business research? To that question we now turn.

The role of second-order cybernetics in business research

12. In considering the role that second-order cybernetics could or should play in the broader world of business research, it is important not to look at the world as a vacuum. A well-established set of research philosophies already exists to guide business researchers. To the extent that these are providing sufficient value, we could easily conclude that introducing a new philosophy – one that violates many of the premises of the existing accepted philosophies – might be a battle not worth fighting.

13. Not surprisingly, the value of business research is a matter of perspective. From the academic researcher's perspective, it has demonstrated extraordinary value. In the US, we typically get paid nearly twice as much as our reference discipline faculty counterparts. A much higher percentage of our doctorates find academic jobs than doctorates from

other fields. Unlike our engineering colleagues, the pressure on us to acquire grants is manageable. Unlike our liberal arts colleagues, we can generally find outside consulting opportunities if we look hard enough. Yes, business research has been good to us; thus, our motivation to adopt new paradigms is decidedly limited.

14. The value of business research looks quite different from the business practitioner's perspective. Without going into details that are presented in Gill (2010) and permitting myself to over-generalize only slightly:

- There are very few examples of our research having had a significant impact on practice;
- Managers frequently do not believe our findings when informed about them;
- In those limited areas where our research has been adopted by practice, the assumptions upon which the research is based often fail;
- We rarely replicate our findings and when we try to, the failure rate is extraordinarily high;
- Most of the research that managers do rely upon comes from sources other than academic researchers, such as consultants.

15. My assessment is that the principal source of these deficiencies is the mismatch between the way we conduct most of our research and the underlying complexity of the domains that we study. In complex domains, effects tend to be the result of complex interactions between large numbers of variables – similar to the interactions between ingredients in a recipe – rather than through the sum of individual main effects, which is what our statistical tools are good at parsing out. Complex domains are also subject to frequent cybernetic cycles, as agents on the landscape react to the environment and the presence of other agents and, in doing so, change the underlying landscape itself. To compound the problem, as agents form self-similar groups as a consequence of the process of adaptation, illusions that suggest significant main effects are present are an expected consequence (Gill 2012: 72).

16. To be effective in highly complex environments, research designs need to be highly localized and need to shed some of the formalisms of the traditional scientific method, such as the hypothesis test (intended to support or refute stable, general propositions). In place of these approaches, the researcher needs to become highly aware of the interactions affecting the local context and must also become expert in the art of observation and the construction of models that reflect the local reality. These are the skills of the discussion case writer. They also parallel the lists Umpleby provides in Tables 2 and 3.

17. To take the matter a step further, as far as highly complex environments are concerned, the entire notion of the independent researcher can be questioned. Where phenomena are highly contextualized, the individual researcher's ability to step in and achieve immediate understanding of what is being observed is limited. Under such conditions, the knowledge accumulated from standing inside is likely to contribute to understanding far more than knowledge of general research principles. In §31, Umpleby introduces Karl Müller's (2016) distinction between exo-mode and endo-mode research. The former describes research conducted by the objective, distant observer, with the latter describing research with high levels of participant involvement. Complex environments are likely to be better served by endo-mode research, driven by the participant-researcher. Moreover, where an external researcher does become involved, the best course would seem to be partnering with a participant from within the system.

18. Umpleby's ICA case study example would seem to present a laudable destination for research conducted in complex environments. Within ICA, the researcher and participant roles became inseparable. Learning was accomplished through local experimentation and sharing of results. Success was not measured through publications, but through observed social change. It should also serve as a cautionary tale for today's researchers, however. It illustrates how effective practitioners can be at solving complex problems… once they have acquired a modest amount of knowledge relating to the design and conduct of research. It is not clear that we have trained our fellow academic researchers to be equally adept at observing, adapting to, and participating in new contexts.

Conclusion

19. Undoubtedly, Umpleby's article has clarified my thinking. I also feel a certain optimism as I look towards the future. Recently, AACSB International – the best known accrediting agency for business schools – has placed a new emphasis on measuring the impact of our research on practice. We have long been aware that the academic journal article that is built around the narrowly defined scientific method is a very inefficient channel for communicating with practice. Moreover, whatever impact is achieved is nearly impossible to detect owing to the distance between the academic and practitioner communities. Achieving and detecting research impact, on the other hand, is among the greatest strengths of the endo-mode research that constitutes second-order cybernetics. Umpleby's ICA example illustrates this in a compelling way; I have seen similar impact, on a smaller scale, in my own case writing experiences. As this new top-down emphasis on impact takes hold, we will perhaps see another major rethinking of our attitudes on what makes for "good" research, paralleling the last dramatic shift, observed in the 1960s.

Author's Response: **Struggling to Define an Identity for Second-Order Cybernetics**

Stuart A. Umpleby

1. I shall discuss the commentaries on my target article using three themes. Two of the commentators emphasized the academic and social context of second-order cybernetics. Two commentators were concerned with how to describe the field both to people outside the field and to those inside. And two commentators were interested in the relationship of second-order cybernetics to management research and practice.

The context of second-order cybernetics

2. **Allenna Leonard** suggests that the lack of acceptance of both first-order and second-order cybernetics is probably related to a "general lack of comprehensive social science research" (§1). By this she means a lack of interest in formulating general theories. She describes a wave of interest in general theories in the 1960s and 1970s (§2). **Leonard**

also points to a "lower sense of common values that … now characterizes our communities and nations" (§1).

3. It is certainly the case that the social sciences lack a common foundation, which the engineering disciplines find in physics and chemistry. And the number of scholars seeking to create a transdisciplinary foundational discipline for the social sciences is limited, as far as I know, to the practitioners of systems science and cybernetics – disciplines that have no established home on US campuses today. Her broader point, that there is a lack of common values in societies, may also be a contributing factor, if people do not assume that establishing a common frame of reference is possible. She is correct that corporations regularly lobby the government to forestall or diminish research that they think will reduce their profits, even if the research would benefit the public (§5). Smoking and climate change are just two examples. And participatory methods are sometimes seen by managers as threatening their prerogatives, even though their use would likely improve the performance of the organization (§4).

4. I also agree with **Leonard** that the incentive systems in universities have in recent years moved strongly toward rewarding research in narrow disciplines rather than transdisciplinary research (§7). Government agencies frequently say that they seek interdisciplinary research proposals, but young faculty members are reluctant to work on research that will not be counted toward promotion. She suggests that the need for more use of systems and cybernetics approaches will become clearer (§9) as the unsustainability of current practices becomes more evident. In the meantime, there is plenty of work to do to develop the new points of view. **Leonard**'s observations help to explain why research that once attracted great interest has diminished in recent decades.

5. **Gastón Becerra** (§10) compares my article to the work of Roland García and Jean Piaget, who claimed that social, cultural, and historical contexts condition the direction that the emergence of knowledge takes. **Becerra** notes these authors describe two types of analysis. Psychogenesis refers to the development of knowledge from childhood to adulthood (§13). Sociogenesis describes the history of science. "Here the epistemic framework refers to a worldview resulting from philosophical, religious and ideological factors that influence the contents of theorizing by enabling or inhibiting our questions" (§10). He quotes García, according to whom…

> a particular cultural pattern enters in a very concrete way into the shaping of science in a particular society at a particular time. It acts as an epistemological obstacle […]. Here the social component is not merely providing directionality to scientific research; it enters deeply into the conceptualization of science. (§10)

These quotations are an excellent description of the different reactions to second-order cybernetics that I have witnessed in Europe and the US. In Europe, second-order cybernetics is welcomed and appreciated. In the US, in the past, it was sometimes attacked, dismissed, and disparaged as if one were denying the most fundamental tenant of a religion.

6. **Becerra** writes that new tools and techniques are needed to integrate multidisciplinary team members (§15). Cybernetics has done this by combining a variety of methods: group facilitation methods such as those of the Institute of Cultural Affairs; causal influence diagrams; process improvement methods; and, when attempting to understand or bring about social change, the use of theories and methods from several disciplines: economics; psychology and anthropology; sociology and political science; and history and

law, organized into a meta-method (Umpleby 2014: 21; Medvedeva & Umpleby 2015). I agree with **Becerra** that additional examples and improvement of methods are needed.

Some debates within the field

7. I understand and share the frustration that **Michael Lissack** and others feel with the word "cybernetics." As a theory of the interaction between ideas and society, the field is unusually large in its range of interests. It essentially encompasses all human knowledge, at least in so far as knowledge is used to achieve human purposes. Currently, not many people have heard of cybernetics, even though most professionals spend several hours each day in "cyberspace." Only parts of the field are known to some academics. And current cybernetics research – in philosophy, social science, and design – is not the part that is most known. Many people in the US interpret the term as a synonym for computer science, which is more confusing than not knowing the field at all. So should we rename cybernetics? **Lissack** suggests a "science of context" (§19). A "science of reflexivity" would come closer to my interests. Unfortunately, no term encompasses all aspects of the field, except the name "cybernetics." The original name has the advantage of providing an address for the literature in the field.

8. Although authors such as Andrew Pickering (2010) and Ronald Kline (2015) said that cybernetics ended in the mid 1970s, the American Society for Cybernetics has held an annual meeting almost every year since it was founded in 1964. Numerous journals have "cybernetics" in their titles, and a gradually increasing number of conferences on the subject are being held each year. Where the field has had difficulty has been in establishing courses and degree programs on university campuses. The institutionalization of cybernetics has been difficult primarily due to its multidisciplinary character. The field touches all the schools on a university campus, but none feel a special responsibility for it.

9. It seems odd that people in business and government frequently call for more communication across disciplines, yet a field that brings together people from many disciplines and has made contributions to many disciplines is not enthusiastically established on campuses. Many books have been written applying ideas from cybernetics to numerous fields – engineering, medicine, management, arts and sciences, design, and philosophy to name a few. There is no shortage of applications. However, in each field of application, other theories and methods already exist. Specialists hesitate to learn new theories and methods when they already have a body of knowledge that is familiar and respected by others within their fields. Not everyone seeks cross-disciplinary understanding.

10. Perhaps the term "cybernetics" should be used simply as an address. Rather than writing about cybernetics or second-order science, we could write about improving the efficiency and effectiveness of science and advocate more attention to how purposeful systems can better achieve their goals.

11. **Peter Cariani** raises the question of whether cybernetics should continue to be a small revolutionary band seeking to make a scientific revolution or a broader field that welcomes interested people from many disciplines. This is a debate that has existed within the American Society for Cybernetics (ASC) since it was founded (Herr et al. 2016). I have always preferred the large-tent strategy. ASC, however, has so far decided to be a small committed band of advocates for a new development in science. What the new direction is changes somewhat with each president, but, since the 1970s, the emphasis has been on some version of second-order cybernetics.

12. **Cariani** points out that my list of constructivist philosophers is incomplete. I appreciate the additional references. He suggests that second-order cybernetics may not be the best banner under which to advance this alternative understanding of science. Perhaps broadening the group involved in the discussion would lead to a new name for the field.

13. **Cariani** provides a succinct but thorough explanation of how practicing scientists, particularly in biology, operate, knowing that the role of the observer is essential. The debate within cybernetics has usually been between engineers and scientists/ philosophers. I agree with **Cariani** that many second-order ideas were present in the earliest writings on cybernetics. But as the field developed, the engineering applications – computers, artificial intelligence, and robotics – received the most attention. The term "second-order cybernetics" was chosen in an attempt to refocus attention on the early interest of Warren McCulloch, Heinz von Foerster, and others in cognition. I agree that there is no need to divide cybernetics, but most people outside the field assume that cybernetics means computers, so some modifier of "cybernetics" has been helpful.

14. Regarding **Cariani**'s comments (§12) on the loss of funding for cybernetics in the 1970s, in conversations that I heard, the subjective approach of cybernetics was considered to be naïve, out-dated, and not worthy of support. This was the view of people both at the University of Illinois and in the Research Applied to National Needs program in the National Science Foundation. The two epistemologies – realism and constructivism – continue to be discussed in the US in two societies, one devoted to systems science and one to cybernetics. There have been efforts over the years to combine the two societies, in part to minimize administrative work. These efforts have failed because of the sharp difference of opinion regarding acceptable epistemologies. In Europe and elsewhere, systems and cybernetics topics can be discussed in one conference, but not yet in the US.

Implications for applications

15. **Thomas Flanagan** emphasizes the need for second-order cybernetics to be connected to practice through methods. He mentions in particular structured dialogic design (SDD), system dynamics modeling, and interpretive structural modeling. I agree that repeated testing is the way to improve and extend a scientific theory. However, I think of second-order cybernetics as being different from the usual theory. Whereas we usually critique scientific practice from the perspective of philosophy of science, cybernetics, because it is based on neurobiology, provides a way to critique present conceptions of knowledge and science. As our conception of knowledge changes, new ways of doing science are considered appropriate.

16. Cybernetics originated in a desire to create a science of the informational domain (i.e., communication and control) in addition to the physical domain (i.e., matter and energy). Since the philosophy of science usually illustrates its examples of how science should be done by citing examples from physics, social researchers, in an effort to create social science, have sought to imitate physics and overlooked cybernetics. The result is a preoccupation in contemporary social science with finding linear causal relations with a high level of statistical significance, often independent of a theory. Meanwhile, schools of management have constructed knowledge in narrow fields – finance, marketing, accounting, and labor relations – without regard to a general theory. And research in universities is evaluated by number of publications in peer reviewed journals, rather than its utility in improving the performance of individuals and organizations. The result is that the literature

on management and social science often lacks clear connections to philosophy, theory, and practice.

17. Second-order cybernetics began with an effort by von Foerster to include the observer in the descriptions created by scientists. This suggestion was resisted, often quite strongly, by those who felt that including the observer would imply self-reference and lead to paradox and inconsistency. Maintaining that the observer could be excluded from research descriptions enabled scientists to claim objectivity and lack of bias. Von Foerster (1971) cited the work of John von Neumann, Gotthard Günther und Lars Löfgren that concluded that self-referential statements do not necessarily lead to inconsistencies, but the desire to avoid political controversy led scientists to adopt the claim that they were doing objective research. My concern is that our current conception of science, that descriptions can somehow be created without observers, is limiting our ability to describe important problems and therefore to devise needed solutions. When people hold different views, creating multiple descriptions is necessary.

18. I believe that cybernetics, as a general theory of communication and control, a general theory of management, and a general theory of an information society, will progress most successfully if it clearly states its connection to philosophy, theory, and practice. So far, second-order cybernetics has tended to emphasize the biology of cognition, i.e., it focused on the individual knower. By connecting second-order cybernetics to the large literature on management methods, particularly group methods, second-order cybernetics can make clear its practical utility and its status as a general theory of management, namely that human groups work more effectively when they use explicit methods for engaging in problem-solving tasks.

19. Some of the best work in management is done by consultants who work most closely with clients and who are not constrained by narrowing what they observe to the issues of interest in an academic field.

20. The Institute of Cultural Affairs (ICA) did not know about second-order cybernetics in the 1950s or even until recent years (§5). The term "second-order cybernetics" was invented in the 1970s. However, the ICA did know about – and disagreed with – Saul Alinsky's approach to community organizing in Chicago (Alinsky 1971), and they had read the work of systems theorists such as Kenneth Boulding and Margaret Mead. My point in the target article was that the work of ICA, both at the local level and in designing and carrying out a global strategy, was compatible with and could be thought of as an illustration of principles from second-order cybernetics.

21. My purpose in describing the work of ICA was to illustrate both a different way of doing research on organizations and a different goal for doing research (§6). Second-order cybernetics originally focused on the role of the observer in doing scientific research and, by extension, the importance of the points of view of different participants in an experiment. Much social science research today focuses on surveying respondents and analyzing data. Facilitating conversations among a group of people who share an interest in an organization is done not to establish a causal relationship among variables but rather to improve the operation of a social system. Facilitated group discussions are a kind of research in that the participants learn what the members of the group are thinking. Decisions are made, acted upon, and after a few months the planning process is repeated. It is an iterative approach – small steps eventually leading to large changes.

22. People at the local level are learning what the group feels is needed and what actions they think will be fruitful. The facilitators are learning what methods seem to work

best and what kinds of problems arise in more than one community. Hence, they can be better prepared when those issues arise in the future. Discussions at the local level also serve as training programs for the participants. After participating in several meetings, a person can move up to leading a small group discussion and later a plenary session. Training sessions for group leaders are also conducted in addition to the planning and organizing meetings. With additional experience, participants are able to lead training programs and later suggest different methods and training programs.

23. Many people in management do not think of management methods as the result of research. The methods are just "how we do things." It is a mistake to think that only academics do research. Process improvement methods constitute a kind of research as well (Umpleby 2002). They are also a way of designing and redesigning an organization.

24. I think of second-order cybernetics not so much as a way of interacting with clients or of developing methods for facilitating group discussions but rather as an argument for why working with people in groups on problems of interest to them is a legitimate form of scientific research (§16). The purpose of science must not only be to publish research results in journals but also to help people achieve their goals by working in harmony with their colleagues and neighbors.

25. Developing management methods (i.e., procedures used in organizations) is often considered to be different from management research, which is thought to involve analyzing data (§17). But individuals and organizations are purposeful systems. Cybernetics is a science of purposeful systems. Developing methods that improve the performance of individuals or organizations is definitely a form of scientific research that should be guided by second order cybernetics. As **Flanagan**'s Figure 1, citing Warfield, illustrates, it is empirical and guided by theory.

26. **Grandon Gill** places my article in the context of two approaches to business research: the case method and classical social science research. He notes that my arguments would support the legitimacy and appropriateness of the case approach relative to social science research (§§2f).

27. I like this framing of the issues. However, I believe there is a third approach. Service-learning has been increasing steadily as a teaching method in recent decades (Umpleby 2011). In service-learning, students work with clients on current problems, rather than examples from a textbook. I have my students do service-learning projects with organizations – a business, a government agency, or a non-governmental organization. I describe the projects as the laboratory part of the course, see http://www.gwu.edu/~rpsol/service-learning. In their project reports, students are expected to describe their activities using as many concepts from the course as possible. In this way, the concepts in the course are connected to their personal experiences and observations.

28. The theory that underlies a cybernetics approach to research is that both individuals and organizations are purposeful systems. The goals of such systems, and how the goals change, is an essential part of understanding and modifying them. Note that the classical approach to science places the observer outside the system being studied. Although this assumption has worked well in the natural sciences, carrying it over to management research conflicts with the phenomenon being studied. Managers are members of the organizations they manage and how the two interact is the subject being investigated. **Gill** notes that not only managers but also the writer of a discussion case study interact frequently with the manager and organization and generally adopt the point of view of the decision-maker (§9).

29. **Gill** wisely and accurately notes that greater involvement by researchers in the organizations they study will require new skills in working within very different cultures (§18). And he expresses his hope that the new emphasis on measuring the impact of research will lead to rethinking our attitudes on what makes for good research (§19).

Conclusion

30. The field of cybernetics, by creating a general theory of communication and control, is a major contribution to contemporary science. It provides a common foundation for the biological and social sciences by pointing out the similarity of circular causal and feedback processes. Reflexive processes, where elements of a social system both observe and participate, have been an important recent addition (Lefebvre 1982, 2006; Soros 1987, 2014). Second-order cybernetics, since the term was introduced in the mid 1970s, has enabled cybernetics to continue to make noteworthy contributions. Work on second-order cybernetics in the past 40 years has led those in the field to believe that it enables a reconceptualization of the scientific enterprise, one that will accelerate the contributions that scientists can make to improving our ability to cope with current and future events.

31. There are many challenges to guide future research. But the primary challenge seems to be explaining cybernetics and second-order cybernetics to the scientific community and to university faculty and administrators. The difficulty of this task provides evidence that cybernetics is a different kind of academic field.

Combined References

Ackoff R. L. & Emery F. E. (1972) On purposeful systems. Aldine-Atherton, Chicago.

Alinsky S. D. (1971) Rules for radicals: A practical primer for realistic radicals. Random House, New York.

Alrøe H. F. & Noe E. (2014) Second-order science of interdisciplinary research: A polyocular framework for wicked problems. Constructivist Foundations 10(1): 65–76. http://constructivist.info/10/1/065

Argyris C. & Schön D. A. (1974) Theory in practice. Increasing professional effectiveness. Jossey-Bass, San Francisco CA.

Ashby W. R. (1952) Design for a brain: The origin of adaptive behaviour. Chapman and Hall, London.

Ashby W. R. (1958) Requisite variety and its implications for the control of complex systems. Cybernetica 1(2): 83–99.

Baecker D. (2013) Beobachter unter sich. Eine Kulturtheorie. Suhrkamp, Frankfurt.

Bateson G. (1972) Steps to an ecology of mind: Collected essays in anthropology, psychiatry, evolution, and epistemology. Chandler Publishing, San Francisco.

Bausch K. & Christakis A. N. (eds.) (2015) With reason and vision: Structured dialogic design. Ongoing Emergence Press, Cincinnati OH.

Bausch K. C. & Flanagan T. R. (2013) A confluence of third-phase science and dialogic design. Systems Research and Behavioral Science 30(4): 414–429.

Becerra G. & Amozurrutia J. A. (2015) Rolando García's "Complex Systems Theory" and its relevance to sociocybernetics. Journal of Sociocybernetics, 13(1), 18–30.

Becerra G. & Castorina J. A. (2015) El condicionamiento del "marco epistémico" en distintos tipos de análisis constructivista [The epistemic framework and the different levels of constructivist

analysis]. In: Ahumada J. V., Venturelli A. N. & Chibeni S. S. (eds.) Filosofía e historia de la ciencia en el Cono Sur. Universidad Nacional de Córdoba, Córdoba: 101–107.
Beer S. (1994) Beyond dispute: The invention of team syntegrity. Wiley, Chichester.
Boden M. A. (2006) Mind as machine: A history of cognitive science. Oxford University Press, Oxford.
Boix-mansilla V. (2006) Interdisciplinary work at the frontier: An empirical examination of expert interdisciplinary epistemologies. Issues in Integrative Studies 31(24): 1–31.
Brün H. (2004) When music resists meaning: The major writings of Herbert Brün. Edited by Arun Chandra. Wesleyan University Press, Middletown CT.
Burbidge J. (ed.) (1988) Approaches that work in rural development: Emerging trends, participatory methods and local initiatives. K. G. Saur, New York.
Burman J. T. (2007) Piaget no "remedy" for Kuhn, but the two should be read together: Comment on Tsou's 'Piaget vs. Kuhn on Scientific Progress.' Theory & Psychology 17(5): 721–732. http://cepa.info/2835
Cariani P. (1993) To evolve an ear: Epistemological implications of Gordon Pask's electrochemical devices. Systems Research 10 (3): 19–33. http://cepa.info/2836
Cariani P. (2009) The homeostat as embodiment of adaptive control. International Journal of General Systems, 38(2): 139–154. http://cepa.info/349
Charter N. & Loewenstein G. (2016) The under-appreciated drive for sense-making. Journal of Economic Behavior & Organization 126B: 137–154.
Christakis A. & Bausch K. C. (2006) Co-laboratories of democracy: How people harness their collective wisdom to create the future. Information Age Publishing, Boston MA.
Clarke B. (2014) Neocybernetics and narrative. University of Minnesota Press, Minneapolis MN.
Cooperrider D. L. & Whitney D. (2005) Appreciative inquiry: A positive revolution in change. Berrett-Koehler, San Francisco CA.
Corona E. & Thomas B. (2010) A new perspective on the early history of the American Society for Cybernetics. Journal of the Washington Academy of Sciences. Summer 2010: 21–34.
de Latil P. (1956) Thinking by machine. Houghton Mifflin, Boston.
Derman E. (2011) Models. Behaving. Badly: Why confusing illusion with reality can lead to disaster, on Wall Street and in life. Free Press, New York.
Dreyfus H. L. & Dreyfus S. E. (1988) Making a mind versus modeling the brain: Artificial intelligence at a branchpoint. Daedalus 117(1): 15–43.
Fals Borda O. (1978) Über das Problem, wie man die Realität erforscht, um sie zu verändern. In: Moser H. & Ornauer H. (eds.) Internationale Aspekte der Aktionsforschung. Kösel, Munich: 78–112.
Flanagan T. R. (2008) Scripting a collaborative narrative: An approach for spanning boundaries. Design Management Review 19(3): 80–86.
Flanagan T. R. (2014) Designing the means for governing the commons. In: Metcalf G. S. (ed.) Social systems and design. Springer, Tokyo: 147–166.
Foerster H. von (1948) Das Gedächtnis. Eine quantenphysikalische Untersuchung. Franz Deuticke, Vienna.
Foerster H. von (1960) On self-organizing systems and their environments. In: Yovits M. C. & Cameron S. (eds.) Self-organizing systems. Pergamon Press, London: 31–50. Reprinted in: Foerster H. von (2003) Understanding understanding: Essays on cybernetics and cognition. Springer, New York: 1–19. http://cepa.info/1593
Foerster H. von (1971) Computing in the semantic domain. Annals of the New York Academy of Sciences 184: 239–241. http://cepa.info/1645
Foerster H. von (ed.) (1974) Cybernetics of cybernetics. Future Systems, Minneapolis MN.
Foerster H. von (1981) Formalización de ciertos aspectos de las estructuras cognitivas. In: Inhelder B., García R. & Voneche J. (eds.) Jean Piaget. Epistemología genética y equilibración. Fundamentos, Madrid: 89–105. Originally published as: Foerster H. von (1976) Objects: Tokens

for (eigen-)behaviours. ASC Cybernetics Forum 8(3–4): 91–96. Republished in: Foerster H. von (2003) Understanding understanding: Essays on cybernetics and cognition. Springer, New York: 261–271.

Foerster H. von (1988) Abbau und Aufbau. In: Simon F. B. (ed.) Lebende Systeme: Wirklichkeitskonstruktionen in der Systemischen Therapie. Springer Verlag, Heidelberg: 19–33.

Foerster H. von (2003a) Cybernetics of cybernetics. In: Foerster H. von, Understanding understanding. Springer, New York: 283–286. Originally published in 1979. http://cepa.info/1707

Foerster H. von (2003b) Cybernetics of epistemology. In: Foerster H. von, Understanding understanding. Springer, New York: 229–246. Originally published in 1974.

Foerster H. von (2003c) Ethics and second-order cybernetics. In: Foerster H. von, Understanding understanding. Springer, New York: 287–304. Originally published in 1992. http://cepa.info/1742

Foerster H. von (2003d) On constructing a reality. In: Foerster H. von, Understanding understanding: Essays on cybernetics and cognition. Springer, New York: 211–228. Originally published in: Preiser F. E. (ed.) (1973) Environmental design research. Volume 2. Dowdon, Hutchinson & Ross, Stroudberg: 35–46. http://cepa.info/1278

Foerster H. von (2003e) What is memory that it may have hindsight and foresight as well? In: Foerster H. von, Understanding understanding. Springer, New York: 101–131. Originally published in: Bogoch S. (ed.) (1969) The future of the brain sciences. Proceedings of a conference held at the New York Academy of Medicine. Plenum Press, New York: 19–64.

Foerster H. von (2014) The beginning of heaven and earth has no name. 7 days with second-order cybernetics. Fordham University Press, New York.

Foerster H. von, Mora P. M. & Amiot L. W. (1960) Doomsday: Friday, 13 November A. D. 2026. Science 132: 1291–1295. http://cepa.info/1596

Foerster H. von & Müller K. H. (2003) Action without utility. An immodest proposal for the cognitive foundation of behavior. Cybernetics & Human Knowing 10(3–4): 27–50. http://cepa.info/2700

García R. (1987) Sociology of science and sociogenesis of knowledge. In: Inhelder B., Caprona D. & Cornu A. (eds.) Piaget today. Taylor & Francis, New York: 127–140.

García R. (1992) The structure of knowledge and the knowledge of structure. In: Beilin H. & Bufall P. (eds.) Piaget's theory: Prospects and possibilities. Routledge, Chicago: 21–38.

García R. (1997) La epistemología genética y la ciencia contemporánea: Homenaje a Jean Piaget en su centenario [Genetic epistemology and contemporary science: Homage to Jean Piaget and his followers]. Gedisa, Barcelona.

García R. (1999) A systemic interpretation of Piaget's theory of knowledge. In: Scholnick E. K., Nelson C., Gerlman S. & Miller P. (eds.) Conceptual development: Piaget's legacy. LEA, Mahwah NJ: 165–184. http://cepa.info/2782

García R. (2000) El conocimiento en construcción: De las formulaciones de Jean Piaget a la teoría de sistemas complejos [Knowledge in construction: From Jean Piaget's writings to the theory of complex systems]. Gedisa, Barcelona.

García R. (2006) Sistemas complejos. Conceptos, método y fundamentación epistemológica de la investigación interdisciplinaria [Complex systems: Concepts, methods and epistemology of interdisciplinary research]. Gedisa, Barcelona.

Gibbons M., Limoges C., Nowotny H., Schwartzman S., Scott P. & Trow M. (1994) The new production of knowledge: The dynamics of science and research in contemporary societies. Sage Publications, London.

Gill T. G. (2010) Informing business: Research and education on a rugged landscape. Informing Science Press, Santa Rosa CA.

Gill T. G. (2011) Informing with the case method. Informing Science Press, Santa Rosa CA.

Gill T. G. (2012) Informing on a rugged landscape: Homophily versus expertise. Informing Science: The International Journal of an Emerging Transdiscipline 15: 49–91.
Glanville R. (1982) Inside every white box there are two black boxes trying to get out. Behavioral Science 27(1): 1–11. http://cepa.info/2365
Glasersfeld E. von (1995) Radical constructivism: A way of knowing and learning. Falmer Press, London.
Glasersfeld E. von (2001) The radical constructivist view of science. Foundations of Science 6(1–3): 31–43. http://cepa.info/1536
Glasersfeld E. von (2005) Radikaler Konstruktivismus – Versuch einer Wissenstheorie. Edited by Albert Müller and Karl H. Müller. Edition echoraum, Vienna.
Glasersfeld E. von (2007) Key works in radical constructivism. Edited by Marie Larochelle. Sense, Rotterdam.
Glasersfeld E. von (2008) Who conceives of society? Constructivist Foundations 3(2): 59–64. http://constructivist.info/3/2/059
Goldman A. & Blanchard T. (2015) Social epistemology. In: Zalta E. N. (ed.) The Stanford encyclopedia of philosophy (Summer 2015 Edition). http://plato.stanford.edu/archives/sum2015/entries/epistemology-social
Gould S. (2010) Triumph and tragedy in Mudville: A Lifelong passion for baseball. W. W. Norton, New York.
Greenwood D. J. & Levin M. (2007) Introduction to action research. Second edition. Sage Publications, Thousand Oaks CA.
Haraway D. J. (1988) Situated knowledges: The science question in feminism and the privilege of partial perspective. Feminist Studies 14: 575–599.
Haraway D. J. (1991) Simians, cyborgs, and women: The reinvention of nature. The Free Association Books, London.
Hawking S. & Mlodinow L. (2010) The grand design. Bantam Books, New York.
Held R. & Hein A. (1963) Movement-produced stimulation in the development of visually guided behaviour. Journal of Comparative and Physiological Psychology 56(5): 872–876.
Herr C. M., Fischer T., Glanville R. & Guddemi P. (eds.) (2016) 50th anniversary retrospective of the ASC. Special issue. Cybernetics & Human Knowing 23(1).
Hollingsworth J. R. & Müller K. H. (2008) Transforming socio-economics with a new epistemology. Socio-Economic Review 3(6): 395–426.
Inhelder B., García R. & Voneche J. (1981) Jean Piaget. Epistemología genética y equilibración. Fundamentos, Madrid. Originally published in French as: Inhelder B., García R. & Voneche J. (1977) Epistémologie génétique et équilibration [Genetic epistemology and equilibration]. Delachaux & Niestlé, Neuchatel.
Juarrero A. (1999) Dynamics in action: Intentional behavior as a complex system. MIT Press, Cambridge MA.
Kahneman D. (2011) Thinking, fast and slow. Farrar, Straus, and Giroux, New York.
Kandel E. R. (2012) The age of insight. The quest to understand the unconscious in art, mind, and brain, from Vienna 1900 to the present. Random House, New York.
Kelly G. (1955) The psychology of personal constructs. 2 Volumes. Norton, New York.
Kline R. R. (2015) The cybernetics moment or why we call our age the information age. The Johns Hopkins University Press, Baltimore.
Knorr-Cetina K. (1989) Spielarten des Konstruktivismus: Einige Notizen und Anmerkungen [Varieties of constructivism: Some notes and comments]. Soziale Welt 40(1/2): 86–96. http://cepa.info/2776
Krippendorff K. & Clemson B. (2016) A merger of two strategic (ir)/reconcilables, 1962–1980. Cybernetics & Human Knowing 23(1): 10–18.
Kuhn T. S. (1962) The structure of scientific revolutions. University of Chicago Press, Chicago.

Kuhn T. S. (1970) The structure of scientific revolutions. Second edition. University of Chicago Press, Chicago.
Lakoff G. (2012) Explaining embodied cognition results. Topics in Cognitive Science 4: 773–785.
Lefebvre V. (1982) Algebra of conscience: A comparative analysis of Western and Soviet ethical systems. D. Reidel, Boston MA.
Lefebvre V. (2001) Algebra of conscience. Revised edition with a second part and a new foreword by Anatol Rapoport. Kluwer, Dordrecht.
Lefebvre V. (2006) Research on bipolarity and reflexivity. With a postscript by Karl Popper. Edwin Mellen Press, Lewiston NY.
Lissack M. (2016) Don't be addicted: The oft overlooked dangers of simplification. She Ji: The Journal of Design, Economics, and Innovation. In press.
Lissack M. & Graber A. (eds.) (2014) Modes of explanation: Affordances for action and prediction. Palgrave Macmillan, Basingstoke UK.
Luhmann N. (1990) The cognitive program of constructivism and a reality that remains unknown. In: Krohn W., Küppers G. & Nowotny H. (eds.) Selforganization: Portrait of a scientific revolution. Springer, Dordrecht: 64–86. http://cepa.info/2712
Luhmann N. (1997) Die Gesellschaft der Gesellschaft. 2 volumes. Suhrkamp, Frankfurt.
Lupton D. (2013) Risk. Second edition. Routledge, London.
Maturana H. R. (1975) The organization of the living: A theory of the living organization. International Journal of Man-Machine Studies 7(3): 313–332. http://cepa.info/547
Maturana H. R. (1978) Biology of language: The epistemology of reality. In: Miller G. & Lenneberg E. (eds.) Psychology and biology of language and thought: Essays in honor of Eric Lenneberg. Academic Press, New York: 27–63. http://cepa.info/549
Maturana H. R. & Varela F. J. (1980) Autopoiesis and cognition: The realization of the living. Kluwer, Dordrecht.
Maturana H. R. & Varela F. J. (1992) The tree of knowledge: The biological roots of human understanding. Revised Edition. Shambhala, Boston.
McCulloch W. (1965) Embodiments of mind. MIT Press, Cambridge MA.
Medvedeva T. A. & Umpleby S. A. (2015) A multi-disciplinary view of social and labor relations: Changes in management in the U. S. and Russia as examples. Cybernetics and Systems 46(8): 681–697.
Miles R. E. Jr. (1978) The origin and meaning of Miles' Law. Public Administration Review 38(5): 399–403.
Munévar G. (1981) Radical epistemology. Hackett, Indianapolis.
Murdoch D. (1987) Niels Bohr's philosophy of physics. Cambridge University Press, Cambridge.
Müller K. H. (2008) Methodologizing radical constructivism: Recipes for RC-Designs in the social sciences. Constructivist Foundations 4(1): 50–61. http://constructivist.info/4/1/050
Müller K. H. (2010) The radical constructivist movement and its network formations. Constructivist Foundations 6(1): 31–39. http://constructivist.info/6/1/031
Müller K. H. (2011) The two epistemologies of Ernst von Glasersfeld. Constructivist Foundations 6(2): 220–226. http://constructivist.info/6/2/220
Müller K. H. (2015) De profundis. Ranulph Glanville's transcendental framework for second-order cybernetics. Cybernetics & Human Knowing 22(2–3): 27–47. http://cepa.info/2699
Müller K. H. (2016) Second-order science. The revolution of scientific structures. Edition echoraum, Vienna. http://cepa.info/1148
Nowotny H., Scott P. & Gibbons M. (2001) Re-thinking science. Knowledge and the public in an age of uncertainty. Polity Press, Cambridge.
Ostrom E. (1990) Governing the commons: The evolution of institutions for collective action. Cambridge University Press, Cambridge UK.
Overton W. F. (1994) Contexts of meaning: The computational and the embodied mind. In: Overton W. F. & Palermo D. S. (eds.) The nature and ontogenesis of meaning. LEA, Hillsdale NJ: 1–18.

Pask G. (1975a) Conversation, cognition and learning. Elsevier, Amsterdam.
Pask G. (1975b) The cybernetics of human learning and performance. Hutchinson, London.
Piaget J. (1970) Genetic epistemology. The Norton Library, New York.
Piaget J. & García R. (1982) Psicogénesis e historia de la ciencia. Siglo XXI, México. English translation: Piaget J. & García R. (1989) Psychogenesis and the history of science: Columbia University Press, New York.
Piaget J. & García R. (1988) Hacia una lógica de significaciones. Centro editor de América Latina, Buenos Aires. Originally published in French as: Piaget J. & García R. (1987) Vers une logique des significations [Towards a logic of significations]. Murionde, Geneva.
Pias C. (2003) Cybernetics: The Macy conferences 1946–1953. Diaphanes, Zurich.
Piattelli-Palmarini M. (1996) Inevitable illusions: How mistakes of reason rule our minds. Wiley, New York.
Pickering A. (2010) The cybernetic brain: Sketches of another future. University of Chicago Press, Chicago.
Reason P. & Bradbury H. (eds.) (2001) The SAGE handbook of action research. Participative inquiry and practice. Sage, London.
Reichardt J. (ed.) (1968) Cybernetic serendipity: The computer and the arts. Praeger, New York.
Rorty R. (1991) Objectivity, relativism, and truth: Philosophical papers. Volume 1. Cambridge University Press, Cambridge.
Rosen R. (1985) Anticipatory systems: Philosophical, mathematical, and methodological foundations. Pergamon Press, Oxford.
Rysiew P. (2016) Epistemic contextualism. In: Zalta E. N. (ed.) The Stanford encyclopedia of philosophy (Summer 2016 Edition). http://plato.stanford.edu/entries/contextualism-epistemology/
Saris W. E. & Gallhofer I. N. (2007) Design, evaluation, and analysis of questionnaires for survey research. Wiley Interscience, Hoboken NJ.
Soros G. (1987) The alchemy of finance: Reading the mind of the market. Simon and Schuster, New York.
Soros G. (2014) Fallibility, reflexivity, and the human uncertainty principle. Journal of Economic Methodology 20(4): 309–329.
Steffe L. P. (2007) Radical constructivism: A scientific research program. Constructivist Foundations 2(2–3): 41–49. http://constructivist.info/2/2-3/041
Suri R. (2008) Growth at the bottom. Delinking poverty-combat from "development." A non-charity solution to global poverty in a short time. WISDOM RISC-Papers, Vienna.
Umpleby S. A. (1974) On making a scientific revolution. In: Foerster H. von (ed.) Cybernetics of cybernetics or the control of control and the communication of communication. The Biological Computer Laboratory, Urbana IL: 130–131. http://cepa.info/2410
Umpleby S. A. (1979) Computer conference on general systems theory: One year's experience. In: Henderson M. M. & MacNaughton M. J. (eds.) Electronic communication: Technology and impacts. Westview Press, Boulder CO: 55–63.
Umpleby S. A. (1987) American and Soviet discussions of the foundations of cybernetics and general system theory. Cybernetics and Systems 18: 177–193.
Umpleby S. A. (1997) Cybernetics of conceptual systems. Cybernetics and Systems 28(8): 635–652. http://cepa.info/1854
Umpleby S. A. (2002) Should knowledge of management be organized as theories or as methods? Janus Head 5(1): 181–195. http://cepa.info/1884
Umpleby S. A. (2003) Heinz von Foerster and the Mansfield Amendment. Cybernetics & Human Knowing 10(3–4): 161–163. http://cepa.info/1876
Umpleby S. A. (2011) Service-learning as a method of instruction. Journal of the Washington Academy of Sciences 97(4): 1–15.

Umpleby S. A. (2014) Second-order science: Logic, strategies, methods. Constructivist Foundations 10(1): 16–23. http://constructivist.info/10/1/016
Umpleby S. A. (2016) Reviving the American Society for Cybernetics. Cybernetics & Human Knowing 23(1): 19–27. http://cepa.info/2837
Umpleby S. A. & Oyler A. (2007) A global strategy for human development: The work of the institute of cultural affairs. Systems Research and Behavioral Science 24: 645–653.
Umpleby S. A. & Thomas K. S. (1983) Applying systems theory to the conduct of systems research. In: Debons A. (ed.) Information science in action: System design. Volume 1. Martinus Nijhoff, The Hague: 381–395.
Vaihinger H. (1924) The philosophy of "as if": A system of the theoretical, practical and religious fictions of mankind. Translated by C. K. Ogden. Routledge, and Kegan Paul, London. German original published in 1911.
Van Fraassen B. C. (1980) The scientific image. Oxford University Press, Oxford.
Varela F. J. & Shear J. (eds.) (1999) The view from within: First-person approaches to the study of consciousness. Imprint Academic, Thorverton.
Varela F. & Singer W. (1987) Neuronal dynamics in the visual cortico-thalamic pathway as revealed through binocular rivalry. Experimental Brain Research 66(1): 10–20.
Vörös S., Froese T. & Riegler A. (2016) Epistemological odyssey: Introduction to special issue on the diversity of enactivism and neurophenomenology. Constructivist Foundations 11(2): 189–203. http://constructivist.info/11/2/189
Warfield J. N. (1986) The domain of science model: Evolution and design. In: Proceedings of the 30th annual meeting of the Society for General Systems Research. Intersystems, Salinas CA: H46-H59.
Warfield J. N. (1987) The domain of science model: Extensions and restrictions. George Mason University white paper, John N. Warfield Collection 35.14. http://hdl.handle.net/1920/3303
Warfield J. N. & Cardenas A. R. (1994) A handbook of interactive management. Iowa State University Press: Ames.
Weick K. (1995) Sensemaking in organizations. Sage, Thousand Oaks CA.
Wiener N. (1948) Cybernetics, or control and communication in the animal and the machine. MIT Press, Cambridge MA.
Wiener N. (1950) The human use of human beings. Houghton Mifflin, Boston MA.
Wittgenstein L. (1967) Philosophical investigations. Translated by G. E. M. Anscombe. Third edition. Basil Blackwell, Oxford. Originally published in 1953.
Yin R. K. (2009) Case study research: Design and methods. 4th edition. Sage, Thousand Oaks CA.

Cybernetics, Reflexivity and Second-Order Science

Louis H. Kauffman

Introduction

1. In this article, I study the questions "What is cybernetics?" and "What is science?" I examine these questions in the form of reflexivity. I shall explain what is meant by a reflexive domain and in the process expand the concept of cybernetics. Cybernetics is concerned with circularity, a circularity that includes the observer (or operator) in the system. The observer is actively in the system. Nevertheless, domains with such circularity remain amenable to rational study. In cybernetics, one attempts to be fully aware of the context of any situation. This means that a reflection on the context, and an inclusion of that awareness of context into the context is always present. What is being evoked is a different sense of the rational than that offered by traditional science. In the face of the circularity of context and observer it is still possible to explore and come to agreements that have every appearance of being scientific facts. I shall describe how it comes about that an arena generated by dynamic interactions and shifting relationships can become a world just like ours, subject to exploration, invention and discovery. One may say that from this point of view, objectivity is an emergent phenomenon! The body of the article is formed in a series of short, numbered paragraphs that I hope will let the reader explore and compare the ideas as they are articulated. The main ideas have already been expressed in this paragraph, but extensions and ways to speak and express these notions occur in the expanded form of the article itself.

2. As cybernetics grew, this notion of circular domains, in which cybernetics occurs, also grew. There are areas of logic and mathematics that have a good correspondence with the sort of circularity that cybernetics needs. I shall speak of reflexive domains.

3. A *reflexive domain* is an abstract description of a conversational domain in which cybernetics can occur. Each participant in the reflexive domain is also an actor who transforms that domain. In full reflexivity, each participant is entirely determined by how he or she acts in the domain, and the domain is entirely determined by its participants. I write $D=[D,D]$ to denote this reflexivity of a domain D (Kauffman 1987, 2001, 2004, 2005, 2009, 2012a, 2012b, 2015; Scott 1971; Varela 1979).[1] The symbol $[D,D]$ denotes all the

1. These references deal with both lambda calculus and the notion of reflexive domains. In this listing I have indicated only those references that explicitly or implicitly deal with reflexive do-

available transformations of the domain D to itself. The equation says that D is identical to the processes that transform it. The point to note is that if a transformation T is defined by the equation $T(D)=[D,D]$, then the transformation T applied to a domain D reveals all the processes that can transform D into itself. *A reflexive domain is itself an eigenform* (see below): $D=T(D)=[D,D]$. Note that a reflexive domain is a context for action, and when we say that the domain D is itself an eigenform, we stand back momentarily from the domain into a larger context that can include it. This means that no domain, even a reflexive domain, is the end of our deliberations. Each domain can be transcended to a new and larger domain. The process is endless and is the source of all our constructions and considerations.

4. An eigenform is a fixed point for a transformation. In the context of this article, an arbitrary transformation is allowed in any mathematical domain with a fixed point either in that domain or in some extension of that domain. This usage is an extension of some technical uses of the term that are special cases of this notion. In the next paragraph, I give a specific example of how an eigenform can arise as the fixed point of a transformation that is simple and syntactical. For our purposes an eigenform is the analog of an eigenvector in analysis or linear algebra, but it is much more general and includes the fixed points that occur in reflexive domains, as will be explained below.

5. I define $T(x)=[x]$. Then I can apply T again and again to an arbitrary x as shown below:

$$\begin{gathered} x \\ [x] \\ [[x]] \\ [[[x]]] \\ [[[[x]]]] \\ [[[[[x]]]]] \\ \ldots \end{gathered}$$

6. If you do this for a long time it begins to look like

$$E=[[[[[[[[[\ldots]]]]]]]]]$$

and this expression has the form of something that does not change if you put one more set of brackets around it. Thus E (above) is an eigenform for the transformation T where

$$T(x)=[x].$$

7. The entity E appears in our perception due to the recursive action of the transformation, and it is a consequence of how one deals with E, seeing it as invariant under T, that makes it into an object for our perception. That E appears as an object is part and parcel of being an eigenform. Thus Heinz von Foerster spoke of eigenforms in the phrase "objects as tokens for eigenbehaviors" (Foerster 2003a). Heinz, in a wonderful turn of perception, turned the mathematical idea on its head. He pointed out that ordinary objects are tokens for eigenbehaviors. *Ordinary objects are invariances of processes performed in the space of our experience.* The "space of our experience" is the context in which we have our experience and it is the experience itself. I make objects by finding fixed points in the recursion of

mains. A reflexive domain is a particular situation where lambda calculus applies.

my interactions. Eigenforms are a touchstone for the relationship of circular and recursive processes and the ground of our apparent worlds of perception. This point of view should be compared with that of Humberto Maturana and Francisco Varela (1987), where they see the construction of conversational domains through the coordination of coordinations of actions of organisms in the course of their evolution.

8. With this in mind, look again at the defining property of the reflexive domain $D, D=[D, D]$, and understand that *the domain itself is an eigenform.*

9. This structural observation has consequences for the applications of cybernetics and applications to the epistemology of second-order science (Kauffman 2015; Müller & Riegler 2014; Umpleby 2014). For if science is to be performed in a reflexive domain, then one must recognize the actions of the persons in the domain. Persons and their actions are not separate. If an action is a scientific theory about the domain, then this theory becomes a (new) transformation of the domain. Theory inevitably affects the ground that it studies. Furthermore, the fact that an entire domain can be seen as an eigenform suggests that one can be an observer of that domain in a wider view of the landscape. Thus physics can be seen as a reflexive domain and one can take a meta-scientific view, allowing physics itself to be one of the objects of a larger domain of in which it (physical science) is one of the eigenforms.

10. How then, do physical and natural science manage to obtain their apparently objective results? The answer lies in circularity and eigenform. Traditional science searches only for those actions that are independent of the observers (persons) involved. This means that traditional science asks to work within a particular subset of available eigenforms. Once this is realized, one can begin to see how to widen scientific operations.

11. Exploration of this theme will proceed in other papers, but the point I want to make here is that the initial place to begin cybernetics and second-order science is in the recognition of circularity and reflexivity.

12. In this article, I shall show that there is no definition of cybernetics that is not a circular definition. This result should be a cause for celebration, for cybernetics is a study of circularity and the fact that it requires a circular definition shows that cybernetics is fundamentally not separate from circularity.

13. The next section discusses definitions and the structure of circular definitions followed by a section showing that basic notions of mathematics would require circular definitions but that mathematics, allowing as it does undefined terms, can give the appearance of avoiding circularity. I continue with a section providing the proof that cybernetics, if it has definitions, must have circular definitions.

14. The subsequent section is a discussion that leads outward to the structure of science and the meaning of eigenform and reflexive domains. I show that our concept of reflexive domain is itself an eigenform and that the central, circular concept of eigenform underlies and overlies all notions of reflexivity. In the Conclusion, I discuss science and second-order science directly. It is my thesis that all science, all attempts to find knowledge, are faithfully modeled by the search for eigenforms in a reflexive domain. A conventionally successful science will find such forms and point to them as the objective results of that science. But a wider look at the situation will reveal that the larger landscape of the reflexive domain has been significantly influenced by these theories. The world has changed as a result of the scientific activity. There is no inviolate ground. The path is being constructed in the act of making it. And yet there are beautiful and important eigenforms to

be found. The quest of science becomes larger and more romantic and more intentional in the embracing of the second-order. I imagine worlds and bring them into being.

Distinction and circularity

15. How do definitions come about?

16. A definition of a term, as it is usually taken, defines that term in terms of other terms. For example, I may say that "a hammer is a device for pounding nails into wood" or "a straight line segment is the shortest distance been two points in a plane space." Definitions in this form are to be found in the dictionary. Some words are genuinely difficult to define. Consider what happens when I look up the world "distinction" in a dictionary. I find that the word distinction is defined as a marking off, an indication of a difference, a discrimination or a distinguishing quality or characteristic.

17. If you already know the meaning of distinction, the dictionary entry can be useful in pinning down some subtleties of usage of the word. But if you did not know the meaning of the word, then you would only find synonyms for distinction in the definition of distinction.

18. The dictionary has engaged in circularity in defining distinction.

19. According to the dictionary:

1. A distinction is a marking off of a difference.
2. A difference is the making of a distinction.

20. This is a circular definition. Each term is defined in terms of the other term. I have been taught that circular definitions are wrong and should be avoided. And yet our dictionaries routinely use circular definitions for the most fundamental terms and concepts such as distinction. I say that the circularity of definitions of distinction is a cause for celebration. For it cannot be otherwise.

21. What is a definition? In order for a definition to be effective it must make a distinction. It must carve a niche in the world of our words. In order for a definition to tell us without circularity what is a distinction, that definition must distinguish distinguishing. This cannot be accomplished without circularity.

Theorem 1. There is no definition of distinction that is not circular.

Proof. If there were a definition of distinction, this would be a distinction that characterizes distinction. Thus the definition would be circular, using the concept of distinction to define distinction. Hence there is no definition of distinction that is not circular. *QED*

22. This theorem does not tell me that I cannot understand distinction. It tells me that such understanding is necessarily a matter of experience. Make a distinction. Draw a circle in a tide-flattened stretch of sand. Bake a cake. Prove a theorem. Sing a song. Attempt to understand understanding. Live a life. Distinction transcends closed worlds of words, and moves into the worlds of feeling and action.

Mathematics

23. Mathematics is based on the idea of a distinction and mathematics is usually regarded as a gold standard for precision. No one said that distinctions could not have precision. I said that they must have circularity, but circularity does not preclude precision.

24. Take numbers, positive integer numbers such as 1, 2, 3. We all seem to agree how to treat them and work with them and they are precise. If I say that I have one apple, you know what I mean. But I may ask what is the number 2? This is a question at a different level.

25. The number 2 is represented by the set TWO = { { }, { { } } } whose members are the two sets { } and { { } }. The first set is empty and the second set has as its member the empty set. Thus these two sets are distinct, and so TWO is a set with two distinct members. TWO represents the number 2.

26. I produced the set TWO and checked that it had two distinct members. So I already knew what two was and used it in the definition of TWO. But TWO is not two! TWO is a set and it is like a ruler. TWO is *the mathematician's standard for the number two.* If you ask a mathematician "Are there two sheep in that field?", he pulls out the set TWO from his waistcoat and compares it with the sheep in the field. He then says to you "I have attempted to make a one-to-one correspondence between the sheep in the field and my set TWO. I discover that there are more sheep in the field than there are members of TWO. Therefore I tell you that there are more than two sheep in that field."

27. I do not need a definition for 2. If someone wants to know if there are 2 things about, I just take out my ruler and check. But could I give a definition of 2?

28. Bertrand Russell gave a definition of number. He writes "[...] a number of a class [is] the class of all classes similar to the given class" (Russell 1938: 115).

29. Thus Russell would say that 2 is the collection of all (possible) pairs. Russell's definition is a deep insight, but it is very wide. How am I to know about all pairs? Let us keep thinking about Russell's idea but leave it aside for now.

30. What if I said that 2 is the set { { }, { { } } }? That will not do. I could call this set TWO, but I cannot call it 2 because *2 is the concept of a pairing* and the set is just a set. So I use the set as a definite example of a pairing, and then I say that something has 2 elements if they can be matched with this set.

31. Now there is an even deeper problem. In defining 2 via this set and the idea of matching, I use the idea of two for the matching. Two things are matched if you put them in correspondence with one another, and so I use the concept of 2 in defining the concept of 2. Matching and pairing mean the same. As I see these features of twoness, I get a precise idea about the concept 2.

32. Precision of concept is something that one can explore, even though there is no way to avoid circularity of definition.

33. There are other issues about number. I have the symbols 1, 2, 3, ... and rules for working with them such as $3+1=4$ and that $9+1=10$. I can specify the rules without circularity and so learn to use the numbers. It takes discussion, teaching and learning to do arithmetic. Later, once one has learned to do arithmetic, it is possible to make definitions and theories and worry about circularity. The worry about circularity has led mathematicians to the remarkable notion of *undefined terms*. In every mathematical theory there are terms that are simply not defined. In this way these terms avoid circular definition.

34. Let us go back to the standard for 2. This is built from sets, and sets are, in their own theory, undefined, but have special operations for building them. The simplest set is the empty set, usually written as { }. This set is distinguished by the fact that it has no members, and is represented as an empty container. As a container, it is a distinction with no contents and no markings other than its boundary. I cannot define a distinction, but I use distinctions all the time and mathematics is built from them. When I go into the realm of distinctions I go into the realm of my own creations and into the realm where I can make the imaginary act that distinguishes myself from what I do. I am a reflexive domain.

The definition of cybernetics and the cybernetics of definition

35. I assert that cybernetics *cannot be defined without circularity*. Recall the theorem of the previous section:

Theorem 1. There is no definition of the concept of distinction that is not circular.

36. I shall use this theorem to prove the next theorem. The point is that cybernetics is so deeply involved in the concept of distinction that it can no more be defined than can distinction be defined without circularity.

Theorem 2. There is no definition of cybernetics that is not circular.

Proof. Cybernetics includes within its purvey the concept of distinction since it studies circular processes and all distinctions are circular processes involving the production of an observer in relation to an observing system and the production of an observing system in relation to an observer. Therefore a definition of cybernetics would entail a definition of distinction. But by Theorem 1 (see above) there can be no definition of distinction that is not circular and therefore there can be no definition of cybernetics that is not circular. *QED*

37. In proving this theorem, I have assumed that cybernetics deals with the fundamentals of circular processes. This is in accord with its history and with its use. Contrast this situation with mathematics. Mathematics studies distinctions, but I do not insist that distinctions themselves are the subject matter of mathematics. Nevertheless, if pressed, I would argue that mathematics is also indefinable!

38. I conclude that a definition of cybernetics could be given in the following form:

Cybernetics is the study of systems and processes that interact with themselves and produce themselves from themselves. This includes cybernetics itself as a system or process that interacts with itself and produces itself from itself.

39. While the reader may object to this definition on grounds that it is not broad enough for a definition of cybernetics, at least the definition is circular.

Cybernetics and reflexive domains

40. I should note that once cybernetics is defined in terms of itself, it becomes what is commonly called "second-order cybernetics." Perhaps there never was a first-order cybernetics but this is a useful distinction. When I distance myself from an object of study,

but nevertheless examine its circularities, it may be natural to call such a study cybernetics of the first-order. Here I am concerned with the second-order, but I will just call that cybernetics.

41. There is a way to do mathematics that is explicitly circular, called lambda calculus (Barendregt 1985; Kauffman & Buliga 2014; Kauffman 2015; Scott 1971). It is called "Laws of Form" (Spencer-Brown 1969) in its full simplicity.

42. First I want to discuss Laws of Form and then pass over to more complex versions.

43. In George Spencer-Brown's book *Laws of Form* (Spencer-Brown 1969), a very simple mathematical system is constructed on the basis of a single sign, the mark, designated by a circle or a box or a right angle bracket. I shall not develop the formalism here. But that sign, in our eyes, makes a distinction in plane in which it is drawn. And the interpretation of Spencer-Brown's calculus has us understand that the sign refers to the distinction that the sign makes (we make that distinction when we are identified with the sign). Thus the sign of distinction in the calculus of Spencer-Brown is self-referential. That is, expressions in the calculus refer to themselves and to other expressions in the calculus.

44. While this calculus is circular and self-referential, it is consistent.

45. By a series of departures and constructions, the calculus of indications of Spencer-Brown can be seen to be a foundation for the construction of all of mathematics, and it gives direct insight into the structure of language and communication. The external observer is intimately involved in this calculus, for it is through that observer that the distinctions take place. Thus the circularity of the calculus engulfs the observer. One is in a cybernetic domain.

46. A more complex form of circularity in mathematics occurs when an expression *explicitly* refers to itself. For example, I could write the self-referring equation $x = 1 + 1/x$ and have an expression x that is defined in terms of itself. More general fixed points indicate other circularities. For example

"This sentence has thirty-three letters."

is a sentence that refers to itself and tells a truth about itself. I can formalize circularity in terms of fixed points. Heinz von Foerster (2003b) was fond of this relationship of circularity and fixed points. He saw that a fixed point, taken in the cognitive world, can connote an object in the perception of an observer. In his terms, the object is a token for an eigenbehavior, the iteration of a transformation whose fixed point is the object. The observer embodies a transformation T that leaves the object fixed in the sense that one has the fixed point equation $T(\text{Object}) = \text{Object}$. This means that no extra distinction is introduced by the application of the transformation T to the Object. Von Foerster called the object so fixed an *eigenform,* and the transformation an *eigenbehavior*.

47. It should be understood that the object, the eigenform, may come into existence through the action of the transformation. *This can happen by an interlock of meanings without any iteration.* By an interlock I mean a direct reference of the sentence to itself, or a mutual reference of two structures. For example, the US dollar bill has a statement written upon it that says "This bill is legal tender." That statement interlocks with the dollar bill upon which it is printed. When we shake hands, each hand grasps the other hand. The hands interlock. In the next example, we have the sentence "I am the one who says I." The sentence prior to the one you are now reading shakes hands with its reader and makes an opportunity for the reader to identify himself or herself.

48. I have the following operator:

$$T(x) = \text{the one who says } x$$

The eigenform for this operator is "I".

"I am the one who says I"
"I = the one who says I"
$$\mathrm{I} = T(\mathrm{I})$$

49. This explication of "I" as an eigenform places its fundamental circularity starkly before us. I is an eigenform of the transformation, but the identity of the I can vary. I can be the one who says I and so can you. The eigenform I is fixed by a linguistic transformation but is handed back and forth in the conversation among a community of I's. This is exactly the condition that Spencer-Brown (1969: 76) describes for his mark when he says "We see now that the first distinction, the mark and the observer are not only interchangeable, but in the form, identical."

50. A community of observer/participators forms a *reflexive domain D*. By this term I mean that each person in the domain is also an actor in that domain. Each one acts upon the others and each can be acted upon by the others and by himself. This is the fundamental condition for the cybernetics of social and economic behavior, and it is also the condition for all scientific behavior.

51. There is a key relationship between reflexive domains and eigenforms. I state this relationship as a theorem.

Church-Curry fixed point theorem. Let D be a reflexive domain and let T be any element of the domain D. Then there is an element X in D such that X is a fixed point for T. That is, $TX = X$.

Proof. Define the following operation on elements of D:

$$Gx = T(xx).$$

That is, G operates on x by letting T operate on the result of x operating on itself.
Since D is a reflexive domain, I can assume that the operation G is itself an element of the domain. Now examine the result of G operating on itself. G operating on itself is the result of T operating on G operating on itself. That is, $GG = T(GG)$. Thus GG is a fixed point for T. *QED*

52. It is quite possible that the reader will find this construction, simple as it is, quite baffling! It is best to discuss it from a number of angles. The key point for a reflexive domain is that participants in the domain can act on themselves. One must take this property to heart. Regard yourself as such a participant and understand that you interact not only with others (for example in a market) but also with yourself. There is a singularity about interacting with oneself that is different from interacting with others. You have to separate yourself from yourself to interact with yourself and this separation is at a conceptual level or an imaginary level, since you are not actually separate from yourself. Think of the G in the theorem as a person. This person acts on another person P by letting the transformation T act on the self-interaction of P. What does it mean to act on the self-interaction of another? It means to go into conversation with that other. Thus G induces T to go into

conversation with P. One can think of G as a coach who gets people to interact with one another. But how does G act on himself?

53. Well G acting on G induces T to be in conversation with G. This means that G's self-interaction is acted on by T. Thus G's self-interaction is indeed the action of T on G's self-interaction. The singularity of self-interaction leads to the fixed point.

54. This singularity of self-interaction is well-known in many puzzles and paradoxes. For example, there is the tale of the village with a barber B who shaves only those who do not shave themselves. This is straightforward just so long as I do not ask "Who shaves the barber?" For if the barber shaves himself then by definition, he cannot shave himself.

55. A deeper relationship with the fixed point theorem is Russell's paradox. For this purpose, let AB denote the relationship "B is a member of A." Thus A acting on B is the act of B becoming a member of A. Then I can define $Rx = \sim(xx)$ where $\sim$ denotes "not." R is a set with the property that *x is a member of R exactly when x is not a member of x*. R is the set of all sets that are not members of themselves. This set R is the famous Russell set. The paradox of the Russell set is that

56. RR is an eigenform for negation. For I have

$$Rx = \sim(xx) \text{ and so } RR = \sim(RR).$$

57. This Russellian eigenform is a problem for those who insist that negation must not have fixed points. Certainly RR is a rogue set from the usual point of view since it apparently can both belong to itself and not belong to itself.

58. This depends upon what is the "usual point of view." Looking from the reflexive domain of persons and selves, I see that RR is very like a person. RR is part of itself, but RR is also the whole of herself and so not a part at all. RR is a member of RR and RR is not a member of RR. In the reflexive domain of persons and selves it is natural enough for RR to be a eigenform of negation.

59. It is an axiom of physical science that one requires results and procedures that can be repeated by others to a sufficient degree of accuracy so that the community of physicists agrees that the results and procedures do not depend on the specific observers (other than that they carry out the actions of the experiment according to the specifications). It should be clear from this well-known description of experimental physical science that an experiment is a transformation E that involves persons P and a result X and I desire an eigenform X so that $E(P,X) = X$ independent of the choice of P (within a community of "competent scientists" P). Why does X appear on the left-hand side of this equation? By this I mean that the experiment is repeatable. You can record X as a result of the experiment and then find that when you perform it again, you get the same result. It is the repeatability that makes a successful experiment into an eigenform.

60. That knowledge should be independent of the observers is related to the repeatability and I will discuss it further below.

61. Physics is defined circularly and it demands certain eigenforms for its successful experiments. These eigenforms are in accord with the description of a world that is objective, a world that is independent of the particular choice of observers. The objective world of physics is not given a priori. The objective world of physics is a construction of the physicists based on the type of results that they deem acceptable as valid physics. Of course it could have happened that physics would fail and there would never be any experiments E of this kind or any results X of this kind. Fortunately for physics, there is a whole world of physical results, and it would appear that there is no end in sight to the progress of physics.

62. I see that one could look at a science that studies eigenforms and reflexive domains and that the participants in such cybernetics are themselves elements/actors in the reflexive domain. One is then led to the question: Which is more fundamental, the circularity of eigenform or the reflexivity of the domains in which these eigenforms occur?

63. It is easier to start with eigenform since I have a sense about how to make an eigenform by iterating a transformation. When one does this iteration one learns curious lessons.

64. One lesson is the astonishing fact that one can make an eigenform just from the form of a transformation as in $X=TTTT\ldots$. Here, X is an eigenform for T since formally $TX=X$. But is this not some sort of magic trick, a sleight of hand? Indeed it is. For I could define

$$T^1(g)=T(g)$$
$$T^2(g)=TT(g)$$
$$T^3(g)=TTT(g)$$
$$\ldots$$
$$T^N(g)=TTTT\ldots T(g)$$

where there are N T's in this composition. Then $T^N(g)$ depends upon g as well as N, for each N. I can imagine $TT\ldots TTg$ with an infinite number of T's in-between the first and the last T, and this would be an eigenform that depended on the choice of g. I would have to imagine an infinity where I can "count down" infinitely as well as counting up.

65. The simplest way to think of this is to imagine that N is a generic large integer so that it is possible to confuse N and $N+1$. Then N acts as though it is infinite, but is actually finite but large.

66. Our usual method produces an infinite limit

$$X=TTTT\ldots \text{ with } TX=X.$$

67. The fixed point does not depend upon the initial conditions. If I take an example such as

$$Tx=1+1/x,$$

then $X=TTT\ldots=1+1/(1+1/(1+\ldots))$ and $X=1+1/X$. I can ask for numbers that satisfy this equation, and there are numerical solutions (the golden ratio, phi=(1+sqrt(5))/2 is a solution to this equation.). This independence of the eigenform on the initial conditions is analogous to an experiment whose results are independent of the particularities of the experimenter.

68. I would like to make a more precise version of this analogy with the possible independence of observation so that one could consider experiments that are both sensitive and insensitive to the conditions of the experimenter. I describe a scientist to be in search of eigenforms that are independent of the action of the observer. Iteration of a transformation can, as explained in the previous paragraph, erase some of the initial conditions that are characteristic of a particular observer. This is only part of the story. To make a complete account of how a given science strives to make its results observer-independent is a project for future workers in this field of second-order science.

69. Eigenforms occur within a reflexive domain, giving rise to recursion, and recursion itself brings forth that reflexive domain. The entire reflexive domain is an eigen-

form. This shows how eigenforms can be dynamic. The entire world of science is itself an eigenform. The world of the stock market is an eigenform. The cybernetics world is an eigenform. These eigenforms are the forms of closure that occur in large interacting systems where the dynamics of individuals leads their identities outward and the individuals become identified with their actions. Just so do words become identified with their actions in Ludwig Wittgenstein (1958: §43): "The meaning of a word is its use [...]." In the same way, Charles Sanders Pierce (Kauffman 2001) had the concept of the individual as a sign for himself. This means that in the dynamics of sign-behavior there arises reference and eventually self-reference and through this self-reference the individual becomes a sign for himself. Word and individuals are both seen as elements of a reflexive domain in the work of Wittgenstein and Pierce. Pierce's concept is directly related to the reflexive domain in which the individual comes to have an identity through the eigenform "I".

70. With this I return to the identification of the mark and the observer in Spencer-Brown (1969), now at the scale of the detailed interactions of individuals with many others.

71. I have the fundamental theorem that every element T of a reflexive domain D has a fixed point X with $T(X)=X$. Eigenforms prevail at the individual as well as the global levels in reflexive domains. The world of reflexive domains and their eigenforms is the world in which cybernetics occurs.

72. How does a reflexive domain come about? In some cases, such as our personhood, one simply lives there and cannot easily give an account of how it happened. I live in the eigenform of my reflexive domain of self-other interaction. It is a mystery how it happens. I do not have a sequential construction for it. On the other hand, consider some given domain D that is not reflexive. By not reflexive I mean that the transformations of D to itself are not in perfect correspondence to elements of D. There are transformations A taking D to D that are not members of D. What should I do? I may include them and form a new domain D^1 But D^1, while larger than D, may still not be reflexive, so I may need to adjoin more elements. The general pattern is $T(D^n)=[D^n, D^n]=D^{(n+1)}$. Then I iterate the construction of enlarging the domain to infinity or to a very large N where D^N and $D^{(N+1)}$ are indistinguishable. At this point I have achieved D^N as an eigenform for T. I have created a reflexive domain. Once I have a reflexive domain, I can avail myself of the Church-Curry fixed point theorem within that domain and all the other amenities of closure that such domains afford. The insight that reflexive domains are themselves eigenforms is directly related to how they can come into existence. The technical details of this limiting process can be seen in the work of Dana Scott (1971).

73. With this point of view, I can think of a reflexive domain as arising in the freedom to assign names to processes, and to allow those processes to become new elements of the original domain of objects (under scrutiny). This sort of freedom is well-known to computer programmers of languages, where an algorithm or procedure is given a name and that name is then at the same level as other original terms in the language. The language expands in time as the programmer makes new "macros" of this sort. In this practical situation one does not have to go to the limit of an infinite tower construction (as described in the previous paragraph) to obtain reflexivity. One only has to recognize that defined processes are subject to being elements of an always-extending language.

74. The construction I made in the Church-Curry fixed point theorem can be accomplished under relatively ordinary circumstances. For example, suppose that I define $Gx=<xx>$ as a *symbolic process*. It is a process that a computer can perform on symbol strings such as x. Then applying G to x means that I duplicate x and place brackets around

the two copies of x. This is an operation that can be performed by a computer. I regard G itself as a string of symbols (with one character). The strange and curious thing is what happens when I apply G to G.

75. With $Gx=<xx>$, I obtain $GG=<GG>$. I obtain a statement of self-reference and, if I take this equality seriously as an action, I obtain the unending recursion (GG is replaced by $<GG>$)

$$GG=<GG>=<<GG>>=<<<GG>>> \\ =<<<<GG>>>>=\ldots.$$

76. All this is compatible with the previous notion that

$$E=<<<<<<<\ldots>>>>>>>$$

is invariant under bracketing: $E=<E>$.

77. In one case, I obtain the eigenform by singular substitution. In the other case, I obtain the eigenform by a limiting process. It should be noted that these examples of strict substitution and transition to infinity are idealized and do not begin the wonderful partial aspects of self-reference that occur in our worlds of language. For example, I may refer to myself without yet knowing what boundaries of the self are intended and indeed with the hope of extending them. I may say "I am a mathematician" or "I am a clarinetist," in each case using the declaration as a performative act that may promote the one who says it into that category. The act of saying can be an essential ingredient in the movement to a state of being. Thus does one say "With this ring I thee wed."

78. It is at this point that the reader will discern that I am in favor of both nouns and verbs. In fact, from our point of view it is indispensible to have both nouns and verbs, and the possibility of changing nouns to verbs and verbs to nouns as the need arises. Due to the long discussions in systems theory about the nature of nouns and verbs, this aspect of our epistemology will be continued in other papers.

79. I shall now see how self-reference can arise in systems that are powerful enough to have reference, and powerful enough to take the names of processes and the names of things and use them in the language on an equal basis (just as one does in human natural languages). Nouns and verbs become interchangeable. Suppose that I have a system S that can make distinctions in its world and give names to the sides of these distinctions.

80. S has a language, and in that language S will write $A \to B$ to mean "A refers to B" or that "A is the name of B." S is endowed with a special process denoted by "#" so that when $A \to B$, S creates a new reference $\#A \to BA$ called the *indicative shift* of the original reference $A \to B$. Thus $\#A$ is the name of "B with a name-tag A attached."

81. I say that $\#A$ is the meta-name of BA. This shifting of names to meta-names is fundamental to the observing system S. The shift enables an object and its name to be superimposed just as I find that persons I know have their name as one of their direct properties for me. It is the shift of naming that allows the system S to achieve self-reference. The system S will create a name for the operation of the shift itself.

82. Let $M \to \#$ denote the name of the shift operation. Note that this is an instance of a process (#) acquiring a name (M) and thereby being associated with a noun. But then this naming of the shift process can itself be shifted, and I find that $\#M \to \#M$ is the result of that shift. In this way, the system S acquires self-reference. If I let $I=\#M$ then I find that I refers to I. If the system could talk, she would say "I am the meta-name of my own meta-naming process!" The structure of this acquisition of self-reference has the same

pattern as the Church-Curry fixed point theorem. Our point in describing it is to show that the self-reference of "I am the one who says I." occurs naturally in a model of a system S that is capable of reference, naming and distinction. Such systems deserve the name of "observing systems."

Conclusion

83. I am now in a position to discuss second-order science in the light of an understanding of reflexive domains. I have discussed how a society of persons can be schematically modeled by a reflexive domain D. The domain contains more than just the human persons. It contains processes that are initiated by them, with each process regarded both as an object in the domain and as a transformation of the domain. The transformative properties of an element can be referred to any other element or person in the domain. I can now look at the practice of science in terms of the landscape of the domain D. I emphasize that the domain D should be seen as "the world" in the sense of Wittgenstein. Here I refer to a world that begins in the context of the early Wittgenstein of the *Tractatus* (Wittgenstein 1922) but continues into the world of the later Wittgenstein of the *Philosophical Investigations* (Wittgenstein 1958). From the point of view of reflexive domains these worlds are not incompatible. What is the case is the present state of action and eigenform. But there is no longer an impenetrable barrier between the descriptions (pictures) of the world and the world itself. Each acts upon the other. Each creates the other. Words acquire their meanings via actions, and distinctions occur in the sharp invariances of certain eigenforms. The world is everything that is the case, and the world evolves according to the theories and actions of the participants in that world. Thus when a scientist in the world proposes a theory, this theory becomes a new element of that world. The theory acts and is acted upon by the world of persons and elements of the world. There is no protection for the world from the effects of such a theory. It is remarkable that physicists have been able to isolate phenomena that do not appear to be affected by the theories of those phenomena.

84. For example, the Maxwell theory of electromagnetism does not appear to affect the behavior of electromagnetic fields. What has been discovered is that the electromagnetic field (as measured by the procedures associated with this physics) is an eigenform of the theory. The physics is defined to be physics when the object of the theory is found to be an eigenform of the theory. But I can take a wider view and examine the effect of the electromagnetic theory in the world at large, not just its effect on the "fields themselves."

85. Since James Clerk Maxwell and the allied discoveries of electrical generators and electromagnetic wave transmission, the entire face of our world has been transformed by the entry of the theory of electromagnetism into that world. This is so striking that it seems hard to believe that it is a common belief that scientific theories are just objective descriptions of the way the world "is." The way the world "is" is an evolving context for interaction and exploration.

86. The key to the apparent objectivity of a science is in its carefully crafted eigenforms that are checked to remain fixed throughout significant periods of time. If one attempts to maintain this sort of objectivity in other forms of social science, the processes can take a much different shape. For example, one can design trading algorithms for the stock market on the basis of seemingly reliable information about the behavior of the market. But these very algorithms, when entered into daily practice in the market itself affect the action

of the market and can even lead to instabilities and behaviors that were nowhere in the original theories. Certainly the same remarks apply to the theories and even the opinions of investors, which when made public can affect the action of the market itself. One may say that this indicates that the market is not a subject for objective science, but this is not necessarily the case. Our definition of objective science is that the actions in the reflexive domain produce relatively stable eigenforms. Thus a new version of stock market economics could arise that searches for such regularities even in the face of the publication of the theories themselves.

87. Other aspects of knowledge certainly have this reflexive pattern. There is a science to learning to play a musical instrument, and it involves principles that do not appear to be changed by the acts of practice. But other aspects such as learning to improvise are clearly a matter of finding a moving eigenform in the course of the action of the play. A simpler example is learning to ride a bicycle. The riding of a bicycle is an eigenform in action and there is no way to find this eigenform except to perform the action. Theories of games that do not submit to exact analysis such as chess or Go evolve in relation to the play of the game. The known theories of chess have seriously affected tournament play, and in turn this tournament play has changed the evolution of the theories.

88. Here is a remarkable story told by the physicist John Archibald Wheeler about a game of twenty questions:

> Then my turn came […] I was locked out an unbelievably long time. On finally being readmitted, I found a smile on everyone's face, a sign of a joke or a plot. I nevertheless started my attempt to find the word. 'Is it an animal?' 'No.' Is it a mineral?' 'Yes.' 'Is it green?' 'No.' 'Is it white?' 'Yes.' These answers came quickly. Then the questions took longer in the answering. All I wanted from my friends was a simple 'yes' or 'no.' Yet the one queried would think and think before responding. Finally I felt I was getting hot on the trail, that the word might be *cloud*. I knew I was allowed only one chance at the final word. I ventured it: 'Is it *cloud*?' 'Yes,' came the reply, and everyone in the room burst out laughing. They explained to me that there had been no word in the room. They had agreed not to agree on a word. Each one questioned could answer as he pleased – with one requirement that he should have a word in mind compatible with his own response and all that had gone before. Otherwise, if I challenged, he lost. This surprise version of Twenty Questions was therefore as difficult for my colleagues as it was for me […] What is the symbolism of the story? The world, we once believed, exists *out there* independent of any act of observation. […] I, entering the room, thought the room contained a definite word. In actuality, the word was developed step by step through the questions I raised […] Had I asked different questions or the same questions in a different order I would have ended up with a different word […] However, the power I had in bringing the particular word *cloud* into being was partial only. A major part of the selection lay in the 'yes' or 'no' replies of the colleagues around the room […] In the game, no word is a word until that word is promoted to reality by the choice of questions asked and answers given. (Davies & Brown 1986: 23f)

Wheeler's allegorical fable was intended to illuminate the conditions of the quantum physicist. In quantum physics, no phenomenon is an actual phenomenon until it is observed and agreed upon by all the physics colleagues. The story just as well illustrates the world of social interaction.

89. My thesis is that all attempts to find stable knowledge of the world are attempts to find theories accompanied by eigenforms in the actual reflexivity of the world into which one is thrown. The world itself is affected by the actions of its participants at all levels. One finds out about the nature of the world by acting upon it. The distinctions one makes change

and create the world. The world makes those possibilities for distinctions available in terms of our actions. Given this point of view, one can ask, as one should of a theory, whether there is empirical evidence for this idea that stable knowledge is equivalent to the production of eigenforms. In this case we have only to look at what we do and see that whenever "something is the case" then there is an orchestration of actions that leaves the something invariant, making that something into an eigenform for those actions. The eigenform thesis is not itself a matter of empirical science. It is a matter of definition, albeit circular definition. Another point of view is that the empirical evidence is all around you. Examine any thing. How does it come to be for you? Investigate the question and you will find that thing is maintained by actions. The action could be as simple as opening your eyes and looking at the cloudy sky. With that action, the cloudy sky comes to be for you. I do not assert that this is the usual scientific explanation of cloudy sky. But if you want to work with such things then it is usually even more transparent. The sharp spectral lines of helium are the result of setting up a very particular experiment that produces them. The experiment, its equipment, the scientists and all that is needed to perform it is the transformation whose eigenform is the spectrum of helium.

90. It is a fruitful beginning to look at present scientific endeavors and to see how they are interrelated and find connections among them, to engage in meta-scientific activity. This can reveal how theories, seemingly objective, actually affect the world through their very being, and how these actions on the world come to affect the theories themselves. In exploring the world, we find regularities. It is possible that these regularities are our own footprint. In the end we shall begin to understand the mystery of the eigenforms that we have created, constructed and found.

Open Peer Commentary:
Remarks from a Continental Philosophy Point of View

Tatjana Schönwälder-Kuntze

1. For me, the great merit of Louis Kauffman's target article is that it brings together – intentionally or not – abstract mathematical and logical formalism with philosophical reflections ranging from everyday-life experience to scientific experience through second-order cybernetics. The article tries to support two interlocked theses within a mathematical framework. The first states that scientific observers always observe themselves, at least in some way. The second says that the observed objects depend on their observers, insofar as they are shaped by the observation (§4). From my point of view, there is nothing to reject. Maybe one could add that the observers are also shaped by the forms and categories with which they observe their objects and subsequently by the knowledge about them – but this is another discussion I do not want to enter into here. Instead, I want to draw attention to some striking similarities between "classical" philosophical approaches, Wittgenstein aside, and the statements in Kauffman's article.

2. Being trained in philosophy of science and formal logic, and being especially familiar with George Spencer Brown's *Laws of Form* (1969), but thinking and doing research in the field of continental philosophy, I want to make two remarks:

a. Kauffman's proposition seems to be an attempt to bring Immanuel Kant's "Copernican turn" into the (self-)reflections and descriptions of science (in its narrow sense);
b. Kaufmann's interpretation of "eigenform" can be seen as Kant's figure of transcendental *conditions of the possibility of the knowledge* of the things we can have experience of.[1]

If (b) obtains, it might help to bridge the – pretended – gap between analytic philosophy, which understands itself as being based on scientific methods, including the conviction that an *independent* "real reality" exists that is *as such* knowable by analysing language, and the "continental" philosophy, which is judged – from the analytic side – as nonscientific because of its "relativism." This judgement derives from its reflexive attitude, called "critical" by Max Horkheimer (1975), Michel Foucault (2007) and others, of taking its own historical dependence and therefore *always* constructed standpoint into consideration.

3. Unfortunately, this commentary is not the place to reconstruct Kant's transcendental philosophy or even his transcendental reflection, which is shown and described in his *Critique of Pure Reason* (1781/1787). Rather, it must be sufficient to explain the Kantian meaning of "transcendental," which is different from its use as "transcendental argument" or "transcendental reason" as, e.g., Ross Harrison (1998: 452) defines it. As pointed out in §2 above, Kant calls the structures ("elements") of our thinking *transcendental,* which make our experience, i.e., primarily the perceptions of things *as different things* in a world, *possible*. These elements are the two "forms of intuition," space and time, and twelve categories called "pure concepts of understanding." After Kant, these categories order spontaneously the manifold impressions that is given by our senses through space and time as phenomena. So how we present the "world" to ourselves while making experience depends on our instruments of observation. Everything we discover, e.g., connections or relations, we discover as such because our minds provide *a priori* a categorical ordering set *arranging* the phenomena this way. This is *in nuce* what Kant refers to, talking about Copernicus: "Hitherto it has been supposed that all our knowledge must [go by] its objects." Instead, Kant proposes to assume "that the objects [have to go by our concepts]" (KrV: B XVI, my translation[2]). Discussing our concept of "causality," Kant is even more explicit when saying that "we are able to extract [...] pure concepts from experience only because we have put them into experience, and so have first brought experience about through them" (KrV: A 196/B 241,[3] Kant 2007: 217). This could be linked to Kauffman's "theory inevitably affects the ground it studies" (§9), but we need to make this more obvious.

---

1. "Ich nenne alle Erkenntnis transzendental, die sich nicht sowohl auf Gegenstände, *sondern mit unserer Erkenntnisart von Gegenständen, insofern diese a priori möglich sein soll,* überhaupt beschäftigt." (Kant KrV B 25, emphasis in the original) – "I call all knowledge *transcendental* which deals not so much with objects as with our manner of knowing objects insofar as this manner is to be possible *a priori*." (Kant 2007: 52, emphasis in the original).
2. "Bisher nahm man an, alle unsere Erkenntnis müsse sich nach den Gegenständen richten [...] Man versuche es daher einmal, [...] daß wir annehmen, die Gegenstände müssen sich nach unserem Erkenntnis richten."
3. "Es geht aber hiermit [den reinen Verstandesbegriffen] so, wie mit anderen reinen Vorstellungen a priori (z. B. Raum und Zeit) die wir darum allein aus der Erfahrung als klare Begriffe herausziehen können, weil wir sie in die Erfahrung gelegt hatten, und diese daher durch jene erst zustande brachten."

4. Transcendental concepts or forms *are* ordering structures of our minds that provide the possibility to make experience of different things, events and connections between them. If they were not given (in one way or other) or if we did not bring them with us, we could not make experiences in the way we do. This is what makes transcendental concepts necessary in the sense of being undoubtable. But these forms not only shape our experience, they also correspond with Kauffman's "eigenform" insofar as they make the "transcendental reflection" possible, i.e., the critical, purifying and scientific reflection and observation Kant himself undertook or realized while writing his book. Kant's self-observation iterates the observing relation between the understanding and its objects by using the same observation and ordering instruments and by dividing itself into an observed part, "the understanding," and an observing part, which is called "reason." Thus the set of ordering instruments found by the reflecting reason to be necessary for experience is also necessary for its own reflective practice, for its own being. Self-observation and object-observation depend on the same structures or forms. With Kauffman (§14), one could say that philosophy, too, has been searching for its "eigenform" in a reflexive field. Even more, one could add that Kant's transcendental reflection as the "objective result" has significantly changed the world of social science since then – at least if we follow Foucault's (1994) interpretation of the philosophical event at the end of the 18th century that might be called the "transcendental turn." Or, as Judith Butler puts it more generally: "But we must also have an idea of how theory relates to the process of transformation, whether theory is itself transformative work that has transformation as one of its effects" (Butler 2005: 2004).

5. What does this kind of observation imply? From my point of view, one can see, in accordance with Kauffman, that philosophical research is also self-referential and reflexive, and that it has not only provided the necessary conditions for making experiences but has also shaped our experiences as well as its own research contexts. From this (Foucaultian) perspective,[4] humanities as an *effect* of the philosophical transcendental turn that reflects *on our own* as human beings as the condition of knowledge *at all* evolved in the self-creating and self-discovering "transcendental era." This self-reflection can be seen to have been re-discovered by mathematics and formal logic more than a hundred years ago. Yet it is well known that in those times, self-referential structures were avoided due to the paradoxes *negative* self-references produce (e.g., Tarski's "theory of types" as a strategy of avoidance). The grounding of mathematics in Spencer Brown's *Laws of Form* (1969) and Kauffmann's reference to it in §§41–44 could be seen as the second step of re-discovering the self-referential structure of science (and thinking in general). Spencer Brown not only deals with negative self-referential figures by interpreting the moving within the paradox as creating time (Spencer Brown 1969: 58–68), he also found a very concise way of making self-referential processes and thus self-referential creations obvious on the basis of a single self-referential sign, ⏋, marking at the same time itself, its other, therefore identity and difference, and the *process* of its own coming into being.

6. But, I am not convinced that the search for "eigenforms" *is necessarily* the driving force of science, though. Of course, it is needed as long as we think we have to search for structures, rules, concepts, laws, etc. that verify themselves. But as soon as we accept that it is us who (want to) find regularities, rules, patterns, etc. in our ordering mind because we draw distinctions by observing our objects and ourselves, we will see that we do not need to focus on what we have already ordered and how we have ordered it, but can also focus

4. The subtitle of Foucault's *The Order of Things* is *An Archeology of the Human Sciences.*

on our fantasy and the power of inventing new ways of thinking. Maybe the truth created by self-referential congruence is not the one and only criterion for valid knowledge.

7. Coming back to my expectation that Kauffman's observations might help to bridge the "gap" between analytical and continental philosophy, I was wondering: if mathematics *and* formal logics could show that science is always self-referential then this would be valid for *every* philosophical standpoint – even for the "analytical" one. If we interpret self-reference in philosophy as the fact that philosophical theories are *always* socio-cultural-historically based, i.e., *shaped* in form and content by their times, too, as Hegel (1986: 15–20), Jean-Paul Sartre (2016), Horkheimer (1975), Foucault (2005) and other contemporary continental philosophers claim, then one of the defining differences between those two philosophical movements will disappear. Even the second difference concerning the status of language, which is seen to represent an independent world, will disappear if we follow Spencer Brown's idea that signs always (also) describe the distinction made by them. To show this in detail requires more space than a commentary, but Kauffman's article is a good starting point.

Open Peer Commentary: **Finally Understanding Eigenforms**

Michael R. Lissack

1. Louis Kauffman's target article is but one of many where the concept "eigenform" (Foerster 1981) gets used. I must confess that despite my serving as the President of the American Society for Cybernetics, and despite nearly two decades' work in cybernetics, I have found this concept to be among cybernetics' most difficult. At its heart, an eigenform is a "stability" – and, like many other examples of circular logic, the meaning of that stability supposedly becomes apparent with use. Until this article by Kauffman, I can woefully confess, such usage failed to ascribe much meaning, at least to me.

2. In an earlier article (2005), Kauffman suggested that the notion of an eigenform was related to the idea that "an object is a symbolic entity, participating in a network of interactions, taking on its apparent solidity and stability from these interactions" (Kauffman 2005: 130). The eigenform is an expression of that stability:

> Heinz von Foerster has suggested the enticing notion that 'objects are tokens for eigenbehaviors.' There is a behavior between the perceiver and the object perceived and a stability or repetition that 'arises between them.' It is this stability that constitutes the object (and the perceiver). (Kauffman 2005: 132)

3. Philosophically, this notion of stability is troublesome. "Stable with respect to what?" becomes an obvious question. The von Foerster answer of "between object and perceiver" sheds little light on this matter. In my view, it leaves out the context in which the network of interactions is situated. To me, there can be no stability except in consideration of the context, but that transforms the notion of an eigenform into at minimum a three-way interaction amongst object, perceiver, and context and, given the possibility of multiple perceivers each with their own context, at maximum an infinite interaction. If, indeed, an eigenform is possibly the product of an infinite network, then how is it possible for that product to be "stable?"

4. Kauffman continues:

> Coalesence connotes the one space holding, in perception, the observer and the observed, inseparable in an unbroken wholeness. Coalesence is the constant condition of our awareness. Coalesence is the world taken in simplicity [...] In the world of eigenform, the observer and the observed are one in a process that recursively gives rise to each. (Kauffman 2005: 133f)

5. At this point, the circular logic tends to defeat me. Stability, as a product of recursion, in the absence of consideration of context, is beyond my reckoning. Kauffman's present article rescues me by restoring the notion of context to the understanding of eigenforms generally. As he states: "a reflection on the context, and an inclusion of that awareness of context into the context is always present" (§1).

6. The key to understanding is located in §3: "A reflexive domain is a context for action" and §4: "An eigenform is a fixed point for a transformation."

7. By this definition, an eigenform is the assertion of a stable coherence (perhaps only momentary) that is required before a distinction can be drawn, a choice made, a boundary or constraint asserted, or a bifurcation cut. The eigenform is the stability against which there is now to be asserted something new or different. When we ask "Different from what? or "Newer than what?" the "what" is the stable eigenform against which we are defining, measuring, drawing the distinction.

8. This meaning of "eigenform" suggests that eigenforms are not generated as the result of a search for stability but rather as a necessary component if change or boundaries (constraints etc.) are to be recognized. Eigenforms become the logically necessary priors for a distinction to be made. Eigenforms afford distinctions, and distinctions are impossible absent their being immediately preceded by an eigenform.

9. The eigenform formulation becomes as follows: if a distinction is to be drawn it can *only* be drawn if one treats the prior condition *as if* it had a coherence, coalescence, unity etc. such that the distinction can be recognized as *distinct*. The prior condition cannot be somehow vague enough that the distinction made is *not* distinct. The context for *all* eigenforms thus must include the need/desire/opportunity for a distinction to be afforded as a possibility. Eigenforms are not merely the product of the interaction networks settling down but instead are the momentary requirement of a process that demands stability so that a new distinction can be made.

10. Some might object that such a conception of eigenforms creates a space for teleology. My rebuttal would be that by definition the eigenform is *not* an attribute of a situation or an object but rather is an assertion of a "state" describing that situation or object by an observer. The observer is drawn to make such an assertion in order to create the preconditions necessary for a distinction to be drawn. In the absence of the perceived possible need/desire for a distinction to be drawn the observer has no "need" to assert the existence of an eigenform. Vagueness and ambiguity are fine when there is no contemplation of distinction.

11. Eigenforms are summoned into existence by the observer contemplating the possibility of choice. Absent that possibility, the eigenform may lie dormant, unneeded, and likely not only unperceived but unperceivable. The perception of the eigenform is afforded by the contemplation of choice and vice versa. Kauffman's recursive form has moved from the contemplation of object by observer (thing to thing) to the logical preconditions if choice is to be afforded to a participant in a given situation.

12. Hans Vaihinger claimed:

> The object of the world of ideas as a whole is not the portrayal of reality – this would be an utterly impossible task – but rather to provide us with an instrument for finding our way about more easily in this world. (Vaihinger 1924: 15)

Werner Heisenberg's observations serve to complement the point:

> The world is not divided into different groups of objects but rather into different groups of relationships. [...] The world thus appears as a complicated tissue of events, in which connections of different kinds alternate or overlap or combine and thereby determine the texture of the whole. (Heisenberg 1963: 107)

Eigenforms are a means by which an observer can for a brief moment stabilize some set of connections so as to enable Vaihinger's wayfinding.

13. Kauffman notes:

> There is no longer an impenetrable barrier between the descriptions (pictures) of the world and the world itself [...] when a scientist in the world proposes a theory, this theory becomes a new element of that world. The theory acts and is acted upon by the world of persons and elements of the world. There is no protection for the world from the effects of such a theory. (§83)

14. In Vaihinger's terms, the invention of a new means of wayfinding changes the very territory through which one wishes to find one's way. Eigenforms are a means to an end, where the end is the ability to make distinctions and to make choices. Eigenforms serve as a constructed picture of what might be reality, much like the picture of a watch drawn by Einstein and Infeld:

> In our endeavor to understand reality we are somewhat like a man trying to understand the mechanism of a closed watch. He will never be able to compare his picture with the real mechanism and he cannot even imagine the possibility of the meaning of such a comparison. (Einstein & Infeld 1966: 31)

15. Eigenforms are *not* a perception of the "real mechanism." They are instead the picture which that the man drew so as to understand, as Richard Rorty pointed out: "Knowledge is not a matter of getting reality right [...] but rather a matter of acquiring habits of action for coping with reality" (Rorty 1991: 1). To the extent that science is a quest for such knowledge, then science is making use of tools which that allow us to cope.

16. Kauffman opens his article by telling us that "the scientific endeavor [is] a search for eigenforms in reflexive domains." When one views eigenforms as the assertions of stabilities that are the prerequisites for making distinctions and choices, then the scientific endeavour is the search for the bases that afford those same distinctions and choices. In such a search, realism and constructivism can find common ground in the transitory "as-if" nature of eigenforms – for the end goal is the product of the choices made, not the process by which they get made.

17. Kauffman has successfully recast eigenforms as intellectual/cognitive elements in the process of distinction making. As such, eigenforms lack an existence separate and apart from the domain/situation they describe and the observer relying upon the existence of that description. Eigenforms are essences – much as von Foerster described. But now, in language I can understand.

Open Peer Commentary: **Eigenforms, Coherence, and the Imaginal**

Arthur M. Collings

1. Beginning with the concepts of eigenform and "reflexive domain," Louis Kauffman's target article explores at a deeply abstract level the question of what constitutes second-order science. Kauffman has written a series of papers over the last 10 to 15 years that take as their point of departure the concept of eigenform, returning again and again to revisit and explicate the ideas of eigenform, reflexivity, and second-order cybernetics.

2. In this commentary, I express fundamental agreement with Kauffman's formulation of second-order science. I also gather a succinct list of principles, to clarify and define the nature of second-order science in relation to reflexive domains.

3. The idea of reflexivity, as Kauffman expresses it, is a formalization of the idea of circularity, defined and described in a process that is necessarily circular. A *reflexive domain* is the name of the space that can encompass this circularity. Kauffman uses a mathematical language, called the lambda calculus, which is simultaneously very simple and highly abstract. Readers who find this formalism difficult may find it useful to think of *the reflexive domain of everyday living* or even more simply *the space in which a conversation occurs*. The keys to note about reflexive domains are:

- the ability to make distinctions exists, and therefore every "element" in the domain has a name;
- every element is also subject to actions, or transformations; and
- transformations in turn may be referenced and thereby given names.

These simple rules describe the domain, and give rise to what Kauffman and Heinz von Foerster call "fixed points." Fixed points are recursions in which, for example, an observer observes itself observing (observing itself observing…). Finally, the reflexive domain itself is subject to reference, naming, and transformation, permitting sentences such as "this discussion would be easier without the mathematics!" or "cybernetics is self-referential."

4. Von Foerster coined the terms *eigenform* and *eigenbehavior* (Foerster 2003a). He conceived that the perception and production of objects by an observer is circular, where the observer's percepts and conceptions iterate upon themselves as *coordinations of coordinations,* in potentially infinite recursion. When the process of eigenbehaviour is stable, when there is a fixed point, von Foerster says that the object is then a *token* for the coordinations that produced it.

> Heinz performs the magic trick of convincing us that the familiar objects of our existence can be seen to be nothing more than tokens for the behaviours of the organism that create stable forms. (Kauffman 2003: 73)

5. What distinguishes this article from his earlier writings on the theme is Kauffman's focus on the question of the production of "objective" knowledge in the practice of science and second-order science. Describing second-order systems, in which the observer too is observed, Kauffman states that…

> domains with such circularity remain amenable to rational study [...] In the face of the circularity of context and observer it is still possible to explore and come to agreements that have every appearance of being scientific facts. (§1)

Context: Science and anti-science

6. In the summer of 2015 and winter of 2016, members of the American Society for Cybernetics (ASC) in cooperation with the Institute for the Study of Coherence and Emergence (ISCE) sponsored small conferences in Boston and Salem, MA to discuss the concept of second-order science. One concern addressed by these gatherings was the American public's lack of knowledge of science, its high level of skepticism about science's reliability and truthfulness, and the premise that second-order science be applicable to some of the problems.

7. Every two years, the National Science Foundation and its controlling body, the National Science Board, publishes a comprehensive statement assessing the state of science in American society, the latest published being *Science & Engineering Indicators 2016.* Depending on their context, these findings are either quite disturbing, as (National Science Board 2016: 49f):

- 24 percent believe the sun goes round the earth
- 50 percent believe that electrons are bigger than atoms
- 41 percent do not know the father's gene determines the sex of the child
- 51 percent state they do not believe that humans evolved from other species
- 25 percent believe to some degree in astrology;

or, upon reflection, somewhat comforting: overall scores reported have remained incredibly constant over the last 22 years of publication (National Science Board 2016: 42), and the belief in astrology is hovering close to the recorded low.

8. Writing in a student-published blog hosted by *Nature*, Ryan Hopkins cites a nearly identical set of facts from *Indicators 2012,* and then offers a plausible explanation:

> So why do political and religious ideologies win when they come into conflict with science? A 2012 study on conspiracy theories offers an explanation. Psychologists found that conspiracy theorists can hold their beliefs in the face of contradictory facts because of a process called global coherence, by which subjects selectively chose to believe only those facts which supported their worldview.[1]

The study Hopkins cites, "Dead and Alive," examines the arising and co-existing of deeply contradictory beliefs held by conspiracy theorists: "Believing that Osama bin Laden is still alive is apparently no obstacle to believing that he has been dead for years," conclude its authors (Wood, Douglas & Sutton 2012: 6). There is no need to delve further into the study, but the key point is of the second order: the writers, who are keen observers, distance themselves from their subjects, maintaining their own coherence.

---

1. "Unbelievable: Why Americans mistrust science," http://www.nature.com/scitable/blog/scibytes/unbelievable

Eigenforms and the imaginal

9. The concept of eigenform applies both to ordinary objects, as well as to non-physical objects (let us call them *extraordinary objects*, or *ideas*). The criteria that is advanced by both von Foerster and Kauffman, stability, applies equally to ordinary or extraordinary eigenforms. The idea, simply, is that eigenbehavior presents sufficient stability that it can be distinguished, named, and referenced. Ordinary objects include most of the artefacts and many of the natural phenomena that we encounter, that we pick-up, hold, touch, drive, and control, while extraordinary objects include concepts, abstractions, archetypes, and myths. Science, too, is an extraordinary idea. Bigfoot, unicorns, the characters in certain films and novels, the pantheon of Greek gods, each have persistence in being named and woven into stories that are repeated again and again. Hallucinations, rumours, lies, falsehoods, and similar non-existent forms are also objects, provided they are persistent.

10. Von Foerster, describing how "objectivity" emerges in the world of objects as we know it, contends that objects arise in the form of agreements that are like circular conversations.

> Under which conditions, then, do objects assume 'objectivity?' Apparently, only when a subject, S_1, stipulates the existence of another subject, S_2, not unlike himself, who, in turn, stipulates the existence of still another subject, not unlike himself, who may well be S_1. (Foerster 2003a: 266f)

11. Kauffman is very clear: although they must be circular, eigenforms can be precise (§§23–32). In my conversations with Kauffman about such extraordinary eigenforms as Bigfoot,[2] unicorns, and generally recurring fictional characters such as Sherlock Holmes, Kauffman pointed out that mathematical objects are evaluated based both on their precision and their internal logical consistency. We also evaluate their more literary cousins, Sherlock Holmes for example, based on a sense of their coherence and consistency; when we encounter the instantiation of such a character, in the performance of a play, we readily determine if the actor is in-character, and whether the performance coheres.

Second-order objectivity, precision, and coherence

12. The view of second-order science that Kauffman describes is deeply grounded in constructivism and second-order cybernetics. I would like to entertain the idea, inherent in the concepts of eigenform and reflexivity as Kauffman states them, that the distinction between scientific inquiry and other forms of constructing knowledge (if we choose to make one) should not be considered fundamental. Rather, it is better to believe that science is one aspect of the imaginal, lived and performed in conversational community, in reflexive domains, neither separate from nor different than other aspects of living – where scientific facts and truths, like all extraordinary objects, are tokens for the stable eigenbehaviors that produce them.

---

2. I live in the Town of Red Hook NY, home of Bard College, 100 miles north of New York City. Red Hook is also home to "Bigfoot Researchers of the Hudson Valley," which claims as its purpose collecting evidence and conducting research to support the contention that Bigfoot individuals are living inhabitants of the Hudson Valley. The organization has received 1 426 likes on its Facebook page to date.

13. This perspective raises a number of issues, not least of which being that the ability to discern truth, reliability, accuracy, and credibility in the claims of others is highly useful. Constructivism and second-order cybernetics reject the idea of the *objective observer* and the correspondence theory of truth. The emergence, as implied by von Foerster's definition of objectivity (quoted in §10 above), of numerous, splintered conversational communities, each with a common set of perspectives and beliefs that diverges from all other groups, can readily be observed in matters of contemporary public discourse, and is bound to create a clamor for objective knowing that cannot be easily satisfied. Kauffman does not explicitly address such a fracturing or multiplicity of worlds of discourse, but the idea is presaged in his use of the plural term *domains*.

14. Given the consideration I have given to the idea of the imaginal, I offer the following three principles (or feed-back loops if one prefers), which I imagine to apply to the practice of second-order science and cybernetics:

- Second-order science is lived and performed within reflexive domains and in conversational community. The coherence, or the logical, emotional, and behavior consistency, of such a conversational community, be it scientific or otherwise (e.g., Thagard 1989), should be determined based on internal rather than external observers. Such coherence, or lack thereof, will be identical to the stability of eigenbehaviors that have produced the agreement of perspective within the conversing community.
- Precision is an indicator (but not necessarily a determining factor) of the degree to which a community discourse is scientific or, more generally, rational. Precision in this sense includes the numeric and statistical, compliance with sound logical inference, and acceptable conformance with well-formulated standards and methods.
- Finally, second-order cybernetics and second-order science are, or should be, committed to perturbing their own coherences and to increasing the varieties of their own perspectives – for example by engaging in transdisciplinary research and discussion, by engaging energetically in dialogue and dissent, and by observing itself reflexively. Such an approach contrasts with extra-rational fringe groups (such as Bigfoot researchers), which maintain coherence by releasing themselves from any particular obligation to comply with method or logic, and also contrasts with establishing and maintaining coherence through a strict enforcement of methodological and linguistic constraints.

Open Peer Commentary: **Conserving the Disposition for Wonder**

Kathleen Forsythe

1. The act of questioning the notion of how a scientific theory relates to the world and affects the world requires that the author and the commentator engage in a reflexive domain of conversation through the writing. The act of reading in itself demonstrates the distinction between objectivity and (objectivity), so often used in a second-order cybernetic explanatory approaches to science. The argument for objectivity as if there were no persons involved can only be made in language by a person that makes a distinction. The making of a distinction requires that a context be cleaved through this act of perception that transforms as a conception that can unfold as a construction of meaning. As the author states in §1, the observer is actively in the system. He further claims that arguments and theories that contain this premise are no less rational if they include the one who is composing and constructing the theory based on observing a world.

2. Distinction by its very nature requires an observation of that which is to be distinguished and, as Kauffman argued in §21: "This cannot be accomplished without circularity [...] Theorem 1: There is no definition of distinction that is not circular" and in §22: "Distinction transcends closed worlds of words and moves into the worlds of feeling and action."

3. I would say these are the words of second-order science that most distinguish this thinking from classical science, in which the observations of an objective world made with reason deny the act of imagination that enabled the perception and the conception of a reality in the first place. There is no acknowledgement of the process in which making a distinction is a property of rationalization that actually requires the imagination of the context in which the distinction becomes possible – that of a person that perceives.

4. As I began to read this article it was with some trepidation. I am a poet and not a mathematician. However, throughout the first paragraph, I was smiling and, as I progressed, I was delighted. For indeed, what Kauffman has done in this article is to deepen Humberto Maturana's extraordinary explanation about the world of objectivity and the world of (objectivity) in parenthesis (Maturana 1988: 28). Kauffman has managed through his article to collapse the distinction into the idea that objectivity itself is an emergent phenomenon (§1).

5. In the heady days of the early 1980s when I was first introduced to second-order cybernetics through conversations with Gordon Pask, Maturana, Ranulph Glanville, Heinz von Foerster, Lou Kauffman and many others, I went through the dissolution of my traditional education to find myself in a space of understanding that the world I was experiencing actually arose through the dynamics of being in it. I also came to understand that I was so structurally coupled that I actually thought there was an objective reality, even though my intuition suggested to me that there were as many worlds to unfold as there were imaginations to conceive them!

6. I knew then, 30 years ago, that the challenge of science as being a description of an objective world, independent of the observers in that world, no longer made sense to me. I, however, did not have the capacity to demonstrate the elegance of the eigenform, nor how

profound the awareness that it can be shown, even in domains such as physics, that the object of a theory can be found to be an eigenform of the theory (§83).

7. Through my poetic vision, I grappled with what I was perceiving and conceiving and, in a moment of revelation, understood the notion that it was in the movement and the moment that perception and conception arose in one flow between domains that both arose and collapsed in the moment and movement of observing them. Maturana named this realization of mine an isophor (personal communication in response to Forsythe 1986). It was from this notion that I began to see that circularity and reflexivity held other ways of seeing. It all depended upon the point of view – whether what we understand as a fixed point or an invariance in nature was actually a distinction made by an observer because "everything said is said by an observer" (Maturana 1987: 65).

8. As I grappled with the seeming illogic that second-order cybernetics suggested, that there was no way to prove that there was a reality separate from myself, I came to appreciate the freedom entailed in seeing myself emerging moment by moment, aware that I am emerging structurally coupled with a rippling wavefront of existence [...] I may think I am an object walking around in an invariant landscape yet I know that all is dynamically and recursively in a dance of which we are only infinitesimally aware.

9. Metaphorically, I began to speak of the imaginary space of conception (Forsythe 1987) in which we live unaware of the shared universe until one day we awake and realize the experiences we are having and give them names. As languaging arises in infants and children, it is these coherences of experience that inform the sounds and context of words that help us to share meaning and bring forth a world in consensual coordinations of consensual coordinations (§7). For the baby, cup-ness emerges when she reaches her hand for the object from which to drink. It is in the experiences of that hollowed vessel containing a liquid and the relational landscape of sound and bodily emotional experience around it with the m/other that an invariance of the process arises in the space of the baby's experience and the baby begins to live in the domain of languaging. I have often wondered at babies who live with people who use different sounds for such an experience and how the baby inevitably sorts out accurately the different spoken languages. My own work with non-verbal children on the autistic spectrum has forced me to look very deeply at the coherences in their observed experience to see where the invariances of process may arise so that we can see that objects do arise in their world even without speech to confirm this.

10. Kauffman refers in §7 to von Foerster's reference to objects and extends this to the "space of our experience" – the context in which we have our experience, and it is from the experience itself I make objects by finding fixed points in the recursion of my interactions. "Eigenforms are a touchstone for the relationship of circular and recursive processes and the ground of our apparent worlds of perception" (§7). Until she wanted to cut, knives as objects did not appear in the perception of my non-verbal 7-year-old learner with autism; when they did, then they also began to appear in her receptive language.

11. My introduction to second-order cybernetics in its early days deeply affected me and I have lived my professional career from this perspective. Being neither a scientist nor a mathematician, I never felt confident enough to make the argument that Kauffman has made, even though I recognized in the emergence of second-order cybernetics that there was indeed a new way to consider science.

12. In particular, in §9 Kauffman says:

> For if science is to be performed in a reflexive domain, then one must recognize the actions of the persons in the domain. Persons and their actions are not separate.

Kauffman is asking us to see that it is in the reflexive domain that conversation occurs. As he says in §3:

> A reflexive domain is an abstract description of a conversational domain in which cybernetics can occur. Each participant in the reflexive domain is also an actor who transforms that domain.

13. Because we humans live in language (as a fish lives in water), we cannot in our observation of an objective world remove ourselves from our relationship to that world (objectively). In the recognition of the reflexivity that exists in the conversational domains through which we bring forth worlds together in language, the role of imagination enters in – even to the most rationally argued scientific theory. For how else could the scientist have conceived an objective world without the capacity to perceive it directly and transform reflexively the perception into the conceptual architecture of languaging?

> In the moment of conception
> We perceive
> the conversation…[3]

And the role of the perception/conception dynamic is precisely the same role as that of the participant in the conversation – to be an actor who transforms that domain.

14. Persons and their actions are not separate. Nor are persons and their perceptions separate. And it is through the act of perceiving that distinctions are made. Scientific theory does not exist without the imagination to conceive it from what one perceives – and perception is the space of our experience from which the invariances of processes give rise to ordinary objects as eigenforms of themselves. The very notion of living systems as autopoietic systems has built in to the definition the dynamic nature that the molecular space is itself generated through the poiesis of experience through structural coupling with its environment. When we extend this to how we know a world, we cannot unglue ourselves from our embodied molecular structure, which requires our perceptual apparatus in order to develop our conceptual capacities that we recognize in and through languaging.

15. My excitement as I worked through Kauffman's mathematics was palpable. If a reflexive domain is itself an eigenform and a reflexive domain is a context for action, then "there is no end of our deliberations," as Kauffman says in §3. For me, this confirmed my aphorism, "Thought without emotion is a womb with no way out." The notion that there can be objective thought without the ground of the state of being of the one who is doing the thinking does not allow for the generative mechanism of creativity. There would be no new distinctions. If we accept that emotions are dispositions for action, then, our actions tell us about what we are experiencing, and we require this perception in order to conceptualize our experience and recognize our actions! This requires the circularity of a reflexive domain and a fixed point of transformation that we hold in our imaginations as we make the transformations. We sometimes call this "thinking" and because thinking requires our bodily engagement, it is what von Foerster referred to as "objects as tokens for eigenbehaviors" (§7, Foerster 2003a). I claim that thinking and rationality require this imaginative

3. From my unpublished *Warrior of the Gentle Passion*, 1991.

act of transformation through which the reflexive domains of conversation in one's own "space of our experience" (§7) give rise to our awareness of our self.

16. I think an epistemology of imagination is fundamental to a second-order science. For the second-order science that is supported in Kauffman's article is one that is grounded in learning as the perception of newness, where newness is a distinction that the one who is learning can make about herself. As Kauffman says in §52, "There is a singularity about interacting with oneself that is different from interacting with others." And it is the newness of distinctions that the scientist makes within his or her own imagination that is used to formulate theory.

17. I have been most appreciative in this article in following Kauffman's arguments that explore those aspects of science that often characterize its inviolability, e.g., the repeatability that is often claimed as proof of an invariant physical universe. Kauffman shows that it is repeatability that makes a successful experiment into an eigenform. In §67 he describes a scientist as being in search of eigenforms that are independent of the action of the observer… indeed, he further expands this to declare that the entire world of science is an eigenform that, by elaboration, continues such a search. The fact that scientists can use their imaginations to map the invariance of transformations in the manner in which the world appears to solidify through its structural coupling does not prove that the physical universe is invariant. If you have followed his argument and have an awareness of yourself as being glued to the world, it arises with you, you will no doubt agree.

18. I think that what is most significant about the arguments put forth is that persons are not separate from their actions and neither is their science. In §82, Kauffman says,

> Words acquire their meanings via actions, and distinctions occur in the sharp invariances of certain eigenforms […] when a scientist in the world proposes a theory, this theory becomes a new element of that world. The theory acts and is acted upon by the world of persons and elements of the world. There is no protection from the world from the effects of such a theory.

Clearly, there are ethical repercussions from seeking to understand a second-order science that includes the observer and, by extension, the environments to which the observer is structurally coupled.

19. Yet it is the homage to creativity and the imagination as the hallmarks of good science that most benefits from this article. As a knowledge architect myself, I understand that knowledge does not exist objectively independent of being brought to life by the person who makes distinctions and demonstrates the conduct adequate for others and him or herself to say that they know. As Kauffman points out in §88, the world is itself affected by the attempts to find stable knowledge of the world because, as he says, "one finds out about the nature of the world by acting upon it."

20. It is this conservation of the disposition for wonder that grounds scientists' pursuit of understanding, discovery and knowledge of the ever-unfolding matrix of existence. Recognizing that we are participants in the processes of the worlds that we unfold and share together though the imaginative distinctions in the reflexive domains of the conversation, as explicated so eloquently in this article, leaves me as awed as any aspect that the universe might be either invariant or completely other than we thought!

Author's Response: **Distinction, Eigenform and the Epistemology of the Imagination**

Louis H. Kauffman

1. In George Spencer-Brown's *Laws of Form* (1969), we have a perfect example of self-reference that occurs apparently without the formal appearance of a fixed point or eigenform. This example is the Spencer-Brown mark, $\urcorner$, which is seen to make a distinction in the plane in which it is drawn and is intended to refer to some given distinction. Since the mark may refer to a given distinction, it can be seen to refer to itself in its own making. Thus we may write an arrow of reference $\urcorner \rightarrow \urcorner$ for this fundamental self-reference of a distinction to itself. We are not yet at the bottom of the rabbit hole. The arrow of reference is also a distinction, and the mark itself can be regarded as a mark of reference from itself to itself. In this sense we can write

$$\urcorner = \urcorner \rightarrow \urcorner.$$

And eigenform appears! Now we have the mark as the solution to the equation

$$X = \urcorner \rightarrow X.$$

The mark is that which is referred to by the mark. The mark is the fixed point of the transformation $T(X) = \urcorner \rightarrow X$. Fundamental self-referential distinction becomes eigenform in the acts of observation, imagination, description, reflection and construction of context. Here I have formalized this transition. These themes appear repeatedly in the comments on the original article and have led me to think all over again about the roles of distinction, eigenform, reflexivity, the imagination of context and the context of imagination. Eigenform is a precondition for distinction and distinction is a precondition for eigenform.

2. In her commentary, **Tatjana Schönwälder-Kuntze** wonders "if mathematics *and* formal logics could show that science is always self-referential then this would be valid for *every* philosophical standpoint – even for the 'analytical' one" (§7). And even the difference between analytical and continental philosophy "concerning the status of language, which is seen to represent an independent world, will disappear if we follow Spencer Brown's idea that signs always (also) describe the distinction made by them" (ibid). Her hope that a reflexive viewpoint could unify a schism in philosophy made me return to the relationship between distinction and eigenform as indicated in the first paragraph above. In the making of a distinction, significant issues arise that are considered in all philosophies. They arise, often in different orders, and appear to give rise to different systems of thought, and yet there is a basic circularity that must be traversed. Distinctions themselves are circular and seemingly have a simultaneous arising of awareness and awareness of a difference that is the distinction. Awareness exfoliates into description and description can become eigenform. It must be understood that eigenform made explicit, made into a fixed point, even for a moment, is descriptive yet can be as timeless as "I am the one who says I." And yet eigenform in the sense of recognition and the eternal or timeless return is no less fundamental than distinction/awareness itself. It is only in the ordering that we put on our experience that one seems prior to the other. Eigenform is a precondition for distinction and distinction is a precondition for eigenform.

3. The apparent ordering of world and perceived world and the structuring of preconditions for perception of form as envisaged by Kant is extended to a circularity in the reflexive view. Just as eigenform is a precondition for distinction and distinction is a precondition for eigenform, so are Kant's a priori categories both the predecessors and the successors of the act of perception. In this way, I agree with **Schönwälder-Kuntze** when she emphasizes that eigenforms are the ordering structures that make experience possible in the first place because they enable "transcendental reflection" (§4). She adds that "Kant's self-observation iterates the observing relation [...] by dividing itself into an observed part, 'the understanding,' and an observing part, which is called 'reason'" (ibid). Referring to §4 in my target article, she claims that "philosophy, too, has been searching for its 'eigenform' in a reflexive field" (ibid). In the reflexive view presented here there is no fixed a priori. Pre and post conditions occur in a circularity wherein one may indeed describe a world that divides itself into a part that is seen and a part that sees in an endless round.

4. This brings us directly to the question of context. In the circularity, the "that which brings forth" the distinction is its context, and it is the distinction that allows us to make a further distinction and discriminate the first distinction from the context in which it has occurred. Often, some context is given beforehand. For example, in a tide-flattened stretch of sand, I wield a stick and draw a circle in that sand, making a distinction between an inside and an outside. The prior context of the sand and stick is a precursor to the appearance of the distinction, and yet the context of the drawn circle in the sand is a different context from the original one. The new context contains the imagination of the marking in the sand as a distinction from an inside to an outside, and it displays the potentiality of the stick as an instrument to bring forth such a pattern in the hands of a maker/observer. In his commentary, **Michael Lissack** speaks of the prior condition or prior context to the emergence of a distinction:

> The prior condition cannot be somehow vague enough that the distinction made is *not* distinct. The context for *all* eigenforms thus must include the need/desire/opportunity for a distinction to be afforded as a possibility. Eigenforms are not merely the product of the interaction networks settling down but instead are the momentary requirement of a process that demands stability so that a new distinction can be made. (§9)

5. I agree with his understanding that the eigenform is not precisely a stability of some system settling down in its interactions, but rather the coming to be of a certain kind of choice of distinction in the interaction of the observer with his observed (not to be taken as separate). The demand for stability "so that a new distinction can be made" is a brilliant phrase on **Lissack**'s part and can be seen so in examples of eigenform. There is no stable self, there is no I except that we create the I, and who are we that we should be able to create ourselves. And yet it comes about in a demand for stability that a new distinction be made. I am the one who says I. Before that saying there is no I. After that saying, the I is imagined and its reality becomes our task in living. There is no I in the prior condition and no Thou either. These are the creations and eigenforms of a context that is capable of transcending itself, including the I that is so declared. The same applies to the eigenforms of science. They are not only the result of empirical inquiry and at the same time they do not partake of a Kantian a priori. In §7, **Lissack** defines an eigenform as "the assertion of a stable coherence (perhaps only momentary) that is required before a distinction can be drawn [...] [It] is the stability against which there is now to be asserted something new or different." I found here a new view of eigenform compared to the one I had held and was

returned to go down the rabbit hole of §1 above to see that context, eigenform and distinction form a circular round, a trinity, with each the progenitor of the others.

6. Let us turn to science. Our scientific and mathematical knowledge is nothing if it is not the record and action of the distinctions we are able to make in the world that forms our momentary context. These distinctions can be seen to be eigenforms, and the hard-won eigenforms of science can be seen to be significant distinctions. In working with the physical world, we must confront that certain of our ideational distinctions such as number come directly into play in ways that seem not to involve the observer directly. Computers factor numbers with the programmer's help, but they do it in a technique that transcends him, and find information for us that we can verify but would never have computed ourselves without the aid of the automatic recursive discriminations of the electronic calculator or even the abacus. Thus we have discovered arithmetic anew in the calculating machine. One should be surprised that these entities do what they do with reliability. The reliability of the machine is itself an eigenform and it is engineered and it is constructed and we have discovered that it works. This is the non-trivial and universally accepted discovery of computer science. It is not an obvious discovery. It might have been quite different and it may become different in the future as we push for ever more competent machines.

7. I have chosen this excursion into computer science in order to bring up a key issue raised by **Art Collings**. He points out that eigenbehavior provides the stability needed to distinguish and reference objects irrespective of whether they are "ordinary" or non-physical "extra-ordinary" ones, such as ideas and science but also "Bigfoot, unicorns, the characters in certain films and novels, the pantheon of Greek gods […] [h]allucinations, rumours, lies, falsehoods, and similar non-existent forms are also objects, provided they are persistent" (§9). This extraordinary range of possible eigenforms calls into question the distinction between scientific endeavor and other forms of imaginative and literary endeavor that produce distinctions and eigenforms. In §13, **Collings** provides a solution for this predicament that focuses on how coherences are dealt with. Science (and with it second-order cybernetics and second-order science) should always be open to perturbing its own coherences and to increasing the varieties of its own perspectives. Other "extra-rational fringe groups," by contrast, want to maintain coherence "by releasing themselves from any particular obligation to comply with method or logic and also contrasts with establishing and maintaining coherence through a strict enforcement of methodological and linguistic constraints." Along with this intent to examine coherences and take the widest possible views, looking for the widest possible agreements, I suggest that we identify scientific discoveries by their combination of imagination and surprise. In each example of what we take to be a good scientific discovery, there is something that is entirely not contrived, something that seems to be the way it is for no reason other than it is that way. This is the case even when the result is mathematical. We can prove the Pythagorean theorem on the basis of the axioms of geometry. The theorem is a fully agreed upon eigenform of Euclidean geometry. But beyond that, there is no further explanation for its truth. Indeed, we did not "make up" this result. It is part of the discovery of geometry. And still **Collings**'s issue can disturb us, for does not Sherlock Holmes in his character have a similar quality of inevitability? We must remember that in the scientific mode there is always the matter of repeatability of experiments and forms of reasoning. This means that the eigenforms produced, the distinctions created are subject to severe forms of testing and criticism of an active kind by the participants in the work. The work is not a production of one individual to be appreciated and assimilated by others. Scientific work is collective work by a community, with the standards of testing that have

evolved. It is these differences that make the textures of scientific work worth examining and distinguish it from other forms of human creation.

8. Finally, we turn to the second-order cybernetic view of the imagination of reality. In §7 **Kathleen Forsythe** writes:

> I grappled with what I was perceiving and conceiving and, in a moment of revelation, understood the notion that it was in the movement and the moment that perception and conception arose in one flow between domains that both arose and collapsed in the moment and movement of observing them.

In §15 she further writes

> The notion that there can be objective thought without the ground of the state of being of the one who is doing the thinking does not allow for the generative mechanism of creativity. There would be no new distinctions.

9. For **Forsythe**, the prior condition is the ground of the state of being, giving rise to and running back to feeling and imagination. All this is present through and through a reflexive way of holding and being. None of this can be excluded in reflexivity or in second order science, which is why an "epistemology of imagination is fundamental to a second-order science" (§16).

10. We can summarize the commentaries by their emphasis on the role of context and the role of eigenform as a form of distinction, possibly momentary and evanescent in the flow of time and process, and yet important as a turning point for the emergence of distinction. This way of holding the theme of eigenform and reflexivity is enlarged by the understanding that each of eigenform and distinction can be seen as the precondition for the other. This is how we have begun this response with a formalization showing how the simple distinction, the mark, can be seen as an eigenform of its own self-reference. The mark is the observed reference of the mark to itself. With this the worlds of eigenforms imagined in the original essay and in the commentaries become worlds of distinctions imagined by communities in the course of becoming the contexts that they themselves observe.

11. Let us return to the self-reference of the mark, the "first distinction." We said in §1 herein ┐⊠┐, indicating the self-reference of the mark. But we did not reach the bottom of the rabbit hole. For the arrow itself is a mark and it is the mark itself that refers to itself and so this expression must be replaced simply by a single mark that is understood to be in the act of self-reference, coalesced with the observer who reads and understands this epistemology of the imagination.

┐

What is created is the imagination of a distinction. Beyond that there is nothing to say.

Combined References

Barendregt H. P. (1985) The lambda calculus. Its syntax and semantic. North Holland, Amsterdam.
Butler J. (2005) The question of social transformation. In: Butler J. (2005) Undoing gender. Routledge, London: 204–231.
Davies P. C. W & Brown J. R. (1986) The ghost in the atom. Cambridge University Press, Cambridge.
Einstein A. & Infeld L. (1966) The evolution of physics: From early concepts to relativity and quanta. Cambridge University Press, Cambridge. Originally published in 1938.
Foerster H. von (1981) Objects: Tokens for (eigen-) behaviors. In: Foerster H. von, Observing systems. Intersystems, Seaside CA: 274–285. http://cepa.info/1270
Foerster H. von (2003a) Objects: Tokens for (eigen-) behaviors. In: Foerster H. von Understanding understanding. Springer, New York: 261–271. Originally published in 1976. http://cepa.info/1270
Foerster H. von (2003b) Understanding understanding: Essays on cybernetics and cognition. Springer, New York.
Forsythe K. (1986) Cathedrals in the mind: The architecture of the metaphor in understanding learning. In: Trappl R. (eds.) Cybernetics and systems '86. Reidel, Dordrecht: 285–292.
Forsythe K. (1987) Isopher: Poiesis of Experience. Working Paper No. 87–2. Center for Systems Research, University of Alberta, Edmonton.
Foucault M. (1994) The order of things. Vintage, London. French original published as: Foucault M. (1966) Les mots et les choses. Gallimard, Paris.
Foucault M. (2005) Was ist Aufklärung? In: Foucault M. (2005) Dits et Ecrits. Schriften Band 4. Suhrkamp, Frankfurt am Main: 837–848. French original published as: Foucault M. (1984) Qu'est-ce que les lumières? Magazin littéraire 207: 35–39.
Foucault M. (2007) What is critique? In: Lotringer Silvère (ed.) The politics of truth. Semiotext(e), Los Angeles: 41–81. French original published as: Foucault M. (1990) Qu'est-ce que la critique? Bulletin de la Société française de Philosophie 84(2): 35–63. (Lecture held in 1978).
Harrison R. (1998) Transcendental arguments. In: Craig E. (ed.) Routledge encyclopedia of philosophy. Volume 9. Routledge, London: 452–454.
Hegel G. W. F. (1986) Differenz des Fichte'schen und Schelling'schen Systems der Philosophie. In: Hegel G. W. F. Jenaer Schriften 1801–1807. Werke 2. Suhrkamp, Frankfurt am Main: 15–51.
Heisenberg W. (1963) Physics and philosophy. Allen and Unwin, London.
Horkheimer M. (1975) Critical and traditional theory. In: Horkheimer M., Critical theory. Selected essays. Edited by M. O'Connell. Continuum Press, New York: 188–243. German original published as: Horkheimer M. (1937) Kritische und traditionelle Theorie. Zeitschrift für Sozialforschung. Reprinted in: Gesammelte Schriften. Volume 4: Schriften 1936–41. Fischer, Frankfurt am Main: 162–216.
Kant I. (1781/1787) Kritik der reinen Vernunft. J. F. Hartknoch, Riga.
Kant I. (2007) Critique of pure reason. Penguin, London.
Kauffman L. H. (1987) Self-reference and recursive forms. Journal of Social and Biological Structures 10(1): 53–72. http://cepa.info/1816
Kauffman L. H. (2001) The mathematics of Charles Sanders Peirce. Cybernetics & Human Knowing 8(1–2): 79–110. http://cepa.info/1823
Kauffman L. H. (2003) Eigenforms – Objects as tokens for eigenbehaviors. Cybernetics & Human Knowing 10(3–4): 73–90. http://cepa.info/1817
Kauffman L. H. (2004) Virtual logic: Fragments of the void – Selecta. Cybernetics & Human Knowing 11(1): 99–107.
Kauffman L. H. (2005) EigenForm. Kybernetes 34(1/2): 129–150. http://cepa.info/1271

Kauffman L. H. (2009) Reflexivity and eigenform. The shape of process. Constructivist Foundations 4(3): 121–137. http://constructivist.info/4/3/121
Kauffman L. H. (2012a) Eigenforms, discrete processes and quantum processes. Journal of Physics: Conference Series 361(1): 012034. http://iopscience.iop.org/article/10.1088/1742-6596/361/1/012034/pdf
Kauffman L. H. (2012b) The Russell operator. Constructivist Foundations 7(2): 112–115. http://constructivist.info/7/2/112
Kauffman L. H. (2015) Self-reference, biologic and the structure of reproduction. Progress in Biophysics and Molecular Biology 10(3): 382–409. http://www.sciencedirect.com/science/article/pii/S0079610715000905
Kauffman L. H. & Buliga M. (2014) Chemlambda, universality and self-multiplication. In: Sayama H., Rieffel J., Risi S., Doursat R. & Lipson H. (eds). Artificial life 14. Proceedings of the 14th International Conference on the Synthesis and Simulation of Living Systems. MIT Press, Cambridge MA: 490–497.
Maturana H. R. (1987) Everything said is said by an observer. In: Thompson W. I. (ed.) Gaia: A way of knowing. Lindisfarne Press, New York: 65–82.
Maturana H. R. (1988) Reality: The search for objectivity or the quest for a compelling argument. Irish Journal of Psychology 9(1): 25–82. http://cepa.info/598
Maturana H. R. & Varela F. J. (1987) The tree of knowledge: The biological roots of human understanding. Shambhala, Boston.
Müller K. H. & Riegler A. (2014) Second-order science: A vast and largely unexplored science frontier. Constructivist Foundations 10(1): 7–15. http://constructivist.info/10/1/007
National Science Board (NSF) (2016) Science & engineering indicators 2016. Chapter 7: Science and technology: Public attitudes and understanding. http://www.nsf.gov/statistics/2016/nsb20161/uploads/1/10/chapter-7.pdf
Rorty R. (1991) Objectivity, relativism, and truth: Philosophical papers. Volume 1. Cambridge University Press, Cambridge.
Russell B. R. (1938) The principles of mathematics. W. W. Norton, New York. Originally published in 1903.
Sartre J.-P. (2016) What is subjectivity? Verso, New York NY. French original published as: Sartre J.-P. (2013) Qu'est-ce que la subjectivité? Les Prairies ordinaires, Paris. (Lecture held in 1961).
Scott D. (1971) Continuous lattices. Programming Research Group, Oxford University. Technical Monograph PRG-7.
Spencer Brown G. (1969) Laws of form. George Allen and Unwin, London. http://cepa.info/2382
Thagard P. (1989) Explanatory coherence. Behavioral and Brain Sciences 12: 435–502.
Umpleby S. A. (2014) Second-order science: Logic, strategies, methods. Constructivist Foundations 10(1): 16–23. http://constructivist.info/10/1/016
Vaihinger H. (1924) The philosophy of "as if": A system of the theoretical, practical and religious fictions of mankind. Translated by C. K. Ogden. Routledge, and Kegan Paul, London. German original published in 1911.
Varela F. J. (1979) Principles of biological autonomy. North Holland, New York.
Wittgenstein L. (1922) Tractatus logico-philosophicus. Harcourt, Brace and Company, New York. http://www.gutenberg.org/ebooks/5740
Wittgenstein L. (1958) Philosophical investigations. MacMillan, New York. Originally published in 1953.
Wood M., Douglas K. & Sutton R. (2012) Dead and alive: Beliefs in contradictory conspiracy theories. Social Psychological and Personality Science 3: 767–773.

Cybernetic Foundations for Psychology

Bernard Scott

The confusion and barrenness of psychology is not to be explained by calling it a "young science"; its state is not comparable with that of physics, for instance, in its beginnings. (Rather with that of certain branches of mathematics. Set theory.) For in psychology there are experimental methods and conceptual confusion. (As in the other case, conceptual confusion and methods of proof.) The existence of the experimental method makes us think we have the means of solving the problems that trouble us; though problem and method pass one another by. (Wittgenstein 1953: 232)

Introduction

1. From 1950s onwards, concepts from cybernetics spread throughout psychology. In particular, they helped give birth to the domain of modern cognitive psychology. Models of "information processing" became ubiquitous and the research interests of cognitive psychologists increasingly overlapped with those of workers in artificial intelligence research, helping spawn the multidisciplinary domain of "cognitive science." Cybernetic concepts also permeated other domains within the broad field of psychology. However, with rare exceptions, the historical origins of the concepts were lost. Also lost was the intent of the early cyberneticians to look for interdisciplinary enrichment and transdisciplinary unity. In this article, I overview the field of psychology as it currently stands, with its many areas of research and application, which, to a large extent, exist as separate specialist domains of activity (for example, the several subdomains that make up biologically and behaviourally based psychology, cognitive psychology, social psychology, developmental psychology, abnormal psychology and the study of individual differences). I then demonstrate how cybernetics, when its contributions are made explicit, can provide both foundations and an overarching unifying conceptual framework for psychology. In order to do so, I make the distinction between first- and second-order cybernetics and briefly define some key cybernetic concepts, including "system," "self-organisation" and "control" (Scott 2011a, 1996). I also make a broad-brushstroke distinction between "process" and "person" approaches within psychology. I go on to show how cybernetic concepts can unify these approaches. I also show how cybernetic concepts can unify individual psychology and social psychology, a unification that also builds useful conceptual bridges with psychology's sister discipline, sociology. I include reference to my personal experiences as

a practitioner psychologist who encountered cybernetics at an early stage in his studies and who has found that cybernetics can indeed provide conceptually satisfying and practically useful foundations for psychology. It can reveal underlying similarities between problem situations and provide tools for modelling those situations. It can facilitate more effective communication between practitioners.

2. The treatment is necessarily terse given constraints on the length of the article. The author may provide a book-length treatment in the future. In the meantime, it is hoped that the article will generate wider discussion of the issues raised. It should also be noted that cybernetics is an abstract discipline. I have not attempted to provide a comprehensive account of its many applications in psychology. There is a wealth of examples in standard texts, though not explicitly named as such. (See for example, Eysenck & Keane 2015).

The story of psychology

3. Standard histories (for example, Miller 1962; Hunt 1993) tell us that psychology emerged from philosophy as a science in the late 19th century, a key moment being the founding, by Wilhelm Wundt, of the first laboratory dedicated to empirical studies of psychological phenomena. An emphasis on the scientific value of empirical data, rather than armchair theorising, combined with the controversies over the validity of data derived from introspection, led to the rise of behaviourism as the dominant paradigm (or "school"), a dominance that lasted until well into the 1950s and early 1960s. Behaviourists aspired to make psychology an objective science. They abjured reference to consciousness and reference to "inner" experience and studied behaviour as objectively observable phenomena, using controlled experimental conditions that afforded replication of findings. For convenience, many studies were carried out using animals, such as rats and pigeons. The main research programme of behaviourists was focused on studying learning. At an extreme, explanations of how and why learning occurred were eschewed in favour of empirically derived "laws" that afforded predictions about when and where learning would occur – for example, under what circumstances a rat could be most effectively induced to learn how to navigate a maze or a pigeon's behaviour shaped so that it responded in predictable ways in response to particular stimuli.

4. Competing paradigms included structuralism, functionalism, Gestalt psychology, depth psychologies (such as psychoanalysis) and humanistic psychology. In the 1960s, inspired by concepts from cybernetics, a new dominant paradigm arose: cognitive psychology. Cognitive psychology addressed issues to do with attention, perception, memory and problem solving, topics that had been addressed in earlier decades and that had amassed a wealth of empirical findings. What the "new" cognitive psychology contributed was new ways of talking about, and modelling, cognitive processes. The central analogy ran like this, "As programs are to computers, so thoughts are to brains." Models of cognitive processes were built that showed the flow of "information" around a cognitive system. Many such models consisted of static images of boxes and arrows. Others adopted a "computational" approach and were written as computer programs. Parallel work in computer science aimed to create "artificial intelligence" programs to solve problems, to serve as "expert systems," process images, interpret natural languages and acquire "knowledge." A new field became demarcated, "cognitive science," centred on the concept that both brains

and computers are "physical symbol systems." This work following this paradigm continues today. I say more about these developments below.

5. For psychology, a seminal text was the book *Plans and the Structure of Behaviour*, authored by George Miller, Eugene Galanter and Karl Pribram (1960). Not only does the book introduce key concepts relevant for the new approaches in cognitive psychology, it also gives an account of the origins of these concepts in the then emerging field of cybernetics. Other texts that highlighted the relevance of concepts from cybernetics for psychology were George (1960) and Pask (1961). As in other fields, as the years passed, researchers took from cybernetics those concepts they found useful for their special areas of interest, ignored or rejected others and very soon forgot their origins.

6. In more recent decades, "cognitive neuroscience" and "physiological psychology" (or, taken together, "biological psychology") have come to the fore, largely due to the ability to map and manipulate activity in the nervous system and the major advances made in understanding these processes, anatomically and physiologically, down to the molecular level, where the interactions of the endocrine system, the nervous system and the immune system can be seen to form a systemic whole. Because of the systemic nature of this whole, in what follows I frequently refer to the "brain/body system" rather than refer to the brain as if the nervous system was all that is of interest.

7. If one considers psychology as a whole field, one can see that over the years there has been a to-ing and fro-ing as paradigms have become more or less dominant or fashionable, with the major shifts having been brought about by the impact of concepts from cybernetics. Mainstream psychology continues to place great emphasis on empirical research. Associated theorising and model building tends to be specific to a domain or subdomain. Overall, there is still conceptual confusion and controversy over what psychology is about: what it should be aiming to achieve and how it should pursue those aims. At a metatheoretic level, there is now an explicit domain of "critical psychology" that questions the assumptions that underlie mainstream practice (see, for example, Sloan 2000). There is also a periodic (and less critical) attempt to examine the epistemological foundations of the several paradigms (see, for example, Chapman & Jones 1980; Leary 1990).

8. To illustrate the unchanging face of psychology as a field consisting of a variety of topic areas and approaches, in the Appendix, I list the contents of standard undergraduate text books: one from the 1960s (Sanford 1966) and two bestselling texts from the 2000s (Hayes 2000; Gross 2010). I, myself, was an undergraduate in the years 1964–1968 and taught undergraduate courses in psychology, on and off, between 1968 and 2000. I was thus a witness to the changes that occurred in those years. One topic not featured in the Appendix that was (and still is) commonly taught as part of undergraduate courses is organisational psychology.

9. In anticipation of the next section, I wish to say a little more about the conceptual confusion that Wittgenstein above refers to. The crux of his critique is that we should look carefully at how we use words to talk about psychological events and processes, as a way of avoiding the ontologising of "mind" and "matter" (for "matter," one could also write "brain") as different kinds of fundamental "substances." This ontologising comes with the adoption of one of the particular metaphysical positions that underly the competing paradigms in psychology. In brief, both functionalism and structuralism employ dualistic parallelism (mental events are correlated with physiological processes); some dualists also advocate a Cartesian mind/brain interaction; mainstream behaviourism is monistically materialist and reductionist (talk of mental events is not permitted); "cognitivists" are on-

tologically monist, materialist reductionists in that they reduce the "mental" to the status of programs executed by a computer.

10. In the unpublished essay "The relevance of Wittgenstein's philosophy of psychology to the psychological sciences"[1] Peter Hacker provides an extended discussion of Wittgenstein's position and its relevance for psychology. As discussed further below, cybernetics in its role of a metadiscipline and a transdiscipline engages in the kind of "philosophical ground clearing" that Wittgenstein (and Hacker) calls for.

Understanding cybernetics and its contributions to psychology: The story of cybernetics

11. I am not aware of any single text that gives a clear and inclusive account of the origins, early years and key later events concerning cybernetics. Here, I will give a very brief summary.[2]

12. The story has several possible beginnings. One common starting point is the publication, in 1943, of the paper "Behavior, purpose and teleology" by Arturo Rosenblueth, Norbert Wiener and Juliann Bigelow and associated discussions that lead up to the Macy conferences on "feedback and circular causality in biological and social systems" held between 1946 and 1953. The paper proposed that the goal-seeking behaviour that could be built into mechanical systems and the goal-seeking observed in biological and psychological systems have a similar form: they are structured so that signals about achieved outcomes are "fed back" to modify inputs so that, in due course, a prescribed goal is achieved (a cup is picked up) or a desired state of affairs (the temperature of a room or of a living body) is maintained. This process is referred to as "circular causality." It was recognised at an early stage that many fields of study contain examples of these processes and that there was value in coming together in multidisciplinary fora to shed light on them, to learn from each other and to develop shared ways of talking about these phenomena. In 1948, Norbert Wiener, one of the participants, wrote a book that set out these ideas in a formal way that not only collected together many of the emerging shared conceptions but did so in a coherent way that not only facilitated interdisciplinary exchanges but also stood as a discipline in its own right: an abstract transdiscipline – the study of "control and communication in the animal and the machine." Wiener called this new discipline "cybernetics." Following the book's publication, the Macy conference participants referred to their conferences as conferences on cybernetics, keeping "feedback and circular causality in biological and social systems" as the subtitle.

13. As the subtitle emphasises, there was an interest in biological and social systems. The participants were interested not only in particular mechanisms, they also looked for the general forms to be found in the dynamics and organisation of complex systems (living systems, small groups and communities, cultures and societies): how they emerge and

---

1. http://info.sjc.ox.ac.uk/scr/hacker/DownloadPapers.html
2. As further reading, I suggest Heims (1991), Glanville (2002), Pickering (2010), Scott (2002, 2004) and Müller & Müller (2007). I also recommend the 2006 biography of Norbert Wiener, written by Flo Conway and Jim Siegelman. One should also consult key texts of cybernetics' founders and early contributors: Wiener (1948), Ashby (1956), Pask (1961), Foerster, Mead & Teuber (1953), Bateson (1972).

develop, how they maintain themselves as stable wholes, how they evolve and adapt in changing circumstances. The term "self-organising system" was adopted by many as a central topic for discussion in later conferences (for example, Yovits & Cameron 1960). Formal models of adaptation and evolutionary processes were proposed.

14. In the years following the Macy conferences, cybernetics flourished and its ideas were taken up by many in many disciplines. Cyberneticians also found common ground with the followers of Ludwig von Bertalanffy, who were developing a general theory of systems (Bertalanffy 1950, 1972).

15. By the 1970s, cybernetics, as a distinct discipline, had become marginalised. A number of reasons have been suggested for this. I believe two are particularly pertinent. The first is that, at heart, most scientists are specialists. Having taken from cybernetics what they found valuable, they concentrated on their own interests. Second, in the USA, funding for research in cybernetics became channelled towards research with more obvious relevance for military applications, notably research in artificial intelligence.[3] Attempts to develop coherent university-based research programmes in cybernetics, with attendant graduate level courses, were short-lived. However, some developments in the field that occurred in the late 1960s and early 1970s are particularly pertinent for the theme of this article.

16. First, it is useful to note that the early cyberneticians were sophisticated in their understanding of the role of the observer. In the later terminology of Heinz von Foerster (see below), their concerns were both first-order (with observed systems) and second-order (with observing systems). It is the observer who distinguishes a system, who selects the variables of interest and decides how to measure them. For complex, self-organising systems this poses some particular challenges. Gordon Pask, in a classic paper, "The natural history of networks" (Pask 1960), spells this out particularly clearly. Even though such a system is, by definition,[4] state-determined, its behaviour is unpredictable: it cannot be captured as trajectory in a phase space. The observer is required to update his reference frame continually and does so by becoming a participant observer. Pask cites the role of a natural historian as an exemplar of what it means to be a participant observer. A natural historian interacts with the system he observes, looking for regularities in those interactions. Pask goes as far as likening the observer's interaction with the system with that of having a conversation with the system. Below, we will see how this insight of Pask was the seed for his development of "conversation theory."

17. Second, the early cyberneticians had the reflexive awareness that in studying self-organising systems, they were studying themselves, as individuals and as a community. Von Foerster, in a classic paper from 1960 "On self-organising systems and their environments," makes this point almost as an aside. He notes:

> [W]hen we [...] consider ourselves to be self-organizing systems [we] may insist that introspection does not permit us to decide whether the world as we see it is 'real,' or just a phantasmagory, a dream, an illusion of our fancy. (Foerster 2003: 3f)

3. For more on this, see Umpleby (2003).
4. The fundamental tenet of systems theory, cybernetics and computer science is that a system's next internal state and its output are a function of its current internal state and its input. These states and inputs and outputs are as distinguished and modelled by the observer.

Foerster escapes from solipsism by asserting that an observer who distinguishes other selves must concede that, as selves, they are capable of distinguishing her. "Reality" indeed exists as the shared reference frame of two or more observers. With elegant, succinct formalisms, Foerster, shows how, through its circular causal interactions with its environmental niche and the regularities (invariances) that it encounters, an organism comes to construct its reality as a set of "objects" and "events," with itself as its own "ultimate object." He goes on to show how two such organisms may construe each other as fellow "ultimate objects" and engage in communication as members of a community of observers.

18. This interest in the role of the observer and the observer herself as a system to be observed and understood led Foerster to propose a distinction between a first- and a second-order cybernetics, where first-order cybernetics is "the study of observed systems" and "second-order cybernetics is the study of observing systems" (Foerster 1974: 1). Foerster also referred to this second-order domain as the "cybernetics of cybernetics."[5] Of relevance for us here is that cybernetics is not only, as noted above, a discipline in its own right that can serve as a transdiscipline, cybernetics can also serve as a metadiscipline that studies not only itself but other disciplines, too.[6] I have discussed these aspects of cybernetics in some detail in Scott (2002).

19. Again, for the purposes of this article, it should be mentioned that others had been thinking along somewhat similar lines to those of Pask and von Foerster. Humberto Maturana in his seminal paper, "Neurophysiology of cognition" (Maturana 1970a), frames his thesis about the operational closure of the nervous system[7] with an epistemological metacommentary about what this implies for the observer, who, as a biological system inhabiting a social milieu, has just such a nervous system. The closure of the nervous system makes clear that "reality" for the observer is a construction consequent upon her interactions with her environmental niche (Maturana uses the term "structural coupling" for these interactions). In other words, there is no direct access to an "external reality." Each observer lives in her own universe. It is by consensus and coordinated behaviour that a shared world is brought forth. As Maturana succinctly points out, "Everything that is said is said by an observer." In later writings (some written in collaboration with Francisco Varela), Maturana uses the term "autopoiesis" (Greek for self-creation) to refer to what he sees as

5. For more detailed accounts of the events that led up to Foerster's making this distinction, see Glanville (2002) and Scott (2004).
6. It is of particular interest that, beginning with Wundt, many psychologists have considered psychology to be the "propaedeutic science" (Greek propaideutikos, i.e., what is taught beforehand) because what it says about human behavior and cognitive capabilities can shed light on how science works and how it can be carried out effectively by practitioners in other disciplines (and, of course, in psychology itself). See, for example, Stevens (1936). In more recent years "the psychology of science" has emerged as an active area of research. See, for example, Gholson et al. (1989) and Feist (2008). Worthy though the aims of this research are, it remains the thesis of this article that they will be best achieved if psychology itself is properly founded using concepts from cybernetics.
7. The nervous system is an example of a circular causal system: it is a sensorimotor system in which what is done (motor "outputs") affects what is sensed (sensory "inputs") and what is sensed affects what is done (Dewey 1896). It is also worth noting (as stressed by von Foerster) that all sensing is a form of acting (sensory cells are primed to send signals to other cells when something happens that may be relevant for the whole system of which they are a part) and all acting includes sensing (by proprioception and kinaesthesia) what is being done.

the defining feature of living systems: the moment by moment reproduction of themselves as systems that, whatever else they do (adapt, learn, evolve), must reproduce themselves as systems that reproduce themselves. In explicating his theory of autopoiesis, Maturana makes an important distinction: the distinction between the "structure" of a system and the "organisation" of a system. A system's structure is the configuration of its parts at a given moment in time, a snapshot picture of the system's state. The organisation of a system is the set of processes that are reproduced by circular causality such that the system continues to exist as an autopoietic unity. In general, a system with this "circular causal" property is said to be "organisationally closed" (Maturana & Varela 1980).

20. The ideas of Pask are particularly relevant for this article. Not only was Pask an early enthusiast of, and contributor to, cybernetics, he also had psychology as his core discipline. As noted above, Pask had an early interest in seeing interactions between an observer and a self-organising system as having the form of a conversation. Central in his research activity was the design of "teaching machines" and "learning environments" that interact with a learner, in a conversational manner, and adapt to the learner's progress so as to facilitate her learning. Pask was familiar with the work of Foerster and Maturana as a friend and colleague and drew on their ideas in creating his theory of conversations. As described below, Pask's theory is a much more fleshed out and elaborated account of human cognition, learning and communication than is to be found in the writings of either Foerster or Maturana.

21. I shall begin my account of Pask's theory by disambiguating the terms "observer" and "observing system" as used in cybernetic writings. Usually, it is clear from the context that "observer" refers to a human observer capable of being a member of a community of observers. The term "observing system" is used more generally to refer to autopoietic systems. A single-celled organism, such as an amoeba, can serve as an example. An amoeba, to maintain itself as a unity, distinguishes itself from its environment. In its interactions with its environment, it adapts. The form of its organisation changes as a consequence of its interactions (its moment by moment structural coupling). As long as these changes do not affect the organisational closure of the system, the system persists.[8] The amoeba becomes "in-formed" about its environment. It has its own perspective on what is its environment, its "environmental niche." There is thus a sense in which to be alive is to cognise. Multicellular organisms with nervous systems that afford rapid transmission and receipt of signals and rapid self-referential operations no doubt have greater cognitive powers. One may speculate that the cognition of a porpoise (say) is qualitatively different from that of a tree.

22. Although much of what Foerster and Maturana have to say is pertinent to humans, arguably it is Pask, the psychologist, who has given us the most comprehensive observer-based cybernetic theory of human cognition and communication. From the earliest stages of his thinking, he was aware that the human self develops and evolves in a social context and that "consciousness" (Latin con-scio, with + know) is about both knowing with oneself

8. It is worth noting that alongside the abstract cybernetic considerations of the systemic property of organisational closure, there is ongoing research in biophysics that seeks to understand the specific mechanisms by which living systems maintain themselves as coherent entities. See, for example, Mae Wan-Ho's review, in which she notes that none of the biophysical theories of the coherence of biological systems, as developed so far, is "as yet complete or fully coherent" (Ho 1995: 733). I suspect the search to understand the "glue" that holds living systems together will continue to be incomplete, just as other theories in quantum mechanics and cosmology remain incomplete.

and knowing with others. Throughout his writings, from the 1960s onwards there is an acknowledgement by Pask of his indebtedness to the Russian psychologist Lev Vygotsky, who argued that, as a child develops, what begins as external speech eventually becomes internalised as an inner dialogue.[9]

23. Pask, at an early stage in his theorizing made a distinction between a cognitive system and the "fabric" or "medium" that embodies it. This distinction is analogous to the distinction between programs and the computer in which they run. However, unlike the cognitivist science community, where the analogy is the basis of the thesis that both brains and computers are "physical symbol systems," Pask is aware that this interpretation of what is a symbol is conceptually naive.[10] He stresses how important it is to take account of the differences between brain/body systems and computing machinery. Brain/body systems are dynamical, autopoietic systems, whose structure is constantly changing, whereas computers are designed to be stable. In Pask's terms, there is an interaction between a cognitive system and its embodiment. A change in the structure of the brain/body system affects cognition. Changes in thinking affect the structure of the brain/body system. It is important to note that Pask's distinction is an analytic distinction, not an ontological one. It affords a way of talking about cognitive processes distinct from physiological processes.

24. In the late 1960s, Pask adopted a new terminology. Brain/body systems and extensions are referred to as "mechanical individuals" (M-individuals). Cognitive systems are referred to as "psychological individuals" (P-individuals). M-individuals (with extensions, such as vehicles, pens and telescopes) are the "processors" that "execute" the P-individuals as cognitive "procedures." Both kinds of system are organisationally closed, self-reproducing systems. As we shall see in later sections, Pask's distinction between the two kinds of individual (or unities) is very useful for the aim of providing psychology with a coherent conceptual framework.

25. In order to avoid some of the confusions a partial or shallow reading of Pask can lead to, I refer to P-individuals as "psychosocial unities" and M-individuals as "biological unities" or "biomechanical unities." Pask himself on occasion referred to conversation theory and his later development of "interaction of actors theory" as theories of the psychosocial (Pask 1996).

Cybernetics in psychology

26. A key feature of cybernetic explanations is their use of models. The cybernetician Frank George proposes that a theory is a model together with its interpretation (George 1961: 52–56), where a model can be anything: marks on paper, a computer program, a mathematical equation, a concrete artefact. The key idea is that a model is a non-linguistic part of the theory. It is a form, a structure, a mechanism that can be manipulated by an observer and that maps onto the "real" system that the theory is concerned with. This is to be contrasted with many so-called "theories" that are to be found in the humanities, where metaphors and analogies are liberally deployed, without formal (non-linguistic) justifica-

9. Vygotsky's work, carried out in the 1920s and 1930s, did not become available in English until 1960 (Vygotsky 1962).
10. See Scott & Shurville (2011) for an extended discussion of this conceptual confusion within the AI/cognitive science community.

tion. Models are to be found throughout the sciences. What makes a model "cybernetic" is the inclusion of circular causality, for example, in a model of a control system, such as a thermostat. Non-cybernetic models feature "linear causality" only, for example, models that show how the magnitude of a variable is a function of the magnitude of another.[11]

27. The mapping between a model and the system modelled has the form of an analogy relation, such as, "A is to B as C is to D," where A and B are parts or states of the model and C and D are parts or states of the system modelled. There may of course be a number of such relations. It is also relevant to note that metaphors are abbreviated analogy relations. For example, the term "The ship of state" is asserting that steering a ship is analogous to governing a nation state. Pask tersely defines cybernetics as "The art and science of manipulating defensible metaphors" (Pask 1975a: 13). Not only does this definition capture the idea of constructing and validating models, "manipulating" carries with it the idea that the observer is in a circular causal relation with the model and the system modelled and the use of the word "defensible" carries with it the idea that the observer is a member of a community of observers.[12]

28. Prior to the advent of cybernetics, psychology's bias was towards reporting empirical findings. As theory, the best that behaviourism could offer was a model of the brain as a kind of telephone exchange where "stimuli" give rise to "responses." Gestalt psychologists used the concept of brain activity being "field"-like in an attempt to explain how perceptual inputs were reconfigured to conform to the "laws of pragnanz" (good form) in perception and problem solving.[13] Now models featuring circular causality can be found throughout psychology, for example, models of perceiving, problem solving, learning, remembering and skilled performance. However, their general form tends not to be highlighted. There is a focus on specific subdomains, rather than an appreciation that the models are part of larger general class.

Unifying "process" and "person" approaches

29. By "process" approaches, I am referring to those that set out to model and understand some particular aspect of human cognition. As mentioned above, models for these processes abound in contemporary psychology, as an examination of standard texts will show (for example, Eysenck & Keane 2015). By "person" approaches, I am referring to those that concern themselves with a human being as a whole, albeit, possibly focusing on some particular set of attributes, such as "personality" or "intelligence." Whole person approaches are sometimes referred to as "humanistic psychology." My proposal here is that cybernetics, because it deals with both the processes that constitute the behaviour of parts of a system and the joint effects that constitute the behaviour of whole systems, can supply a conceptual framework that unifies the two approaches. I have written about this

---

11. For more on cybernetic explanations and cybernetic modelling, see Klir & Valach (1967) and Scott (2000).
12. For more on the use of analogies in science, see Hesse (1966). For more on the use of analogies in cybernetics, see Pask (1963).
13. In "hands-on" studies of the brain (neuropsychology), more sophisticated models were constructed, as in the classic work of Donald Hebb (1949), whose models are clearly "cybernetic" in the sense used here.

possibility elsewhere (Scott 2001d, 2011b, 2011c) and have drawn on two main sources, Pask and von Foerster.

30. In the field of "cognitive science," which subsumes artificial intelligence research and certain approaches to cognitive psychology and the philosophy of mind, there have been several attempts to build a "unified cognitive architecture." See, as examples, Newell's SOAR (Newell 1990),[14] and Anderson's ACT-R (Anderson 1983).[15] Both systems are built from components. Both systems take inspiration from (and can be seen as embodying) theories of human cognition. Both systems are "artificial intelligences" in their own right. In SOAR, every decision is based on current sensory data, the contents of working memory and knowledge retrieved from long-term memory, where long-term memory contains procedural knowledge, semantic memory and episodic memory. ACT-R's main components are: perceptual-motor modules, two kinds of memory module (declarative and procedural), buffers that access modules and a pattern matcher that matches buffer contents to the possible actions ("productions") stored in procedural memory. Further details are not relevant for the argument being made.

31. In contrast, von Foerster makes clear that the components of a unified cognitive architecture are inseparable:

> In the stream of cognitive processes, one can conceptually isolate certain components, for instance (i) the faculty to see (ii) the faculty to remember (iii) the faculty to infer. But if one wishes to isolate these faculties functionally or locally, one is doomed to fail. Consequently, if the mechanisms that are responsible for any of these faculties are to be discovered, then the totality of cognitive processes must be considered. (Foerster 2003: 105)

32. More generally, von Foerster criticises "the delusion, which takes for granted the functional isomorphism between various and distinct processes that happen to be called by the same name." In this context, he mentions the misapplication to computing machines of the terms "memory," "problem solving," "learning," "perception" and "information" (Foerster 2003: 172).

33. Theorising in any discipline needs foundations: somewhere to begin the telling of explanatory stories. In psychology, it has been common practice to begin with elementary building blocks, such as "habits," "expectations," "stimulus-response bonds," "memory states," "drives," "thoughts," "instincts," "cognitive processes," "feelings." I believe that von Foerster provides a cybernetic foundation for psychology with his concept of a "self-organising system," as set out in his 1960 paper "On self-organising systems and their environments." A self-organising system "eats energy and variety from its environment" (Foerster 2003: 6). The rate of change of redundancy in the system is always positive. The system is always becoming more ordered. The observer is continually obliged to update her reference frame.[16] He points out that, reflexively, the observer is just such a system. A classic example from the human domain is a human infant exploring its environment. Of

---

14. See also http://soar.eecs.umich.edu
15. See also http://act-r.psy.cmu.edu
16. Foerster (2003: 281) refers to Varela, Maturana and Uribe as the inventors of the idea and to their joint paper (Varela, Maturana & Uribe 1974) as the first statement of the idea in English. Elsewhere (Foerster 2003: 251), he notes that the general form of the closed system of recursively applied operations that constitutes autopoiesis was described by Maturana before it was named (Maturana 1970a, 1970b).

course, metabolic requirements mean it has to rest once in a while as energy and variety are assimilated and accommodated.

34. In later years, von Foerster refined the concept of a self-organising system, citing the concept of autopoiesis as a useful way to speak about an organism as an autonomous entity: "Autopoiesis is that organization which computes its own organization"; "Autopoietic systems are thermodynamically open but organizationally closed" (Foerster 2003: 281). I believe von Foerster's definitions are a very useful way of uniting the earlier and later literatures.

35. In his studies of human learning and cognition, which lead to the development of his conversation theory (CT), Pask took von Foerster's concept of a self-organising system and made it a cornerstone of his theorising about the dynamics of learning, arguing that humans have a "need to learn." He refers to his interest in the whole system aspects of human cognition as "macrotheory."[17] In contrast, he refers to his (and colleagues') accounts of how human subjects construct particular cognitive structures as "microtheory." Pask (1975b) refers to the processes that are the parts of a cognitive system by the general term "concept." Pask's usage of the term is quite unusual as his concepts are dynamic processes. In mainstream cognitive science, concepts are typically thought of as relatively static representations.[18] Pask defines a concept as a procedure that recalls, recognises, constructs or maintains a relation. A concept may be likened to a program or operator that solves particular problems. "Relation" is used here as an empty slot or label for that which is being acted upon by the process as input or product.

36. Recursively, there are concepts whose domain of application, whose input and products, are other concepts. This affords the construction of hierarchies of concepts. Thus, there can be problem-solver concepts, the task of which is to construct and select from amongst lower-level putative problem solvers, guided by feedback from the problem domain about the success or not of their application. Thus learning is an evolutionary process. One of Pask's very elegant definitions of learning is that it is the construction of a hierarchy of problem solvers (Pask 1975b). Micro and macro aspects of his theorising are married in the idea that "conceptualisation," the process of creating and recreating concepts, is an ongoing dynamic activity. A Paskian P-individual is a system of concepts that is self-reproducing. Particular hierarchies of concepts are seen to be temporary constructions and re-constructions within an overall heterarchical, organisationally closed system of processes.[19,20]

37. In CT, in an effective learning conversation, the role of the teacher (human or machine) is to facilitate the learner's construction of new concepts. This is done by providing the learner with descriptions and demonstrations of what is to be learned, as part of an

17. Macrotheory is crucially concerned with giving some account of "awareness" and "consciousness" as being concerned with seeking variety and the consequent reduction of uncertainty. It is not possible here to address these topics satisfactorily, see Pask (1981) and Scott & Bansal (2014).
18. Walter Freeman (2000) gives an elegant description of the differences between representationalist accounts of cognition and dynamic and "enactive" accounts from the perspective of contemporary findings in neuroscience. His arguments in favour of dynamic approaches are cognate with Pask's theorising.
19. Within mainstream representationalist cognitive science, there have been attempts to develop theories of concept system dynamics. See, for example, Barsalou (2012). Arguably, these accounts are unsatisfactory because they lack the concept of an organisationally closed unitary system.
20. For further discussion of these core ideas of CT, see Scott (2009).

ongoing conversation. In return for these affordances to help in her learning, the learner is invited to say what she is aiming to learn and how she intends to go about it (what strategy for learning she has, if any). Periodically, the learner's understanding of new concepts is assessed by requiring her to "teach back" what she has learned.[21] With respect to this ongoing cycle of learner and teacher interactions, Pask not only views the two participants as self-organising systems in interaction, he also views the learning conversation itself as an emergent self-organising system, a P-individual (psychosocial unity) in its own right. As a generalisation, Pask then argues that all conversations are, at heart, learning conversations. In conversations, whatever else the participants are doing, they are learning about each other.

Unifying individual and social psychologies

38. What is also innovative and unifying in Pask's conversation theory (CT) is the voiding of the distinction between the human individual and the social processes that are constitutive of him/her and that he/she constitutes. Pask agrees with George Herbert Mead, Leo Vygotsky, Martin Buber and von Foerster that the psychological individual is dialogical in form, *is* a social process, *is* constituted by an inner dialogue, *is* an inner conversation. As a good cybernetician, Pask abstracts from specific cases and voids the distinctions and thus argues that all conversations, all dialogues, all social processes are psychological individuals. They are all organisationally closed, self-producing, collectives of concepts (psychosocial unities). Thus, in ontogeny, individuals and collectives are co-evolving psychosocial unities. For an extended discussion of this view, see Scott (2007). We can now see the usefulness of making a distinction between M-individuals (biomechanical unities) and P-individuals (psychosocial unities) in that the two types of unity need not necessarily be in one-to-one correspondence. A single M-individual (a brain/body system, for example) may embody several P-individuals (the inner conversation). A single P-individual (the outer conversation that unifies a collective) may be embodied in several M-individuals.

39. CT is useful for providing a conceptual framework that helps in understanding the dynamics of interpersonal perception and the pragmatics of human communication (see Scott 1987, 1997). As a reflexive theory of theory building (learning), CT accounts for its own genesis. Top down, it accepts that theories are the consensual constructions of communities of observers engaged in conversation, including conversations about conversation. As such, it is cognate with the "discursive" approach in the humanities and social sciences (also known as social constructionism).[22] Bottom up, its foundations lie in the cybernetics of self-organising systems and their interactions as described above.

40. It is also worth noting that the CT concept of a psychosocial unity provides an alternative, cybernetics-based, concept of a social system to that developed by the sociolo-

---

21. For more details about CT's application in the design of a conversational learning environment, see Pask, Scott & Kallikourdis (1973).

22. As examples, see Gergen (1999) and Gergen, Schrader & Gergen (2009). The latter is a collection of readings; authors of contributions include Rom Harré, John Shotter, Steve Duck, Erving Goffman, Harold Garfinkel and Ludwig Wittgenstein.

gist Niklas Luhmann (1995). Luhmann distinguishes three kinds of "autopoietic" system:[23] biological, "psychic" and social. Pask's unification of the individual and the social distinguishes just two kinds of organisationally closed system: the biological and the psychosocial (M-individuals and P-individuals).[24]

Future directions

41. There are two areas in which I believe an observer-focused cybernetics can continue to contribute to psychology and the cognitive and social sciences at large. One is conceptual clarification; the other as a foundation for and a reframing of the education of psychologists.

42. As so ably pointed out by Hacker (op. cit.), conceptual confusion abounds in psychology, cognitive science and the neurosciences, not least in talk about "consciousness" as an ontological essence or of brains and computers having the same ontological status as "physical symbol systems." Hopefully, second-order cybernetics will continue to do its job of conceptual ground-clearing, and the ongoing empirical and theoretical research into "minds," "brains," "individuals" and "societies" will be better conceived and more fruitful.

43. Arguably, the education of psychologists should begin with an understanding of complex adaptive systems and the specific concept that humans are self-organising, autopoietic wholes that in their ontogeny and social interaction develop organisationally-closed cognitive and affective systems and become psychosocial unities (psychological individuals).[25] It should then set out, in broad-brushstroke form, the unifying conceptual framework I have sketched out above.

Conclusion

44. I have proposed observer-based cybernetic foundations (with complementary first and second-order aspects) and a unifying conceptual framework for psychology and have argued for the value of my proposals based on the experience of how cybernetics served me. As an undergraduate, encountering cybernetics transformed my approach to studying and understanding psychology. It gave psychology a conceptual coherence that, previously, I had found lacking. In later years, as my understanding of cybernetics deepened, I continued to use second-order cybernetics as a foundation and framework for my work as an experimental psychologist (summarised in Scott 1993) and my later work as a practitioner in educational psychology (Scott 1987) and educational technology (Scott 2001a). The transdisciplinary and metadisciplinary nature of second-order cybernetics empowered me to read widely (and, on occasion, deeply) in other disciplines (logic, mathematics, comput-

---

23. Luhmann takes this term from Maturana and Varela to refer to systems that are self-reproducing and organisationally closed. His use of the term is controversial. See, e.g., Buchinger (2012) and the associated open peer commentaries.
24. Pask and Luhmann are compared more systematically in Scott (2001b) and Buchinger & Scott (2010).
25. Elsewhere I have outlined a curriculum for "cybernetic enlightenment," which sets out some of my proposals in more detail (Scott 2014).

er science, philosophy, linguistics, the natural sciences, the social sciences).[26] Second-order cybernetics helped me learn how to learn. It helped me to appreciate readily the concepts and methods that inform other disciplines and their applications. I hope my account here will encourage others to explore, or to continue to explore, what second-order cybernetics has to offer.

Appendix: Contents of standard undergraduate textbooks

Contents listing for Sanford (1966)

Part One: Introduction 1. Knowing the human being. 2. Theories of people.
Part Two: Biological Foundations of Behaviour 3. The developing organism. 4. Biological basis for integrated behaviour.
Part Three: Methods in Psychology 5. Tests and measurements in psychology. 6. Experimental design and psychological statistics. 7. Intelligence.
Part Four: Segments of the Psychological Process 8. Motives. 9. Emotions. 10. Sensation. 11. Perception. 12. Basic processes of learning. 13. The management of learning. 14. Higher mental processes.
Part Five: Behaviour of the Whole Organism 15. Personality. 16. Adjusting. 17. Neurosis, psychosis and psychotherapy. 18. Social psychology.

Contents listing for Hayes (2000)

1. Perspectives in Psychology.
Section 1: Cognitive Psychology 2. Perception and Attention 3. Memory 4. Language and Literacy 5. Thinking and Representation.
Section 2: Individuality and Abnormality 6. Intelligence 7. Theories of Personality 8. The Medical Model of Abnormal Behaviour 9. Alternatives to the Medical Model.
Section 3: Physiological Psychology 10. Brain Development and Clinical Neuropsychology 11. Consciousness 12. Sensation and Parapsychology 13. Emotion and Motivation.
Section 4: Social Psychology 14. Self and Others 15. Understanding Others 16. Social Influence and Social Action 17. Attitudes, Prejudice, and Crowd Behaviour.
Section 5: Developmental Psychology 18. Learning and Skill Development 19. Cognitive Development and Social Awareness 20. Social Development 21. Lifespan Developmental Psychology.
Section 6: Comparative Psychology 22. Introducing Comparative Psychology 23. Animal Behaviour 24. Animal Communication 25. Methods and Ethics in Psychology.

Contents listing for Gross (2010)

The Nature and Scope of Psychology: What is this thing called psychology? Theoretical approaches to psychology. Psychology as a science.

26. A propos of this, the developmental psychologist, Jean Piaget (1977: 136) writes, "Thus cybernetics is now the most polyvalent meeting place for physicomathematical sciences, biological sciences, and human sciences."

The Biological Basis of Behaviour and Experience: The nervous system. Sensory processes. Parapsychology. States of consciousness and bodily rhythms. Substance dependence and abuse. Motivation. Emotion. Learning and conditioning. Application: health psychology.
Cognitive Psychology: Attention and performance. Pattern recognition. Perception: processes and theories. The development of perceptual abilities. Memory and forgetting. Language, thought and culture. Language acquisition. Problem solving, decision-making and artificial intelligence. Application: cognition and the law.
Social Psychology: Social perception. Attribution. Attitudes and attitude change. Prejudice and discrimination. Conformity and group influence. Obedience. Interpersonal relationships. Aggression and antisocial behaviour. Altruism and prosocial behaviour. Application: the social psychology of sport.
Developmental Psychology: Early experience and social development. Development of the self-concept. Cognitive development. Moral development. Gender development. Adolescence. Adulthood. Old age. Application: exceptional development.
Individual Differences: Intelligence. Personality. Psychological abnormality: definitions and classification. Psychopathology. Treatments and therapies. Application: criminological psychology.
Issues and Debates: Bias in psychological theory and research. Ethical issues in psychology. Free will and determinism, and reductionism. Nature and nurture.

Open Peer Commentary: **Wielding the Cybernetic Scythe in the Blunting Undergrowth of Psychological Confusion**

Vincent Kenny

Introduction

1. It is an admirable and generous project to attempt to reconstitute what is known as "psychology" by using the keen insights from the framework of second-order cybernetics. However, before Bernard Scott embarks on a book-length version of his target article, and before he sets off to use the glue of cybernetics to stick together psychological components that might not have much, if anything, to do with one another, I would like to add some fundamental issues to his list of "conceptual ground-clearing." These issues mostly relate to the fragmental chaos of that which is called "psychology."

"Psychology" does not exist

2. "Psychology" uses very different operations of distinctions with very different cultural and language preferences and with very different socio-cultural and political intents. For example, the dominant values of psychology for the past century were to "predict and control," which, every so often, emerge publicly as critical moral issues in situations such

as the long-running scandal of the American Psychological Association members secretly colluding with the CIA in the torture of prisoners (American Psychological Association 2015a). Historically, we have seen in Russia the use of psychiatric diagnosis for the incarceration of political prisoners (Lader 1977), and many other socio-political abuses of the "prediction and control" mentality (Breggin 2008; Johnstone 2000; Mills 2013; Tyler 2013).

3. There is no universal – or even dominant – consensual agreement as to *which* operations of distinction are to be used to bring forth any *given* version of a "psychology" nor for which purposes. This raises serious questions about "values" in psychology, and reflects Scott's statement that "there is still conceptual confusion and controversy over what psychology is about" (§7).

4. It is no less than extraordinary that the report on the APA's long-term involvement in torture should raise questions as to what "psychology" is. To quote the chairman of the report, former federal prosecutor David Hoffman, who says that their investigation "will help define the meaning of psychology," he feels it necessary to warn that when the psychology profession allows for the possibility that psychologists will intentionally inflict pain on defenceless people "…faith in the profession can diminish quickly" (American Psychological Association 2015b: 72).

5. The absence of any consensual agreements has had the effects of generating countless versions of what "psychology" is understood to be. Donald Bannister and Fay Fransella observe:

> In the past, the carving up of the field into mini-psychologies has allowed a 'live and let live' policy. Each psychologist has been free to stake his own claim and produce work which had no implications, nice or nasty, for the endeavours of those in other territories. (Bannister & Fransella 1971: 56.)

6. With such unilateral attitudes, it is clear that "psychology" is not a coherent entity, and is not a "unity." "Psychology" does not exist in the way that we understand physics, chemistry, etc. to be existing – as an accumulated body of knowledge that is reliable and experimentally replicable. Instead, there is a proliferation of "mini-theories" about highly selective areas of human experiencing ("memory," "motivation," etc.) which has been described as "the sickness of chapter-heading psychology which has made a textbook convenience the limits of our imagination" (Bannister & Fransella 1971: 15).

7. Scott himself illustrates this in his observations of the "unchanging face of psychology as a field" (§8) – that is, that the textbooks still continue today with the same "chapter-heading" limitations.

8. "Psychology" is not a unity to be recomposed as an "entity" because it has never existed as such and has never been organised into a coherent system of knowledge.

9. This, then, is to do with the traps "psychology" makes for itself by its use of idiosyncratic operations of distinctions that are then obscured by the object they have brought forth.

Irrelevance of psychological research

10. What has been produced over a century of research amounts to little more than a mass of trivialities having little or nothing to do with people's actual experience of living

their lives. That is, the dominant form of "chapter-heading psychology" – which reduces the person to convenient segments – has entirely failed to deal with its proper subject.

11. Apart from the stark irrelevance of academic research psychology to the actual living experiences of people, there has not been very much connection between the world of research on the one hand, and clinical psychology and psychotherapy practice on the other hand (Tavris 2004). That is, clinicians have not found much help from their research colleagues in their daily task to be of some helpful relevance to people in dealing with their ongoing states of suffering. Robert Joynson (1974: 34) observes that: "... the psychologist's findings seem either to be a mere repetition of what ordinary good sense already knew, or, regrettably, a distinctly inferior brand of information."

Cooking the books

12. Going back to the last century in British psychology, we find the infamous case of Sir Cyril Burt, whose research work on the heritability of intelligence was shown to have been fabricated. Leslie Hearnshaw, a fellow psychologist and his official biographer, concluded that most of Burt's data from after World War II were unreliable or fraudulent (Hearnshaw 1979).

13. Most recently, we have seen the case of Diedrik Stapel, who was shown to have fabricated his research to the extent of rendering at least 30 of his publications fraudulent, and throwing much doubt on many other publications (Stapel 2014).

14. There have been continuing alarms sounded about the status of research in psychology and the level of unreliable research findings (Ioannidis 2005; Simmons, Nelson & Simonsohn 2011). An anonymous survey of 2 000 psychologists by Leslie John, George Loewenstein and Drazen Prelec (2012) finds that "questionable practices may constitute the prevailing research norm." And Tom Farsides and Paul Sparks wonder:

> Consider the roll call of those who have in recent years had high-status peer-reviewed papers retracted because of confirmed or suspected fraud: Marc Hauser, Jens Förster, Dirk Smeesters, Karen Ruggiero, Lawrence Sanna, Michael LaCour and, a long way in front with 58 retractions, Diederik Stapel. [...] Could most of what we hold to be true in psychology be wrong (Ioannidis 2005)? (Farsides & Sparks 2016: 368).

The replication crisis

15. Apart from the sheer volume of faked results, we also have the enormous problem of replicability – or rather of the *non-replicability* of experimental results in "psychology." This problem was already flagged by Hans Eysenck 60 years ago when he warned that...

> the root of many of the difficulties and disappointments found in psychological research, as well as the cause of the well-known difficulties in duplicating results [...] lies in this neglect of individual differences[...]
>
> Hundreds of extremely able psychologists spent time, energy, and a considerable amount of money [...] apparently quite pointlessly; must this sort of thing be repeated endlessly before we learn the lesson that individual differences [...] may not be pushed aside and forgotten when experiments are designed which purport to reveal universal truths?" (Eysenck 1966: 26)

16. It is just as well that Eysenck is not around anymore to learn that indeed it seems that psychologists *are* condemned to repeat the same errors endlessly, getting nowhere with the replication issue 60 years later! Psychologists are supposed to know something about "learning," but if they do they certainly do not know how to reflexively apply their learning schedules to themselves.

17. Farsides and Sparks also comment on the very serious problem of replicability in psychological research.

> Few successful attempts have been made to rigorously replicate findings in psychology. Recent attempts to do so have suggested that even studies almost identical to original ones rarely produce reassuring confirmation of their reported results. (Farsides & Sparks 2016: 370)

18. Commenting on the fact that the Reproducibility Project has revealed that only 36 per cent of findings in psychology appear to stand up to a replication attempt:

> [F]or any recently published significant result in a leading psychology journal, there is only a one in three chance that the research, if repeated, would produce a statistically significant replication. […] Furthermore, the effect size of the repeated study is likely to be less than half of that originally reported. (Morris 2015: 858)

Instead of "giving psychology away" as George Miller (1969) once tried to do – and who would want it even for free? – must we now think of just "throwing it away"?

"Guild-ing the wily"

19. So arising from the fact that "psychology" is not what it pretends to be, and that it has produced very little of relevance to people's daily experiencing, one of the major tasks in hand is to deal with the guild-like functioning of "psychological societies" or "associations," usually organised at a national level, who perpetuate the illusion of "psychology" as a "science."

20. In describing what he calls "disabling professions," Ivan Illich notes that they go beyond the type of powers operated by guilds and unions in that:

> They claim special, incommunicable authority to determine not just the way things are to be made, but also the reason why their services are mandatory. Many professions are now so highly developed that they not only exercise tutelage over the citizen-become-client, but also determine the shape of his world-become-ward.
>
> A profession, like a priesthood, holds power by concession from an élite whose interests it props up. (Illich 1977: 16f)

He continues:

> Professionals assert secret knowledge about human nature, knowledge which only they have the right to dispense. They claim a monopoly over the definition of deviance and the remedies needed […] Public affairs pass from the layperson's elected peers into the hands of a self-accrediting élite. (ibid: 19f.)

21. So here we have the serious problem of a professional psychological élite that exerts the powers conceded to it to demarcate territory, controls who can work within this territory, establishes and imposes price-lists, and operates to expropriate the competencies

and self-governing understandings and skills that have always existed in the common-sense domain of human living.

22. The extensive fragmentariness of "psychology" is hidden and obscured by the operations of psychology organisations, which create the social illusion that there does in fact exist a coherent scientific body of work that legitimises their claims to "expertise" (when the opposite is the case). These are self-declared "experts" with little scientific or other basis for their claims to the right to expropriate that which belongs to the common citizen (McCann, Shindler & Hammond 2004).

23. Criticising the tendency among "psychology professionals" to prefer the sensation of certainty arising from the exercise of control over others rather than face the uncertain task of creating personal significance in one's living, Kelly observes:

> We would rather know some things for sure, even though they don't shed much light on what is going on. Knowing a little something for sure, something gleaned out of one's experience is often a way of knowing one's self for sure, and thus of holding on to an identity, even an unhappy identity. And this in turn, is a way of saying that our identities often stand on trivial grounds. If I can't be a man I can, at least, be an expert. (Kelly 1977: 7)

Conclusion: A genuinely *psychological* psychology

24. So what is necessary in any new psychology is a return to the common understandings of the "layperson" as opposed to perpetrating the pretence that "psychological science" has ever, or will ever, "discover" something that will replace the common understandings. "Psychological science," after more than a century of research, has significantly *underwhelmed* the layperson because whatever they have "discovered" is either something already well known (the layperson's understandings were correct) or they discovered something so banal that it had little or no relevance to the way people live their lives. Instead of perpetuating "segmented man" in the guise of "physiological psychology," "neurological psychology," etc., we need a psychology that is capable of being "*psychological*"!

25. We need new ideas of what a psychology could be and what kind of psychology is needed for the forms of mentality/experiencing of *this* century. We need to throw away the stale, outmoded notions of over 100 years ago. People today have very little in common with people of the Victorian era. What it means to be "human" has radically changed, along with the kind of world within which we must find novel adaptations. This applies especially to those born after the commencement of this millennium. Anyone who works in psychotherapy and who deals with young people will know exactly the profound irrelevance of trying to apply psychological models from the last century to their experience of living today.

26. Cybernetics would be better used not as a "glue" for sticking together things that cannot belong together – into a "PsychoFrankenstein" – but rather to create a completely different understanding of what a genuinely useful psychology could be for this new era.

Open Peer Commentary: **To What Extent Can Second-Order Cybernetics Be a Foundation for Psychology?**

Marcelo Arnold-Cathalifaud & Daniela Thumala-Dockendorff

1. Bernard Scott's target article is among the few works of psychologists who discuss a possible unification of psychology in a reflexive way. Scott does not refer to an "imagined identity," nor does he argue from a historical background or from considerations following in the footsteps of pioneers of contemporary specialties or "schools." In other words, he is not trying to create nostalgia or feelings of loss. Instead, his work is strongly proactive. The focus is therefore on proposing adherence to a theoretical construction – second-order cybernetics – warning that it could set strong demands and cause changes in the current mode of the discipline of psychology.

2. Scott's arguments, while brief and focused, incorporate many aspects that evidence long work and reflection on these issues. In this commentary, we assume the position of the Devil's advocate in order to encourage further discussion on the subject. To do this, we have selected some indications and critical aspects of the text, especially those that can lead to a productive exchange of opinions in a diverse, yet constructively oriented, academic community.

3. We highlight that Scott's diagnosis, although disciplinarily focused on psychology, represents a general condition in human and social sciences. Scott argues that up to now, these disciplines have seemed to be content with developing increasingly sophisticated methods, whose applications accumulate specialized but disconnected knowledge. Certainly, and agreeing with Scott, all these disciplines could be seen to be in a pre-paradigmatic phase, as well as lacking internal unity, not only among its specialties but also within them. Although this generalization may be correct, we can still envision some cases that depart from this pattern. For example, linguistics and economics seem to have more internal consistencies than other social disciplines. How could they come to this? Perhaps Scott could shed light on this question and venture some comparisons.

4. Moreover, a possible "blind spot" in Scott's diagnosis can be found when he implicitly and arbitrarily assumes a positive value for the conceptual and theoretical unification of disciplines such as psychology. Although this may sound acceptable (and could even be partly shared by us) it is not enough to justify the need for something that has not prevented psychology from becoming an autonomous discipline. Even more so, how is it possible to explain that the coexistence of organizational and disciplinary spaces of many "psychologies," some of them almost isolated from each other, has not fractured psychology? In other words, is that unity useful? Or is it just a matter of values and preferences of those who attempt to give coherence to their choices? Finally, we wonder, do the most "mature" sciences such as physics or biology enjoy unity? In short, the "obviousness" of the need for a coherent conceptual discipline, as well as the benefits that would result from having general theoretical models, are arguable. Scott seems to have a perspective, but this perspective would need to be cleared up.

5. The main argument of the article is that second-order cybernetics has the characteristics to unify the scattered field of interests and applications of modern psychology. The author notes some progress when he documents that certain cybernetic concepts have prematurely permeated different fields of psychology, starting from those with a cognitive

orientation. He argues that these assimilations have been used in research and applied areas regardless of their foundations or their subsequent developments. Undoubtedly, this is correct but it could be argued that cybernetics, as well as systems theory, does not have a unified conceptual body. In fact, cybernetics and systems theory are full of contradictions and open disputes (Cadenas & Arnold 2015). Finally, is Gordon Pask's cybernetics (to which Scott refers) not just *a version* of cybernetics? How could a single version satisfy the need for a unified psychology?

6. The guidelines for distinguishing between systemic and constructivist perspectives such as "second-order observation," "self-organized systems" and general indications about "observer systems" are very powerful and yet problematic. Can psychology integrate them? Or would it need to ignore the differences between the notions of "self-organized systems" and "autopoietic systems"? Specifically, would it need to ignore the differences between the classical distinctions of "circular causality" of second-order cybernetics and the notions of "operative closure," "structural determination" or "structural coupling" developed by Humberto Maturana and his colleagues (for example, Varela, Maturana & Uribe 1974; Maturana 2002), especially when these notions are transferred from machines and organisms to human and social systems? What is Scott's perception of the emergence of these new levels of complexity? Or does he only propose a metaphorical use of such advanced second-order cybernetic notions?

7. When we take Scott's perspective to a particular field of psychology, for instance clinical psychology and, in particular, to psychotherapy, more specific questions arise. Many diverse psychotherapeutic models and schools are widely recognized, some of which have shown more clinical effectiveness than others. However, the persistence of this diversity of approaches (and its increase) shows the complexity implied in the distinction between psychological problems and their treatments. From most orthodox versions of behaviorism to the most orthodox versions of psychoanalysis, the approaches cover multiple visions of

a. mental health and mental illness,
b. the therapeutic objectives that must be reached,
c. notions of change and its possibilities in therapy, and
d. the role of the therapist and methods of intervention, among others.

Can second-order cybernetics provide a sufficiently broad and integrative framework and at the same time be specific enough to guide the generation of knowledge and psychotherapeutic practices?

8. Considering Niklas Luhmann's perspective (1984, 1986), it is interesting that by recognizing the unity of "psychic systems" as "autopoietic systems," this sociologist has also somehow demarcated a disciplinary field for psychology and related disciplines into something that we call "psycho (auto) poiesis" (Thumala-Dockendorff 2010; Arnold 2010) as an emerging and distinguishable unit. Luhmann's demarcation involved a conceptual re-specification of the notion of autopoiesis that up to now has been very controversial (Arnold, Urquiza & Thumala 2011). In this sense, the question arises of whether Scott, when comparing the works of Luhmann and Pask, puts both at the same level. And in what way does he appreciate their similarities and differences?

9. Finally, in our opinion, even considering our own proximity to Luhmann's theory, we believe that the complexity of the matters we intend to deal with is too great to be confined to a single theoretical observation program irrespective of its sophistication. However,

we agree with Scott that systemic, cybernetic and constructivist notions are priceless contributions that deserve our full attention and dedication to develop them and apply them to disciplines such as psychology, especially due to the normative character of many of its applications. It remains to be seen how the author will continue his approach by applying it to the development of knowledge in different fields of psychology. As Scott's proposal is well-founded and opens interesting questions and possibilities, the conversation remains open.

Open Peer Commentary: **The Importance – and the Difficulty – of Moving Beyond Linear Causality**

Robert J. Martin

1. We owe a debt of gratitude to Bernard Scott for opening a conversation on the failure of the foundations of psychology to move beyond linear causality toward circular causality and other concepts central to cybernetics.

Linear causality: The underlying paradigm of science

2. The concern with linear causality in this commentary is not with the idea in itself, but with the exclusion of circular causality and other cybernetic concepts from science. Science, including psychology, is rooted in the eighteenth and nineteenth century reductionist idea that the universe is a giant clock. The assumption was that if you understood the constituent parts of the clock, you understood the clock and you could predict the operation of the clock. Keep in mind that "clock" at this time does not have to mean only a clock: watch and clock makers built a variety of automata using clockwork mechanisms. One example is the orrery, a model of the solar system that both modeled the relationships of the planets and, like a clock, predicted their positions in real time.

3. Physics was seen to reveal cause and effect in a game of billiards, chemistry was seen to reveal the secrets of molecules and how to make them, biology was seen to reveal the relationship between stimulus and response (for example, as demonstrated by making the legs of a recently dead frog twitch when an electrical charge is applied). While these examples are now viewed as early steps leading to the complex disciplines of physics, chemistry, biology, psychology, and so on, one thing remains: the assumption that everything can be understood as a physical mechanism and that the role of science is to uncover or reveal the underlying causality that underlies all phenomena. It is this assumption that second-order cybernetics questions.

4. Surrounded by physical mechanisms that modeled the universe, it seems understandable that the metaphor of physical mechanisms would have come to influence or even dominate the thinkers who lived in these centuries. Of course, mechanistic models could not explain human thinking, so, perhaps in an effort to make room for human thinking, Descartes proposed a dualism of mind and body. Rejecting the notion of a non-material mind, modern science has identified the mind with the brain and proceeds on the assumption that the mind/brain can be understood in reductionist fashion as a physical mechanism.

5. Differences are not physical entities. Anthropologist Gregory Bateson, one of the founders of cybernetics, was deeply troubled by this reductionism. Bateson (1972) argued

that science excludes human beings (as living systems). By this I believe he meant to point out that the assumption that everything can be understood as a physical mechanism prevents the study of those processes that characterize human beings and that are usually identified as "mind" in the dualism of mind/body. On the one hand, Bateson was against any supernatural explanation of mind; on the other hand, he was against any assumption that mental processes – or any processes that require computation of a difference – could be reduced to physics or chemistry. Bateson argued that relations are not material; they arise through the computation of a difference, and those differences (which are not things) are what make possible the formation of a hand from a genetic instruction, the ability of a tree to reach toward the light, or the ability of human beings to think. Bateson was not denying that these things are based on physical processes; the point is that they cannot be explained by physics or chemistry, but only by higher-order complexity.

6. Predictions are not explanations. In the example of the orrery, prediction could also be considered an explanation: this is how and when the planets move. In the case of behaviorism, as Scott points out in §3, "[…] explanations of how and why learning occurred were eschewed in favour of empirically derived 'laws' that afforded predictions about when and where learning would occur […]" Explanations were eliminated in favor of predictability.

Why a paradigm shift is worthwhile but difficult to achieve

7. Granted that circular causality and other concepts need to be folded into the foundations of psychology, the foundations of cognitive science, and even into science in general, we need to ask the question: What interferes with this happening? The following interlocking answers suggest possibilities.

8. Use conditions understanding. When we use technology, we probably think of that technology in terms of use through controlling a switch, button, lever, knob, or key rather than in terms of understanding the underlying mechanism. Even a technician may not need an adequate understanding and description. In other words, the adequacy of a description depends upon the user's goal and the user's view of what serves as an adequate description.

9. Technological cultures use the tools and metaphors appropriate for the task at hand. People use the understanding they need to accomplish the goal they have in mind. In cybernetics we regularly refer to the thermostat as an example of feedback, but we need to keep in mind that just as we think of the thermostat – and feedback – as an example of circularity, most people continue to think of the thermostat as a cause and effect mechanism. The point is that we live in a technological culture where we are surrounded by circular systems that are thought of as linear causal mechanisms that we, the users, control, and this dominant metaphor of technological cultures influences our thinking at a deep level (see Martin 2015).

10. Implicit and unexamined paradigms are powerful. Much of what an individual knows is implicit; much of what an individual knows is performative and implicit; that is, it reveals itself in how she behaves. Even as a scientist, much of what ones does is performative and implicit. For example, in designing research, writing a grant, writing a paper, and so on, what one knows is embodied in the skills one uses to produce the artifacts (such as this commentary) that come from the performance of one's skill set. As I pursue my work, I follow the strategies I have learned through study and apprenticeship within my profession.

11. One can understand and agree with a concept without incorporating the implications of that concept into the performative aspects of how one does research. We are immersed in a worldwide culture that recognizes circularity in specific processes but typically treats them as something that can be controlled through an understanding of linear causality. Recognizing circularity in learning, cognition, problem solving, etc., does not by itself change a rootedness in linear causality. Scott (§28) points out that psychology has moved in the direction of accepting circularity in processes such as perception, memory, problem-solving. However, the issue is that psychology has not embraced the concept of human beings as self-organizing systems whose circularity is intrinsic and foundational, not peripheral.

12. The rise of information processing took place when the possibility of writing digital computer programs capable of solving problems (e.g., playing checkers or chess) arose: Can we write a software program to do X? The need for circular processes in writing software is understood: that is what a software program is – a series of loops that produce a result through iteration. Still, the underlying thinking is not necessarily (or even usually) based on the understanding of cognition as a process that is circular, and the metaphor that "the brain is a computer" is a return to the idea that learning, thinking, and remembering are identical to computer processes such as storage, input, output, and retrieval.

13. The gold standard – and the prestige that follows from meeting the gold standard – belongs to the laboratory and the field experiment, with their emphasis on linear cause and effect. Even in those many studies designed to provide correlations, the underlying idea is still to answer questions such as: Which method works best? For example, which method of teaching reading works best? Does more homework produce better scores? Is this antidepressant better than a placebo?

14. It is important to understand that in §1 Scott is not talking only about revising the historical foundations in order to create a coherent discipline (though he is certainly wants that to happen). In §44 Scott is also concerned with how research is carried out in the present. The study of cognition cannot advance until we see living systems and cognition as more than the result of cause-and-effect processes that incidentally include circular loops (e.g., practice in learning). If researchers are embedded in a system of proposal, funding, research, and publication that rewards proposals and papers that follow only certain established paradigms, those paradigms will tend to preserve themselves through many generations of researchers.

15. Findings that point toward constructivist concepts are thought of as anomalies. Decades of research on perception have revealed that color vision does not have an isomorphic correspondence with electromagnetic frequencies (Gregory 1970). Decades of research also show many other examples of how we see what we expect to see, not what is supposedly there (Eagleman 2015). If these research findings were not considered anomalies – features of human perception rather intrinsic properties of cognition – could scientists still think of themselves as representing an external reality?

16. Our preference as human beings, whether individually or in groups, has been to try to control others, and our environment, both social and physical. Incorporating circular causality and self-organization would threaten the belief that humans can control others and nature through appropriate understanding of linear cause and effect. This belief is embedded at all levels of thinking in our cultures: parenting, the law and justice, business management, science, and the technology of modern life – children, employees, criminals, one's toaster, one's automobile, and the earth itself – can be understood and controlled.

17. Given all of the barriers to the acceptance of non-linear causality within psychology, it is all the more surprising that there has been a small but increasingly important thread of psychology that is built on the concepts of constructivism and circular causality based on the work of Jean Piaget. What I would like to add to Scott's history is that during the same period that behaviorism became the dominant model of learning and, later, information processing became the dominant model of cognition, Piaget was evolving a theory of cognition based on ideas of self-organization (the mind organizes itself through a series of assimilations of and accommodations to the environment). Piaget's fieldwork and the theory that flows from his fieldwork is a constructivist theory of learning and cognition that incorporates the ideas of circularity and self-organization, though without using those terms. Piaget was very clear that the mind organizes itself – and this organization creates our understanding of the world. Piaget (1974) titled one of his books: *To Understand Is To Invent*. Ernst von Glasersfeld – who coined the term and the notion of radical constructivism, and was also a Piaget scholar and a director of doctoral theses using Piaget's constructivist assumptions, incorporates the notion of the mind organizing itself according to its goals (we might use other terminology today) into radical constructivism (Glasersfeld 1984). In the last four decades, the importance of Piaget and neo-Piagetians has increased, though Piaget continues to be thought of as a stage theorist in the area of cognitive development rather than a researcher and theorist of cognition and learning.

18. An important idea in Scott's target article is his consideration of concepts of procedures. Scott mentions Gordon Pask, but not Piaget, in his consideration of concepts as procedures. Piaget also thought of concepts as procedures – procedures in a constant process of change and development. This is not to take away anything from Pask, but to point out that in Piaget, there is no conceptual difference between motor schemes and conceptual schemes: both change through interaction, both can be thought of as operators. In fact, this is Piaget's central idea: we know the world not through representation but through operation on it using motor and concept schemes – and conceptual schemes develop through using motor/conceptual schemes to process the environment. I think we can make a better argument for including cybernetic and constructivist concepts into the foundations of psychology by pointing out that these ideas have also been implicit in the work of Piaget and Lev Vygotsky.

Conclusion

19. The history of humanity is a history of changing technologies, including bookkeeping, mathematics, and writing, which allow for greater control of time, people, and resources. In all of these efforts, assumptions of control through cause and effect are almost universal. A causes B. Push the button and your car starts, the bomb drops, the vending machine delivers a candy bar, and so on. Never mind that A does not cause B, it merely initiates a sequence of events; this is how we human beings prefer to understand "reality." In the nineteenth century, the idea of science as a rational way to discover how nature worked established the idea that the discovery of cause and effect relationships could be made into objective process from which neither humans or nature could escape. Thus were cemented together two key components of modern human life: grafting the objective findings of science and the objective results of technologies onto the subjective experience of cause and effect: I do something and I see a result. The assumption of cause and effect is a phenomenon that permeates almost all theory and all practice in modern life.

20. By the end of the twentieth century, psychology's concern with learning and cognition had three main very different approaches to learning and cognition: the reductionist approaches of behaviorism and information-processing, and the non-reductionist, constructivist approaches of Piaget and Vygotsky and those influenced by them. Behaviorism has expanded to include mental behaviors, especially in the field of mental health, and virtually all behaviorism includes cognitive behavioral strategies. Information processing/artificial intelligence has developed enormously more powerful software, but has not changed its stance vis-a-vis cognition as a self-organizing system. Piaget's and Vygotsky's work has been embraced by some in the education community, especially those concerned with teaching science, because a constructivist approach works better than the traditional "transmission of information" approach that continues to dominate the rest of education. Behaviorism and information processing could profit from including the concepts of self-organization and circular causality in their core understanding. Piaget's work already incorporates the concepts of self-organization and circular causality and the construction of knowledge, but researchers influenced by Piaget could benefit from using specific terms such as self-organization and circular causality to inform their teaching and research. Piaget, Heinz von Foerster, Pask, Humberto Maturana, Glasersfeld, and other constructivists and second-order cyberneticists have important insights, but their insights tend not to affect the research done by cognitive scientists.

21. In closing, I agree with Scott that psychology needs to incorporate circular causality and other concepts from cybernetics – both first- and second-order (§§43f) – but this may not create a paradigm shift. Many areas of psychology are aware of circular processes, as Scott (§28) points out. The problem is that psychology (i.e., the structure of psychology as represented by various groups with power and influence, not just individual psychologists) needs to see clearly that the reductionist model that underpins much of the research interferes with incorporating circularity, self-organization, and other cybernetic concepts into the underlying psychological understanding of human beings as self-organizing, circular systems.

22. A more flexible way of thinking about human beings can continue to use designs and methodologies that allow researchers to move forward. For example, while using the traditional tools of psychological research, positive psychology also uses concepts related to constructivist and cybernetic concepts. These researchers do empirical research that accepts and investigates the circular nature of remembering. Our anticipation of satisfaction and our memory of satisfaction in participating in an event have been found to be more closely related to one another than they are to our satisfaction at the time of the event; in other words, our memories are more closely related to our internal expectations than to external events – we do not remember what we experienced, we remember what we expected to experience.

23. Finally, conversation is essential in developing these ideas. I came to understand the ideas in Scott's article and in my commentary by writing about them – a conversation with myself – and by listening to others having conversations with them. The important task is that professional communities undertake to reflect on the assumptions that underlie their practice.

Acknowledgement

Extensive editorial suggestions by biologist Dr. Suzanne Martin are gratefully acknowledged.

Open Peer Commentary: Obstacles to Cybernetics Becoming a Conceptual Framework and Metanarrative in the Psychologies

Philip Baron

The sidelining of cybernetics

1. Bernard Scott (§§5,15) posits two reasons for cybernetics becoming side-lined during the 1970s.

a. Scientists used only what they deemed fit for the paradigms within which they were working and thereafter concentrated on their own interests, ignoring the roots of their specialisation and use of cybernetics.
b. Funding models in the USA have favoured research geared to military exploits over research in psychology – although cybernetics was also a proponent in the military research domain. Scott (2012: 75) believes that in the 1970s the "new cybernetics" literally went unnoticed in circles outside of the systems movement.

Scott's (§15) two reasons are fair; however, with more than 40 years elapsing since the heralding of the "second order," there still remains limited explicit intermingling of cybernetics in mainstream psychology, which is exactly what Scott (§§7, 10) is concerned about. There is no denying that many cybernetic principles can be found dappled across the psychologies, but as Scott (§1) notes, there is almost no awareness that such principles and other evolved derivatives have originated from cybernetics. Cybernetic approaches are thus rarely found as a complete curriculum in psychology studies in either American or European universities. I propose four additional reasons for this status quo of the marginalising of explicit cybernetics in psychology disciplines. I believe these obstacles are still at play and act as boundaries to Scott's ideal of cybernetics providing both a conceptual tying together of competing psychology approaches as well as becoming a meta-view for psychology. Also presented are some findings from the South African context in order to balance the Americentric and Eurocentric context.

Obstacle 1: Teacher and learner challenges

2. Second-order cybernetics is challenging for students to grasp and for educators to teach (Baron 2015). This may be due to the epistemological shifts that are usually required upon embracing cybernetics, not as a model that can be objectified, described, and then applied as something separate from the observer but as both a theory and a lived experience, addressing both the sophia and phronesis of knowledge (Baron 2014; Glanville 2015). Students (and others) grapple with the idea that cybernetics may be something that can be used to frame everything else, while still being personally connected to this very frame.[1] Scott (§43) rightly argues that education in psychology should begin with topics on autopoietic

1. This statement is based on the feedback from visitors (mainly students) who have visited www.ecosystemic-psychology.org.za This website is a resource for people who are interested in eco-

wholes and complex adaptive systems; however, would it be reasonable to expect learners to grasp these principles when they have not yet learned cognitive and social psychology? How does one present these cybernetic topics when the learners do not yet have knowledge of human mental processes, memory, and perception? Language, too, is a major feature in cybernetics, both in the manner in which much cybernetics text is written – the specificity of words, phrases, and their intended meanings – and in the topics of cybernetics research on communication systems (conversation theory, for example). This is especially challenging when the learners have a different mother tongue than the teacher's, which is often the case in large multicultural universities, in particular in the South African context. The educators and practitioners themselves would need to re-think their teaching and learning, as many of them may have already become entrenched in a particular paradigm of knowledge, the same paradigms that Scott (§4) believes have not acknowledged cybernetics.

3. Educators and scientists who are interested in cybernetics may find the observer-dependent realities, non-purposeful drift, structural determinism and coupling, entropy/negentropy, equifinality and equipotentiality all troubling aspects to incorporate into their research and hence their teachings. This forms part of the first obstacle. Thus, it may be beneficial to create a guide for educators on how to present these cybernetic topics, as well as an introductory book for learners in a format that is at a low level and not intimidating ("conceptual ground-clearing," §42), such as the popular mainstream book brand *For Dummies*. The *introductory* titled book on cybernetics is Ashby's (1956) *An Introduction to Cybernetics*, which is an important text, but may be too mathematical for a new student in the psychologies.[2] Scott (§2) does mention that he may provide a book on the topic he proposes regarding cybernetics as a unifying framework for psychology, but being an avid researcher in education himself, he may also consider something of the order of: *Cybernetics for Dummies: A Guide for Teachers and Learners*. This can assist in overcoming the adoption of explicit cybernetics into mainstream psychology curriculums, and may solve the problem of favouritism of some cybernetic topics while other equally valuable topics go ignored, possibly owing to their perceived complexity. Humberto Maturana and Francisco Varela (1992) did well in simplifying their work and opening it up to a wider audience with their book titled: *The Tree of Knowledge*.

Obstacle 2: Traditional universities

4. The second barrier, dependent on the first, rests on the structure of traditional university curriculums. Scott (§§26, 28) notes that psychology studies are populated with various theories and their models, which is part of the reason for Scott's quest to delineate the commonalities across the different psychology approaches in the first place. However, in keeping with the traditional university style of compartmentalising knowledge areas by separating disciplines, presenting topics independently from other topics without addressing their connections,[3] and often disregarding individual learning styles, many students

---

systemic psychology and cybernetics in therapeutic psychology. The site traffic averages 49 page views/day.

2. There is another, lesser-known book with same title as Ashby's. This translated book was written by Viktor Glushkov and published in 1966 as document No. FTD-TT-65–942, Air Force Systems Command, Foreign Technology Division, Wright-Patterson, Airforce Base, Ohio.
3. Pask (1976: 101) noted this point with regards to mechanics and electricity in university curriculums.

thus also address their coursework in the same serial manner – learning to compartmentalise their studies. This is further exacerbated when the educators specifically create assessments that ask questions in an outcomes-based approach that further isolates the parts of a single curriculum, often required for the auditing bodies who want to measure the learners' performance against a pre-determined scale for each course outcome or knowledge area. Students get accustomed to the disconnect between themselves and their study areas, missing the point that knowledge and knowing are not synonymous, for knowing requires a knower and is tied to context and epistemology (Glasersfeld 1990). Thus, in the uncommon event of explicit cybernetics being a topic within a certain module of a university degree/diploma, it simply forms the next topic placed adjacent to the others in the list of knowledge areas in which the learner must achieve competence.[4] Further, the same method of "applying" each paradigm/theory to a psychology case study, for example, now takes place with cybernetics as the tool, resulting in abundant confusion. If educators were versed in conversation theory as Scott (§37) describes, tools such as Teachback, analogy learning, etc. could be used in a widespread fashion, assisting in steering the learners and thus engaging with the different styles of learning that each learner demonstrates. However, this is particularly difficult in distance-learning universities, where verbal conversations are a luxury.

5. In undergraduate years, there may be an introductory module providing an overview of the main approaches in psychology theorists of the last century. For example, in South African public universities, a personology course would consist of the depth psychology approaches (Sigmund Freud, Carl Jung, Alfred Adler, Erich Fromm, and Erik Erikson), the behavioural and learning theory approaches (BF Skinner, Julian Rotter, Albert Bandura, and Walter Mischel), the person-orientated approaches (Abraham Maslow, Carl Rogers, George Kelly, and Viktor Frankl), and then lastly the alternative approaches (Eastern, African, and ecosystemic) (see Meyer, Moore & Viljoen 2008). What is notable is that the ecosystemic approach reflects the explicit cybernetic approaches, which is unfortunately presented as a separate section in this particular personology module. Thus, in addressing Scott's (§§43f) goals, the textbook would need to be re-written from Scott's (§§1,7,10,43) view of tying the cybernetic tenets that are implicitly used within some of the neighbouring approaches and concluding with cybernetics as a meta-view, instead of simply being a separate independent approach. The mega university in question is called the University of South Africa (UNISA),[5] which has student numbers of over 300 000, with 89 000 of these enrolments in the humanities (DoHET 2013: 4f). Two out of the three top universities in Africa (the University of Cape Town, the University of the Witwatersrand, and the University of Pretoria) have systems theory explicitly as part of their curriculums, however with a limited scope. In these two instances, the explicit use of cybernetics (first-order) is within family therapy or group therapy praxis. Thus, Scott's (§§29, 44) attempt at grounding the competing psychology paradigms within cybernetics, whether process- or

4. Pask (1976: 96) was concerned about how modules are structured for students to learn serially/operationally. This topic is still relevant even 40 years later.
5. While this is an African university, it reflects international Western trends in curriculum structure. This particular university is also one of the only universities on the African continent to offer their clinical psychology master's degree from an ecosystemic approach (cybernetic); yet from browsing the undergraduate curriculums, one would not assume this fact. This in turn means that only learners who achieve the master's degree would have had an opportunity to engage in a cybernetic approach to clinical psychology.

person-orientated, is not without merit, also allowing for an appreciation that many models are part of a larger class with the goal of addressing whole systems. A review of traditional university psychology curriculums and the prescribed texts may need to go hand in hand in overcoming this barrier of introducing cybernetics as a conceptual framework in the psychologies.

Obstacle 3: Linear causality, research methodology, and technological efficacy

6. My third proposed barrier rests on the well-established linear research methodology. Sigmund Koch (1976: 485) stated, "at the time of its inception, psychology was unique in the extent to which its institutionalisation preceded its content and its methods preceded its problems." Empirical research is a major activity within psychology as Scott (§§3, 4, 7, 28) notes, however Scott (§42) would like it to be informed by second-order cybernetics. With abundant psychology research and what Scott (§4) refers to as "a wealth of empirical findings," one wonders how rich these enquiries are when mostly undertaken according to an epistemology that has not accounted for observer-dependent realities and contexts. Dorothy Becvar and Raphael Becvar (2006) call for re-research, alluding to the idea that when research findings are understood from a frame of reference that does not account for its own worldview, this research should be viewed tentatively – not being as rich as initially thought. Scott (§7) mentions critical psychology and the attempt to review the assumptions of mainstream psychology, which is important; however, there is still a barrier in that Western thought idolises the individual and one's ability to control and manipulate one's environment. Linear causality is central to the Western mind and the dominance of positivism, prizing mechanisation and the objectification of measurements, which are often the goals of funding models. In terms of psychotherapy, efficacy that relies on standardisation has found its way into psychotherapy with the "manualisation" of process to provide a cheaper intervention (Soldz 1990; Werbert 1989). In terms of therapy practices, there is an increasing need for psychology as a profession to demonstrate that its interventions yield tangible and measureable results to clients and their families, as well as to human rights groups in light of inhumane practices of some psychiatric institutions or abusive traditional healing practices in some low- to middle-income countries (Kagee & Lund 2012: 103).[6] Cybernetic approaches in research methodology have the scope for an ethical approach; however, comparing measureable success against other approaches almost requires a different set of measurement criteria, or at least an understanding of systems thinking.

7. The move to technological efficacy (see Ellul 1964) brings forth an epistemology of highly controllable, linear, predictable, and structured systems that do not readily adapt for humanness. Artificial intelligence supported by cognitive science does have a history of cybernetics as an important proponent in this endeavour, as highlighted by Scott (§30). However, in his unpublished manuscript "Metadesign,"[7] Maturana stresses the term "consensual" in explaining existence: consensual living, consensual emotions, consensual co-ordinations, consensual behaviours, and consensual conversations. It seems that humanity, in the presence of machines, still has a lot to achieve for consensual existence to occur. The structure of the majority of technology is not readily consensual (Baron 2013). With humans at the receiving end of technology, humanity may become conditioned to

---

6. See also the "chain-free initiative," http://www.emro.who.int/mental-health/chain-free-initiative, accessed 30 December 2013.
7. http: //www.inteco.cl/articulos/metadesign.htm

what Jacques Ellul (1964: 324) termed "the law of technique." The deep integration of technology into the day-to-day living of people has resulted in major shifts in how people communicate and achieve their daily goals. This technological efficiency may adjust our worldview, and should not be thought of as something neutral (Heidegger 1977: 4). It is not surprising that one of the most influential humanist psychologists noted, "In our technological society, people's behaviour can be shaped, even without their knowledge or approval" (Rogers 1980: 140). Thus, the linearity of technology programming, too, may be a barrier to embracing circular causality in research.

Second-order cybernetics is, however, an important approach to research. Some anthropologists have recently realised the importance of acknowledging research methodologies from their sister disciplines in addressing past mistakes, especially in terms of ethnographic works that arrive at conclusions that upon revisiting do not hold their ground (Lembek 2014). This "new insight" into *new* observer-dependent research methodologies further depicts the lack of adoption of cybernetics in other disciplines too, now expanding Scott's scope. One of the earliest advocates of the second order was an anthropologist herself – Margaret Mead – who advocated the importance of alignment in both the theorising and the praxis of research for the fruits of cybernetics to be realised (Mead 1968). Thus, while Scott focuses on the psychologies, one wonders if his argument also applies to other disciplines.

Obstacle 4: Personal preference, ethics, and responsibilities

8. The last obstacle rests on personal preference. Scott may do well to provide a coherent conceptual framework for the psychologies, but there may be an audience – as always – who choose not to acknowledge cybernetics, or who dismiss it simply as a type of post-modernism. Scott (§18) would like to see cybernetics as a metadiscipline, which is ideal, but how does one achieve a meta-narrative with groups of people who have not yet understood even the early cybernetic principles? This is indeed a dilemma. Second, there is a high degree of unpredictability in state-determined systems that require continuous updates for participant observers, as Scott (§16) describes, citing Maturana, Gordon Pask, and Heinz von Foerster. This may not be a well-liked position for researchers to subscribe to. As Scott states (§17) "cyberneticians had the reflexive awareness that in studying self-organising systems, they were studying themselves." The associated ethics and responsibilities that arise out of second-order cybernetics may be overwhelming. This is an unsettling no-man's land for many scholars and students, who in turn opt out of this challenging reflexive epistemological domain.

Conclusion

9. Scott (§7) notes that if the field of psychology is looked at as a historical whole, there has not yet been any single paradigm that stands as a dominant victor, rather competing paradigms are at play in different areas. This means there is still scope for cybernetics to "re-enter" and take a seat at the table of dominant approaches in psychology, gaining its position as the metadiscipline while also not excluding other knowledge systems. However, there are boundaries that need to be addressed in order for the step to embracing cybernetics to take place.

Open Peer Commentary: **The Social and the Psychological: Conceptual Cybernetic Unification vs Disciplinary Analysis?**

Eva Buchinger

1. It was Aristotle who said that "it is a characteristic of man that he alone has any sense of good and evil, of just and unjust" and that "the individual, when isolated, is not self-sufficing; and therefore he is like a part in relation to the whole" (Aristotle 1920: 29). One could argue that this is a perfect and still up-to-date expression of the inseparability of the social and the psychological dimension of human beings.

2. Psychology and sociology nevertheless have been developed as distinct academic disciplines with an established link in the form of social-psychology (Goethals 2007; Ross, Lepper & Ward 2010). What are, therefore, the benefits of another approach to interrelating these two disciplines? In the section "Unifying individual and social psychologies" of his target article, Bernard Scott argues that the benefits can be found in better supporting the understanding of the dynamics of interpersonal perception and human communication (§39). He argues further that a concept of social systems based on the conversation theory of the cybernetician and psychologist Gordon Pask could provide an alternative to the theory of social systems developed by the sociologist Niklas Luhmann (§40).

3. Scott's excellent analysis of the role of cybernetics in psychology and his inspiring thoughts concerning the unification of process and person approaches indeed support the better understanding of the interlinking of the social and the psychological. He further gives a clear illustration of how Pask's P-individuals are conceptualized as psycho-social unities (§§25, 38) that are hosted by either one M-individual (human being, biomechanical unity) or by several M-individuals. Of special interest is the latter idea, which indicates that – complementary to an inner dialog – an outer dialog exists (outer conversation that unifies a collective) that is also a P-individual, but of another nature. The same holds for Pask's definition of conversation as "concept sharing" along a certain togetherness and thereby distinguished from communication as "signal transfer which may, or may not, be conversational" (Pask 1980: 999). The expression "certain togetherness" makes sense because

a. there must be enough togetherness supported by institutions (e.g., a dining table, café, market, organization), but
b. too much togetherness gives rise to individual or social malaise.

The second point is the result of digitalization – the growth of data storage and computation, in Pask's wording (Pask 1980: 1000) – which rapidly creates an information environment. Pask concluded that communication that only resembles conversation will be amplified in this signal-overloaded information environment and will therefore "appear as major hazards in the future" (Pask 1980: 1001). All these considerations are based on a second-order cybernetic understanding[1] that refers to the work of Ross Ashby, Norbert Wiener

---

1. First-order cybernetics is understood as the exclusion of the observer within an observation and second-order cybernetics as the inclusion of the observer in what is observed. Second-order therefore means that the observer is "inside the box" (Brand, Bateson & Mead 1976: 38), respectively "a person who considers oneself to be a participant actor" (Foerster 2003: 289), and is contrary to "the first order of classical black boxes and negative feedback" (Pask 1996: 355).

and Heinz von Foerster,[2] among others, which are astonishingly up-to-date concerning the recent debates about the digital revolution. Scott's proposition of a unifying conceptual framework on the basis of conversation theory is therefore quite promising.

4. But one should be aware of the complexity of the world society, which is the focus of Luhmann's social theory and which is not covered by conversation theory. From a sociological point of view, this can be seen as an expression of strengths of progressing within the disciplinary path. While elaborating the societal micro-macro link as part of his social systems theory, Luhmann emphasized the difference between interaction and society by focussing on the latter (Luhmann 1977, 1982a, 1982b):

a. Interaction forms the basic type of social system, which emerges whenever present individuals perceive one another (face-to-face interaction). They communicate verbally and/or non-verbally with those that are present (with the option to speak about those that are absent).
b. Society represents the comprehensive system of all communicative interactions.

Very simplified, it can be said that after a first evolutionary transformation from segmentation to stratification, another evolutionary transformation from stratification to functional differentiation led to the world society in which we live now. Functional systems (which are societal subsystems) co-evolved with symbolically generalized communication media. Here are some examples:

- Money is the communication media of the societal system economy, which operates on the basis of the binary code pay/not pay.
- Truth is the media of the scientific system, with the code true/false.
- Power belongs to the political system and the code is government/opposition.

Thus, an economy is a self-referential system based on all the communication elements that fall into the scheme of pay/not pay; all communication belonging to the code (scientifically) true/false constitutes the functional system of science; and all communication-elements belonging to power and government/opposition generate the political system. As a consequence, operational closure results in particular system rationalities: scientific rationality, economic rationality, political rationality, etc. Modern societies neither have one center (one top) nor one rationality integrating the particular rationalities of the different societal systems. "It is a society without an apex or center" (Luhmann 1990a: 31) Therefore, modern society is characterized by an enormous degree of complexity and it is the explanation of this (complex, functionally differentiated) world society that is Luhmann's objective (Luhmann 1982b, 2012: xiii).

5. An integration of Pask's conversation and Luhmann's interaction approach could benefit from Pask's rich account of learning as the evolution of concepts within a conversation as well as from Luhmann's advanced elaborations on the complex dynamics of modern society (Buchinger & Scott 2010: 118). Such integration could be based on already-established conceptual links, since Luhmann himself was very much influenced by cybernetic considerations. For example, his notion of resonance was inspired by von Foerster (as discussed in Buchinger 2012: 23), his notion of self-reproduction by Ashby,

2. See Pask (1970; 1996). For sources see, for example, Rosenblueth, Wiener & Bigelow (1943), Wiener (1948); Shannon & Weaver (1949) and Ashby (1952).

Humberto Maturana and Francisco Varela (Luhmann 1990b, 1995: 34, 369), and his notion of mutualistic-dialogical unities by Pask (Luhmann 1995: 38).

6. The progress in each discipline thereby provides the ground for the integration. For overall scientific advancement, both are needed, disciplinary specialization on the one hand and conceptual integration (or spill-over between disciplines, or inter-/transdisciplinary fields) on the other.

Open Peer Commentary: **Second Thoughts on Cybernetic Unifications**

Tilia Stingl de Vasconcelos Guedes

1. The problem of conceptual unification is usually considered critical for a mature academic discipline. The main point in Bernard Scott's target article is to demonstrate the possibility of using concepts of second-order cybernetics to provide a foundation and a unifying conceptual framework for psychology. In fact, he argues that cybernetic concepts have influenced psychology since the 1950s. He also claims that the origins of psychological concepts based on cybernetic ideas were lost and the interest of cyberneticians in working on transdisciplinary unity did not last (§1). Scott's article is an attempt to offer means and ideas for this unity. According to Scott, one can build conceptual bridges between psychology and sociology – by merging individual and social psychology. Scott's motivation comes from his personal experiences as a practicing psychologist, in which he understands cybernetics as a provider of useful tools for modeling specific situations.

2. In this commentary, I want to demonstrate that the journey to unification in an actor-based discipline such as psychology has not ended with Scott's article but has only just started. In my view, three important yet missing aspects must be addressed in order to reach the stage of unification, i.e., the problem of levels, the problem of multiple-level dynamics, and the problem of being sufficiently different.

The problem of levels

3. Fields such as psychology, sociology or ethnography deal with individuals and societies, both in static and in dynamic aspects. Scott recommends that these individuals should be described in a dual manner, namely by a differentiation of mechanical individuals, or M-individuals, and psychological individuals, or P-individuals. Both are considered as self-organized and organizationally closed systems. This distinction between M-individuals or M-unities (§25) as processors for the procedures of P-individuals or P-unities may have its unifying merits, but they are, in my view, insufficient and under-critical as unifiers.

4. In recent decades, we can observe that there has been an enormous literature in the cognitive neurosciences, which have made spectacular advances in promoting the level of neurons and neural network interactions as the basis for studying psychological phenomena such as perception, emotions, memory, etc. Can we still use Scott's M and P differentiations at the neural level as well? Can we use M- and P-unities at the level of neurons and their interactions? In other words, can we use the split between M- and P-unities across different levels, starting with the neural level and single neurons as acting unities to higher

neural levels up to the micro-level of individuals? If the answer turns out to be negative, then the proposed cybernetic framework would be inadequate to reach its unification goal.

5. The same argument can be applied upwards towards higher levels of aggregation such as groups, organizations or even regions or states. Using a similar distinction from sociology or economics, micro-psychology can be understood as an actor-based configuration whereas macro-psychology deals with unities as composites of individuals. Can we also use Scott's M and P-separation for these macro-unities or are we bound to the individual or micro-level alone? Again, a negative answer would demonstrate the basic restrictions of Scott's approach.

6. Finally, the question arises of whether Scott's approach is able to deal with problems of the subconscious as well, which, at first sight, the framework of M and P unities does not seem to address.

7. But even a positive answer that the distinction between M- and P-unities can be used across all levels from the basic neural level up to the level of macro-psychology, including the level of the subconscious, would result in a very serious new challenge, which can be classified as the problem of the dynamics in multiple-level configurations, as discussed in the next section.

The problem of multiple-level dynamics

8. As my second critical comment to Scott, I would like to address the issue of the analysis of multiple-level configurations, which also is a general problem in many disciplines outside the social sciences.

9. After 1945, we can observe the appearance of a large number of new actors operating beyond the traditional local, regional or national level. Transnational enterprises, transnational public and private organizations and very large-scale transnational systems such as the financial markets emerged. This lead to a highly complex multiple-level configuration where relevant actions occur at all different levels simultaneously. Rogers Hollingsworth and Karl Müller (2008) remind us that the dynamics across a multiplicity of levels is still an unsolved analytical challenge.

10. The governance of the contemporary world and the interconnections among governance, democracy and knowledge are far more complex than most observers recognize. No single level is decisive in shaping the world in which we live. Moreover, the levels are nested and linked with each other. So one of the great remaining challenges is to comprehend the nature of this nestedness, to understand how governance, democracy and knowledge are linked together not only at each of these levels but also how these processes are linked together across different levels. As societal institutions are increasingly nested in a multi-level world, we are all faced with the perplexing problem of how to govern ourselves (Hollingsworth & Müller 2008: 417). Unfortunately, in his article, Scott has provided no clue as to how to deal with the issue of multiple-level dynamics, which, as stated above, forms a problem for his approach if he offers a positive answer to the differentiation of M- and P-unities across different levels.

The problem of being sufficiently different

11. Finally, a third aspect of Scott's program for the cybernetic unification of psychology needs to be urgently addressed: the status of previous and current unification attempts

in psychology. For example, sociology has seen many unifying approaches, starting from the days of Max Weber at the beginning of the 20th century, culminating with Talcott Parsons for a short period in the 1950s, and ending in its current configuration with a multiplicity of different unification approaches by, to name a few, Pierre Bourdieu, Michel Foucault, Anthony Giddens, Jürgen Habermas, Niklas Luhmann and Richard Münch. Two important phenomena can be observed. First, due to their inherent differences, these unification approaches further fragment sociology rather than unify it. Second, the multiplicity of different foundations has not affected the methodology of empirical social research, which continues in its normal operations of quantitative and qualitative analyses. Thus, in sociology, any new unification attempt increases the number of available alternatives and leaves the empirical work of sociologists largely unaffected. From this, one can conclude that each new attempt moves the unification of sociology a step further *away*.

12. Will the same happen to psychology? Given the failed attempts in sociology, what can Scott's new cybernetic approach offer that makes it different from previous unification attempts, and how can his second-order cybernetics framework also affect the empirical work of psychologists in order to make a difference with regard to the normal practices of psychologists?

13. The work on a unifying foundation for a discipline such as psychology based on second-order cybernetics literally begs for a high degree of self-reflection and considerations of the impact that this kind of unification could generate for the community of psychologists. Perhaps in academic disciplines such as sociology, economics and psychology, one can only start a discussion about the relative advantages or disadvantages of various unification approaches without being able to reduce them to a dominant paradigm in the sense of Thomas Kuhn.

14. In any case, even without Scott's contribution, psychology has already begun to use cybernetic concepts to provide solution-focused, fast, effective ways to deal with daily issues – many of these methods are based also on cybernetic ideas, with ongoing research considering concepts of second-order cybernetics, especially in the field of systemic therapy (see, for example, Schlötter 2005; Varga & Sparrer 2016; Vorhemus 2015). Even though the systemic work represents a specific sector of psychology, its explanations are based on cybernetic thoughts being accepted as the foundation of systemic methods. How do these highly practical methods fit Scott's call for unification?

15. In this respect, Scott's article may be considered only one of several starting points for a long journey towards unification, rather than its finishing line.

Open Peer Commentary: Cybernetics and Synergetics as Foundations for Complex Approach Towards Complexities of Life

Lea Šugman Bohinc

Introduction

1. With a background in psychology and psychotherapy, working as a university teacher in three (inter)disciplinary fields, i.e., social work, psychotherapy, and education, and teaching a variety of courses in bachelor's, master's, and doctoral programmes, I have been rewarded with experience of learning about a field through learning about the history of its ideas more than once. That is why I have been excited to read Bernard Scott's target article and to learn about "his" story of cybernetics, which fulfils two promises:

- it shows how second-order cybernetics can provide a much-needed foundation for constructing a new meaning or order in the conceptually messed-up discipline of psychology;
- it brings together the so-far more-or-less divided branches of individual and social psychology by unifying person and process aspects.

2. I have always been interested in constructing patterns that connect (rather than disconnect or divide) the social and natural sciences and the theories developed within their disciplines. During more than twenty years of teaching, my students of social work, psychotherapy, and education (e.g., teaching, social education, and special and rehabilitation pedagogy) have repeatedly expressed confusion with the variety of (often contradictory) research findings, theories, concepts, work models, methods, or skills in their use, best practice examples, etc. Students have usually found it relieving, informing, and empowering to make sense of the distinctions and similarities between different approaches when interpreting them through the lens of first- and second-order cybernetics. Defining those premises always demands negotiating the meaning of the concepts used by the students. This eventually leads them to arrive at new understandings, such as new individual interpretations and new agreements, including the agreement to disagree (Pask 1987: 18f). These understandings arise among two or more "locally synchronised" participants (Pask 1980: 999) while maintaining or increasing their (interpersonal or polyvocal interpretative) distinctions (Pask 1987: 23).

3. In what follows, I will introduce two of the connecting patterns that have proved useful for me and that I believe to be useful for Scott's intention to unify psychology.

"Third-order" cybernetics

4. For me, the epistemology of second-order cybernetics is more than just a constructivist theory of knowledge, a philosophical world view, a viewpoint, or even a science. I interpret it as a set of assumptions at the basis of individual and collective patterns of cognitive acts (Maturana & Varela 1992: 173f), in which "every act of knowing brings forth a world" (Maturana & Varela 1992: 26). As such, cognitive acts are processes leading

to certain products (e.g., thinking processes lead to certain thoughts and acts or decision making processes result in a decision or act) that then serve as a starting point for a new process (of thinking, decision making). These processes are recursive and the relation between processes and products is complementary. Bringing high sensitivity to the influences of structural societal factors, such as power imbalance relations, into the reflection on how people construct their worlds, I have joined authors such as Rudi Dallos, Ros Draper, and Amy Urry (Dallos & Draper 2010; Dallos & Urry 1999). These authors complemented the notion of second-order cybernetics with a "third-order" one – not as a different epistemology but as an additional recursion or a variation of the constructivist paradigm, with important implications for the life quality of human population and social justice. It is what Scott, in §39, refers to as "social constructionism." In my opinion, the role of dominant public discourses performed by social elites and unconsciously internalised (i.e., ingrained in their inner dialogues, which progressively become saturated with those dominant narratives) and lived out by the public, as reflected in the work of Michel Foucault (1980) and other critical theory authors, has to be explicitly articulated and emphasised when we talk about the epistemology of second-order cybernetics. Foucault, in a personal communication with Hubert Dreyfus and Paul Rabinow (1983: 187), asked: "We know what we think; we think we know what we do; but do we know what what we do does?" Tracking down the effects of our cognitive acts for other people's lives, Michael White (2011: xxviii) could identify one of their origins: the expert knowledge twisted together with the new forms of power thus creating the "politics" (in the general sense of relating to other people and directing their behaviour by using our power) of widely accepted and unquestioned "expert power" in any field of expertise. The reflection on the consequences of the structural factors for how we, as humans, socially construct our realities needs to be complemented with a reflexiveness on how our expert knowledge, along with our gender-, class-, culture-based identities, enters and effects our conversation with others in whatever interactional context.

5. So, rather than a worldview, I prefer to understand the epistemology of second- and third-order cybernetics as an attitude towards oneself, the others, and life or world as one constructs them, reflected in one's continuous endeavour to make his or her thoughts, decisions, feelings, values, etc. congruent with actions and vice versa. That is how, according to Gordon Pask (in Scott 2001c), we construct conceptual ("knowing why") and procedural ("knowing how") knowledge (Scott 2001c) in different contexts of dialogical practice. That is how we make it our experiential knowledge about the world, constructed and lived through as our "lived experience" (White & Epston 1990: 9). Again, it is our conversation partners (students or clients or service users, to use the traditional linear terminology, as well as our colleagues, family, and community co-members etc.) who feed their understanding of our related congruence or incongruence back to us. It is they who teach us about the consequences of our acts in interaction with them. The precondition for that to happen is an established relationship context, mutually perceived as safe and trustworthy. To acknowledge the meaning of lived experience is to acknowledge the complexity of any conversation participant (i.e., one or more of his or her constructed selves or voices or "psycho-social individuals," as Scott refers to Pask in §§24f, 36–38, and 40) in both their individual (autonomous) and social (relational) self (Flaskas 2002: 91).

Synergetics

6. Generally speaking, conversation participants can be interpreted as self-organizing systems in which nonlinear interactions (at the microscopic level) might result in emergent new patterns (at the macroscopic level), such that an observer can interpret them as a (new) self-organizing system in itself (§§13, 37). In the history of ideas, cybernetics as a transdisciplinary science of patterns and complex systems is referred to as one of the main origins for another transdisciplinary science of self-organizing processes, i.e., synergetics (see, e.g., Haken 1983, 2006, 2009).

7. Since 2005, I have been using synergetics as an experimentally supported theory for describing and dealing with complex living and non-living systems through a perspective of interpretive (macroscopic) common principles. The "generic principles of synergetics" (see Schiepek et al. 2005a, 2005b) make it possible to overcome the traditional split between natural and social science by integrating them into one unified conceptual and methodological framework (Šugman Bohinc, 2016). The main focus of synergetics is the exploration of the conditions in which a complex system spontaneously, i.e., in a self-organizing and qualitative manner, changes the pattern of its operation as a result of non-linear interactions among the system's elements, which can themselves be complex systems. Understanding the circumstances that can potentially stimulate a complex system to adapt to those very circumstances by reorganizing its operational patterns can increase the chances of creating an encouraging environment for processes of (desired, needed, agreed upon) change. In other words, it is useful to understand the changes in the environment that, when interpreted in a certain way, serve as stimulation for the complex system to reorganize its operational patterns and thus adapt to the new circumstances. Synergetics has been used and proved meaningful and successful in many disciplines, such as psychotherapy (e.g., Schiepek et al. 2005a, 2005b; Schiepek, Tominschek & Heinzel 2014) and social work (Sommerfeld et al. 2005), as well as in psychology, education, organizational sciences, economy, linguistics, etc. and, of course, in different fields of biology (e.g., in ecology), physics etc.

8. Personally, I have used synergetics in the last two research projects on which I have collaborated. One deals with the change processes in teaching and learning as well as giving support and help to children in a school setting, the other deals with multi-challenged families in their communities. Furthermore, I teach synergetic theory of complexity to my students of social work, systemic psychotherapy, and education. The more experienced they are in their profession, the more they find this conceptual and procedural framework meaningful and useful. The bachelor's students usually interpret it as very abstract and difficult to grasp; the master's students, who already have work experience, report that their intellectual and practical knowledge finally becomes integrated, and holistically as well as critically reflected. The doctoral students make use of synergetics in their research design and interpretive synthesis of analysed data whenever they are dealing with processes of change in the functioning of complex systems.

Conclusion

9. Understanding the interconnectedness and embeddedness of biopsychosocial complex systems in other complex systems, e.g., cells within organs within a brain/body within a family within a community within state administrative systems as well as socially con-

structed norms, roles, identities, etc., enables us to bridge the dichotomies that have been developed in the last century, such as micro and macro context, individual and social, theory and practice, evidence-based and practice-based research, etc. emerging in psychology, as Scott claims in many sections throughout his article. Our understanding that interactions (e.g., conversations) of complex, self-organizing biopsychosocial systems, such as human beings, produce emergent complex, self-organizing systems of a different order can serve as a bridge over the conceptual distinctions and divisions brought forth in the development of psychology and other social and natural sciences along with the very split between those two categories of scientific research. The notion of constructivist (social constructionist) epistemology and the common principles offered by the transdisciplinary sciences of complexity, such as cybernetics and synergetics, have laid the foundations for a more unified and integrated approach to the complexities of life. It can be a self-reflective and self-reflexive, socially critical and responsible approach to participating in conversations that would lead to creating complex answers to the complex challenges of our time.

Author's Response: **On Becoming and Being a Cybernetician**

Bernard Scott

1. I am happy that my commentators are generally supportive of my proposal that cybernetics can provide a unifying framework and foundations for psychology. (As a point of clarification, when I refer to "cybernetics," I mean the complementary union of both first- and second-order cybernetics.) However, I quite understand that, as pointed out by several commentators, this proposal will not be acceptable to everyone. Many domain specialists in any discipline lack an interest in the more holistic issues of foundations and conceptual unification. They have other priorities. Many have cognitive styles (by habit or heritage) that are not conducive to this sort of contemplation (for more about individual differences in cognition and learning, see Scott 1993).[1] But I do believe that many can benefit if my proposal is adopted. As noted in my article, my early exposure to cybernetics certainly helped me. As an undergraduate student of psychology, I was an indifferent and poorly-motivated student in the midst of what I saw as a mess of a discipline, in which my teachers, espousing different paradigms, were incapable of constructive conversation with one another.[2] Cybernetics enabled me to make sense of this mess and inspired me to become an enthusiastic scholar. It is thus no surprise that **Vincent Kenny**'s impassioned account of the sorry state of psychology resonates with me. I see a properly-founded and articulated

1. We also continue to have wide gaps between the two cultures of the sciences and the humanities. Many in the latter camp are quite "illiterate" when it comes to science, mathematics, logic and technology. Arguably, popular writings on these topics are helping to bridge the gaps.
2. **Marcelo Arnold-Cathalifaud** and **Daniela Thumala-Dockendorff** (§3) assert that linguistics and economics are more internally consistent than psychology. Of course, this depends on how one defines these fields. I certainly see competing paradigms, especially if one adds the psychological and sociological dimensions, without which the disciplines are very limited to the point of irrelevance and sterility.

cybernetic psychology as the "psychological psychology" he seeks. I am certainly not advocating any kind of "glue" (§1). I see my proposed foundations and conceptual framework not only as unifying but also as filters that sift out and reject dross.

2. I was inspired, eventually, to regard myself as being a cybernetician. Heinz von Foerster stated that

> we need a theory of the observer. Since it is only living organisms which would qualify as being observers, it appears that this task falls to the biologist. But he himself is a living being, which means that in his theory he has not only to account for himself, but also for his writing this theory. (Foerster 2003: 247)

Thus the aim of second-order cybernetics is to explain the observer to herself. He also stated that "Life cannot be studied *in vitro*, one has to explore it *in vivo*" (ibid: 248). I took these ideas to heart. As a transdiscipline, cybernetics empowered me to cross disciplinary boundaries. This was exhilarating. I also understood other transdisciplines (systems theory, general semantics, synergetics) to be quite cognate with cybernetics and, at a high enough level of abstraction, despite differences in terminology, to have conceptual structures homomorphic or isomorphic with those of cybernetics.[3]

3. In response to **Arnold-Cathalifaud & Thumala-Dockendorff** (§5), I should like to point out that I see all "versions" of cybernetics as having a core commonality. It is obvious that every scholar or practitioner will have her own narrative and ways of doing things and that these may be undergoing changes with experience and further study and reflection. What I detect with cybernetics is a commonality that evolved amongst a community of scholars, where differences in emphasis, terminology and areas of interest and practice mask underlying agreements and similarities of form. To emphasise what I say in my article, I count amongst this community certain central figures: Norbert Wiener, Warren McCulloch, Ross Ashby, Gregory Bateson, Stafford Beer, Gordon Pask and Humberto Maturana. There are, of course, precursors, not least Jean Piaget, who embraced cybernetics when he encountered it. I thank **Robert Martin** for highlighting the significance of Piaget's contributions (§§17f), which I did not stress in my target article but which I perhaps should have. Certainly, his work has been a central influence in the development of conversation theory (see also below). I also thank **Martin** for his more general endorsement of my proposals and for his additional elaborations of the significance of the concept of circular causality.

4. I agree with **Philip Baron** (§§2–8) that there are challenges to trying to take my proposal forward. I have already noted that not all students take to holistic thinking and, of course, there are many institutional barriers. Discussions about how best to place cybernetics within educational curricula have been going on since shortly after its inception. The (now defunct) Department of Cybernetics at Brunel University, where I studied for my PhD, had postgraduate students only, arguing that one needed to have a strong disciplinary base before embarking on transdisciplinary studies. I myself am a supporter of Jerome Bruner's concept of the "spiral curriculum":

> A curriculum as it develops should revisit the basic ideas repeatedly, building upon them until the student has grasped the full formal apparatus that goes with them. (Bruner 1960: 13)

---

3. On the application of homomorphism and isomorphism to conceptual structures, see Pask, Kallikourdis & Scott (1975).

> We begin with the hypothesis that any subject can be taught effectively in some intellectually honest form to any child at any stage of development. (ibid: 33)

As a teacher at primary school level, I introduced my pupils to the concept of circular processes as part of encouraging them to gain some understanding of the ecosystem.

5. I also agree with **Baron** (§3) that a "dummy's guide to cybernetics" could be useful.[4] In 2010, with excellent technical support, I produced a multimedia "Dummy's Guide to Learning Design" for the British Armed Forces (sadly, not available to the wider public), in which I embedded cybernetic concepts. Courses on learning to teach and learning to learn can readily include explicit reference to cybernetics. Diana Laurillard's influential book, *Rethinking University Teaching: A Conversational Framework for the Effective Use of Learning Technologies* (Laurillard 2002), although it does not explicitly mention cybernetics, uses Pask's conversation theory as the source if its core model for teaching and learning. Nigel Ford's *Web-Based Learning Through Educational Informatics* (Ford 2008) makes even more extensive use of the theories and research findings of Pask and his research team. Ford states that his disciplinary background is in "information science," which of course can be considered as a part of the broader field of cybernetics.

6. Not everyone who studies cybernetics becomes a cybernetician who studies "the cybernetics of cybernetics." There are many scholars of cybernetics who look on only from their main area of practice and position themselves in the first instance as being historians, philosophers, architects, biologists, sociologists, psychologists and so on. In doing so, I believe they miss the point, the sense of what it is to be a cybernetician and a member of the cybernetics community.

7. Some of the commentators invite me to comment on issues and disciplines, such as sociology, that are beyond my immediate concern with psychology. I agree that these topics are of interest and that cybernetics has a role to play in conceptual clarification and unification. It is relevant to note that there is a very active community of international scholars concerned with "sociocybernetics," see https://sociocybernetics.wordpress.com. I thank **Eva Buchinger** for her discussion of Luhmann's cybernetic macrotheory of functional social systems. It is beyond the scope of my article to comment much further here, except to note that the P-individual concept can be readily extended to include the recursive nesting and the dynamics of interaction of social actors at different levels.[5] I also note that, as **Buchinger** emphasises, sociology, as a discipline, departs from psychology and social psychology when sociologists choose (as do Talcott Parsons and Niklas Luhmann) to study social systems that, by definition, have an autonomous existence beyond the level of individual human beings. One can, of course, draw on cybernetics in making these studies (as do both Parsons and Luhmann). In contrast to the social systems of sociologists, P-individuals at the social system level have their existence in the conversations (both internal and external) of particular human beings, not least those who hold ultimate responsibility and are accountable for the form those social systems take (kings, presidents, ministers of state, heads of institutions, leaders of professions and so on).

---

4. In the 1970s, Frank George, Professor of Cybernetics at Brunel University, wrote *Cybernetics* (George 1976) as part of a "Teach Yourself" book series that was similar in intent to the "Dummy's Guide" books. Of its time, it does not include reference to second-order cybernetics.
5. I refer the reader to the collection of my papers *Explorations in Second-Order Cybernetics* (Scott 2011a), in which I discuss aspects of the cybernetics of social systems in several chapters (5, 10, 15, 21, 22, 23, 26, 27, 28, 30, 32, 34, 35 and 36). See also Scott (in press).

8. It is also worth noting that Luhmann follows Parsons in basing his concept of a psychic system on the controversial theories of Sigmund Freud and his followers.[6] I find Freud's concept of "the unconscious"[7] as a repository of repressed desires, hopes and fears particularly troublesome. Studies of brain dynamics and the processes of learning and skill acquisition show that many cognitive processes occur without conscious awareness. This is discussed in detail in Scott & Bansal (2014), which presents a cybernetic theory of consciousness and "the unconscious," understood as an ongoing evolutionary process of conceptualisation and internal and external conversation.

9. In answer to a question from **Tilia Stingl** (§4), the P-individual concept cannot be applied at the neuronal level.[8] P-individuals are psychosocial unities that emerge within human communities. I discuss the ontogeny and ontological status of P-individuals in some detail in Scott (2007) and Scott & Shurville (2011). The works of Piaget, Lev Vygotsky and George Herbert Mead play central roles in these accounts, alongside references to the ideas of Pask, von Foerster and Maturana.

10. In her question concerning levels and interactions of different systems, **Stingl** (§9) refers approvingly to an article by Rogers Hollingsworth and Karl Müller (2008). Interesting though this article is, by their own declaration, the "new paradigm" they promote ("Science II") is monistic. They contrast this with the ontological Cartesian dualism found in "Science I" (ibid: Tables 1 and 6). The complex systems and networks they refer to, whatever their origins in particular disciplines, are just that: complex systems and networks. For them, it is a virtue that the "natural" and the "social" can be studied with similar models and methods and that, because of this, the distinctions between disciplines can be voided. This is in contrast to the P-/M-individual distinction, which is a theoretical, analytic way of distinguishing the "social" and "symbolic" from the "natural" and the "mechanical."[9]

11. In cybernetic terms, the different disciplinary studies Hollingsworth & Müller refer to are all first-order: they are studies of observed systems. Interestingly, cybernetics (first- or second-order) is not mentioned by name in their account of the history of work on complex systems, nor do they reflexively acknowledge that their own academic endeavours are a part of an evolving, complex, self-organising system of academic activity and that, as participant observers, they are engaged in bringing about changes in scientific discourse of the same kind as they claim are happening. I am reasonably sure that this circularity is virtuous. This can be usefully contrasted with the Cibercultura y Desarrollo de Comunidades de Conocimiento research programme of El Centro de Investigaciones Interdisciplinarias en Ciencias y Humanidades (CEIICH) at La Universidad Nacional Autónoma de México (UNAM), http://www.ceiich.unam.mx/0/20Ciberc.php, in which the self and other observation of the observers of observers of observed systems plays a central role.

---

6. Psychoanalysis was not included in the undergraduate psychology syllabus that I studied, as it was considered not to be open to refutation and thus not scientific. See Popper (1963) for a very influential critique of psychoanalysis. In general, psychoanalysis plays only a small part in mainstream psychology. In contrast, it frequently plays a major role in literary criticism.
7. The term has entered popular culture, along with other Freudian concepts ("ego," "id," "superego" and so on). Stingl (§6) seems to use the similar term "the subconscious" uncritically.
8. Incidentally, whilst Stingl repeatedly attributes the P-/M-individual to me, it is, of course, as I hope is clear in my article, originally due to Pask.
9. Pask (1979) is a forceful critique of the limitations and dangers of what he refers to as "systemic monism."

12. **Stingl** also asks for comment on recent work in psychotherapy that is informed by cybernetic concepts (§14). As described in Scott (1987) and as evident in the commentary by **Lea Šugman Bohinc**, there is a long tradition of the use of cybernetic concepts in psychotherapy. I see this as an excellent justification for taking my proposals seriously.

13. In her very informative commentary, **Šugman Bohinc** refers to a third-order cybernetics. In the literature, there are several attempts to invoke higher levels of cybernetics. One can certainly do this. However, it is important to recognise that, as **Šugman Bohinc** does, higher levels, whilst having explanatory usefulness, do not add anything new epistemologically. This point was made by von Foerster (2003: 301). The key step is the transcendence to a new domain, the second-order domain, in which reflexivity is introduced. **Šugman Bohinc** refers to her interest in power relations as revealed in discourse and social interaction. I see this concern as one that is central in second-order cybernetics and I thank her for raising this topic, which features as a major theme at conferences on sociocybernetics. Some of my own thoughts about this can be found in Scott (2006).

14. Having read the commentaries, I am even more persuaded that my proposals concerning cybernetic foundations and a unifying conceptual framework for psychology have merit. I acknowledge that the proposals face institutional barriers and may have limited uptake amongst students and practitioners of psychology. However, for those who do take the proposals on board I see great benefits, not least the insights and understandings provided by second-order cybernetics concerning the human condition, which I believe should be promulgated widely. I am further persuaded that I should broaden the scope of my proposals to include the social sciences more widely. Accordingly, I am now considering writing an introductory text with the provisional title *An Introduction to Cybernetics for the Social Sciences*, in which I will bear in mind that…

> social cybernetics must be a second-order cybernetics – a *cybernetics of cybernetics* – in order that the observer who enters the system shall be allowed to stipulate his own purpose […] [I]f we fail to do so, we shall provide the excuses for those who want to transfer the responsibility for their own actions to somebody else. (Foerster 2003: 286)

Acknowledgements

I thank the editors for their hard work in putting this volume together and I thank the authors of the open peer commentaries for taking on the task of reading and commenting on my article.

Combined References

American Psychological Association (2015a) New APA policy bans psychologist participation in national security interrogations: Association takes strong stance in response to findings of independent review. Monitor on Psychology 46(8): 8. http://www.apa.org/monitor/2015/09/cover-policy.aspx

American Psychological Association (2015b) Report to the special committee of the board of directors of the American Psychological Association: Independent review relating to APA ethics guidelines, national security interrogations and torture. http://www.apa.org/independent-review/revised-report.pdf

Aristotle (1920) Politics. Translated by Benjamin Jowett Oxford University Press, Oxford. Originally written between 335–323 BCE.

Arnold M. (2010) Constructivismo sociopoiético. Revista Mad 23: 1–8.

Arnold M., Urquiza A. & Thumala D. (2011) Recepción del concepto de autopoiesis en las ciencias sociales [English Translation]. Sociológica (México) 26(73): 87–108.

Ashby W. R. (1952) Design for a brain: The origin of adaptive behaviour. Wiley, New York. The second edition was published in 1960.

Ashby W. R. (1956) Introduction to cybernetics. Wiley, New York.

Ashby W. R (1956) An introduction into cybernetics. Chapman Hall, London.

Bannister D. & Fransella F. (1971) Inquiring man: The theory of personal constructs. Penguin, Harmondsworth.

Baron P. (2013) A conversation with my "friend" technology. Cybernetics & Human Knowing 20(1–2): 69–81.

Baron P. (2014) Overcoming obstacles in learning cybernetic psychology. Kybernetes 43(9/10): 1301–1309.

Baron P. (2015) A challenge to objective perception in hearing and seeing in counselling psychology. Kybernetes 44(8/9): 1406–1418.

Barsalou L. W. (2012) The human conceptual system. In: Spivey M., McRae K. & Joanisse M. (eds.) The Cambridge handbook of psycholinguistics. Cambridge University Press, New York: 239–258.

Bateson G. (1972) Steps to an ecology of mind. Intertext Books, London.

Becvar D. S. & Becvar R. J. (2006) Family therapy: A systemic integration. Allyn & Bacon, Boston MA.

Bertalanffy L. von (1950) An outline of general systems theory. British Journal for the Philosophy of Science 1: 134–165.

Bertalanffy L. von (1972) The history and status of general systems theory. In: Klir G. (ed.) Trends in general systems theory. Wiley, New York. http://cepa.info/2701

Brand S., Bateson G. & Mead M. (1976) For god's sake, Margret: Conversation with Gregory Bateson and Margret Mead. CoEvolutionary Quarterly (10): 32–44.

Breggin P. (2008) Brain-disabling treatments in psychiatry: Drugs, electroshock, and the psychopharmaceutical complex. Springer, New York.

Bruner J. (1960) The process of education. Harvard University Press, Cambridge MA.

Buchinger E. (2012) Luhmann and the constructivist heritage: A critical reflection. Constructivist Foundations 8(1): 19–28. http://constructivist.info/8/1/019

Buchinger E. & Scott B. (2010) Comparing conceptions of learning: Pask and Luhmann. Constructivist Foundations 5(3): 109–120. http://constructivist.info/5/3/109

Cadenas H. & Arnold M. (2015) The autopoiesis of social systems and its criticisms. Constructivist Foundations 10(2): 169–176. http://constructivist.info/10/2/169

Chapman S. J. & Jones D. M. (1980) Models of man. The British Psychological Society, Leicester.

Conway F. & Siegelman J. (2006) Dark hero of the information Age: In search of Norbert Wiener, the father of cybernetics. Basic Books, New York.

Dallos R. & Draper R. (2010) An introduction to family therapy: Systemic theory and practice. Third edition. McGraw-Hill, Maidenhead.

Dallos R. & Urry A. (1999) Abandoning our parents and grandparents: Does social construction mean the end of systemic therapy? Journal of Family Therapy 21: 161–86.

Dewey J. (1896) The reflex arc concept in psychology. Psychological Review 3: 357–370.

DoHET. Department of Higher Education and Training (2015) Statistics on post-school education and training in South Africa: 2013. DoHET, Pretoria. http://www.saqa.org.za/docs/papers/2013/stats2011.pdf

Dreyfus H. L. & Rabinow P. (1983) Michel Foucault: Beyond structuralism and hermeneutics. Second edition. Chicago University Press, Chicago.

Eagleman D. (2015) The brain. Pantheon Books, New York.

Ellul J. (1964) The technological society. Translated by J. Wilkinson. A. A. Knopf, New York.

Eysenck H. J. (1966) Personality and experimental psychology. Bulletin of the British Psychological Society 19: 1–28.

Eysenck M. & Keane T. (2015) Cognitive psychology: A student's handbook. Psychology Press, London.

Farsides T. & Sparks P. (2016) Opinion: Buried in bullshit. The Psychologist 29: 368–371. https://thepsychologist.bps.org.uk/volume-29/may-2016/buried-bullshit.

Feist G. (2008) The psychology of science and the origins of the scientific mind. Yale University Press, Boston MA.

Flaskas C. (2002) Family therapy beyond postmodernism: Practice challenges theory. Brunner-Routledge, Hove.

Foerster H. von (1960) On self-organizing systems and their environments. In: Yovits M. & Cameron S. (eds.) Self-organizing systems. Pergamon Press, London: 31–50. Reprinted in Foerster H. von (2003) Understanding understanding. Springer, New York: 1–19. http://cepa.info/1593

Foerster H. von (eds.) (1974) Cybernetics of cybernetics. BCL Report 73.38. Biological Computer Laboratory, Dept. of Electrical Engineering, University of Illinois, Urbana IL. Republished in 1995 by Future Systems, Minneapolis MN.

Foerster H. von (2003) Understanding understanding. Springer, New York.

Foerster H. von, Mead M. & Teuber H. L. (1953) A note from the editors. In: Cybernetics: Circular causal and feedback mechanisms in biological and social systems, transactions of the eighth conference, 15–16 March 1951. Josiah Macy Jr. Foundation, New York NY: xi–xx. http://cepa.info/2709

Ford N. (2008) Web-based learning through educational informatics. Hershey, New York.

Foucault M. (1980) Truth and power. In: Gordon C. (ed.) Power/knowledge. Selected interviews and other writings. Pantheon Books, New York.

Freeman W. H. (2000) Brains create macroscopic order from microscopic disorder by neurodynamics in perception. In: Arhem P., Blomberg C. & Liljenstrom H. (eds.) Disorder versus order in brain function. World Scientific, Singapore: 205–220. http://cepa.info/2702

George F. H. (1961) The brain as a computer. Pergamon Press, Oxford.

George F. H. (1976) Cybernetics (Teach yourself). Hodder and Stoughton, London.

Gergen K. (1999) An invitation to social construction. Sage, London.

Gergen K., Schrader S. & Gergen M. (2009) Constructing worlds together: Interpersonal communication as relational process. Pearson Education, Boston MA.

Gholson B., Shadish W. R., Neimeyer R. A. & Houts A. C. (1989) Psychology of science: Contributions to metascience. Cambridge University Press, Cambridge MA.

Glanville R. (2002) Second order cybernetics. In: Encyclopaedia of life support systems. EoLSS Publishers, Oxford. Web publication. http://cepa.info/2708

Glanville R. (2015) Living in cybernetics. Kybernetes 44(8/9): 1174–1179.

Glasersfeld E. von (1984) An introduction to radical constructivism. In: Watzlawick P. (ed.) The invented reality: How do we know what we believe we know? W. W. Norton, New York: 17–40. http://cepa.info/1279

Glasersfeld E. von (1990) An exposition of constructivism: Why some like it radical. Journal for Research In Mathematics Education 4: 19–29 & 195–210. http://cepa.info/1415

Goethals G. R. (2007) A century of social psychology: Individuals, ideas, and investigations. In: Hogg M. A. & Cooper J. (eds.) The SAGE handbook of social psychology. Sage, London: 3–23.

Gregory R. L. (1970) The intelligent eye. McGraw-Hill, New York.

Gross R. (2010) Psychology: The science of mind and behaviour. 6th Edition. Hodder Education, London.

Haken H. (1983) Synergetics,: An introduction. Springer, Berlin.

Haken H. (2006) Information and self-organization. A macroscopic approach to complex systems. Third enlarged edition. Springer, Berlin.

Hayes N. (2000) Foundations of psychology: An introductory text. Third edition. Cengage Learning EMEA, Andover.

Hearnshaw L. S. (1979) Cyril Burt: Psychologist. Cornell University Press, Ithaca NY.

Hebb D. O. (1949) The organisation of behaviour. Wiley, New York.

Heidegger M. (1977) The question concerning technology and other essays. Translated by William Lovitt. Harpes & Row, New York.

Heims S. J. (1991) Constructing a social science for postwar America: The cybernetics group, 1946–1953. MIT Press, Cambridge MA.

Hesse M. B. (1966) Models and analogies in science. University of Notre Dame Press, Notre Dame IN.

Ho M. W. (1995) Bioenergetics and the coherence of organisms. Neuronetwork World 5: 733–750.

Hollingsworth J. R. & Müller K. H. (2008) Transforming socio-economics with a new epistemology. Socio-Economic Review 3(6): 395–426.

Hunt M. (1993) The story of psychology. Doubleday, New York.

Illich I. (1977) Disabling professions. In: Illich I., Zola I. K., McKnight J., Caplan J. & Shaiken H. (eds.) Disabling professions. Marion Boyars, London: 11–39.

Ioannidis J. P. A. (2005) Why most published research findings are false. PLoS Medicine 2(8): E124.

John L. K., Loewenstein G. & Prelec D. (2012) Measuring the prevalence of questionable research practices with incentives for truth telling. Psychological Science 23(5): 524–532.

Johnstone L. (2000) Users and abusers of psychiatry: A critical look at psychiatric practice. Routledge, London.

Joynson R. B. (1974) Psychology and common sense. Routledge & Kegan Paul, London.

Kagee A. & Lund C. (2012) Psychology training directors' reflections on evidence-based practice in South Africa, South African Journal of Psychology 42(1): 103–113.

Kelly G. A. (1977) The psychology of the unknown. In: Bannister D. (ed.) New perspectives in personal construct theory. Academic Press, London: 1–19.

Klir G. & Valach M. (1967) Cybernetic modelling. Iliffe Books, London.

Koch S. (1976) Language communities, search cells, and the psychological studies. In: Arnold W. J. (ed.) Nebraska symposium on motivation. University of Nebraska Press, Lincoln NE: 447–559.

Lader M. (1977) Psychiatry on trial. Penguin, Harmonsworth.

Lambek M. (2014) Recognizing religion: Disciplinary traditions, epistemology, and history. Numen: International Review for the History of Religions 61(2/3): 145–165.

Laurillard D. (2002) Rethinking university teaching: A conversational framework for the effective use of learning technologies. Second edition. Routledge, London.

Leary D. E. (ed.) (1990) Metaphors in the history of psychology. Cambridge University Press, Cambridge MA.

Luhmann N. (1977) Differentiation of society. Canadian Journal of Sociology 2: 29–54.

Luhmann N. (1982a) Interaction, organization, and society. Translated by Stephan Holmes and Charles Larmore. In: Luhmann N., The differentiation of society. Columbia University Press, New York: 69–89. German original published as: Luhmann N. (1975) Interaktion, Organisation, Gesellschaft. In: Soziologische Aufklärung 2: Aufsätze zur Theorie der Gesellschaft. Westdeutscher Verlag, Opladen: 9–20.

Luhmann N. (1982b) The world society as a social system. International Journal of General Systems 8: 131–138. http://cepa.info/2814

Luhmann N. (1984) Soziale Systeme: Grundriß einer allgemeinen Theorie. Suhrkamp, Frankfurt am Main. English translation: Luhmann N. (1995) Social systems. Stanford University Press, Stanford CA.

Luhmann N. (1986) The autopoiesis of social systems. In: Geyer F. & van der Zouwen J. (eds.) Sociocybernetic paradoxes. Sage, London: 172–192. http://cepa.info/2717

Luhmann N. (1990a) Political theory in the welfare state. Translated by John Bednarz Jr. Berlin-New York, Walter de Gruyter. German original published as: Luhmann N. (1981) Politische Theorie im Wohlfahrtsstaat. Olzog, Munich.

Luhmann N. (1990b) The autopoiesis of social systems. In: Luhmann N., Essays on self-reference. Columbia University Press, New York: 1–20. Originally published in 1986. http://cepa.info/2717

Luhmann N. (1995) Social systems. Translated by John Bednarz Jr, with Dirk Baecker. Stanford, Stanford University Press. German original published as: Luhmann N. (1984) Soziale Systeme. Suhrkamp, Frankfurt am Main.

Luhmann N. (2012) Theory of society Vol. 1. Translated by Rhodes Barrett. Stanford University Press, Stanford. German original published as: Luhmann N. (1997) Gesellschaft der Gesellschaft. Suhrkamp, Frankfurt am Main.

Martin R. J. (2015) The role of experience in the ASC's commitment to engage those outside the cybernetics community in learning cybernetics. Kybernetes 44(8/9): 1331–1340.

Maturana H. R. (1970a) Neurophysiology of cognition. In: Garvin P. L. (ed.) Cognition: A multiple view. Spartan Books, New York: 3–24.

Maturana H. R. (1970b) Biology of cognition. BCL Report No. 9.0. University of Illinois, Urbana, Illinois. Reprinted in: Maturana H. R. & Varela F. J. (1980) Autopoiesis and cognition. Reidel, Dordrecht: 5–58. http://cepa.info/535

Maturana H. R. (2002) Autopoiesis, structural coupling and cognition: A history of these and other notions in the biology of cognition. Cybernetics & Human Knowing 9(3–4): 5–34. http://cepa.info/685

Maturana H. R. & Varela F. J. (1980) Autopoiesis and cognition. Reidel, Dordrecht.

Maturana H. R. & Varela F. J. (1992) The tree of knowledge: The biological roots of human understanding. Revised edition. Shambhala, Boston MA.

McCann J. T., Shindler K. L. & Hammond T. R. (2004) The science and pseudoscience of expert testimony. In: Lilienfeld S. O., Lynn S. J. & Lohr J. M. (eds.) Science and pseudoscience in clinical psychology. Guilford Press, London: 77–108.

Mead M. (1968) Cybernetics of cybernetics. In: Foerster H. von, White J., Peterson L. & Russell J. (eds.) Purposive Systems, Spartan Books, New York NY: 1–11. http://cepa.info/2634

Meyer W. F., Moore C. & Viljoen H. G (2008) Personology: From individual to ecosystem. Fourth edition. Heinemann, Sandown.

Miller G. A. (1962) Psychology: The science of mental life. Harper and Row, New York.

Miller G. A. (1969) Psychology as a means of promoting human welfare. American Psychologist 24: 1063–1075.

Miller G. A., Gallanter E. & Pribram K. (1960) Plans and the structure of behaviour. Holt, Rinehart and Winston, New York.

Mills C. (2013) Decolonizing global mental health: The psychiatrization of the majority world. Routledge, London.

Morris P. E. (2015) The reproducibility project and the BPS. The Psychologist 28: 858. http://thepsychologist.bps.org.uk/volume-28/november-2015/science-or-alchemy
Müller A. & Müller K. H. (eds.) (2007) An unfinished revolution? Heinz von Foerster and the Biological Computer Laboratory, BCL, 1958–1976. Edition echoraum, Vienna.
Newell A. (1990) Unified theories of cognition. Harvard University Press, Cambridge MA.
Pask G. (1960) The natural history of networks. In: Yovits M. C. & Cameron S. (eds.) Self-organising systems. Pergamon Press, London: 232–261.
Pask G. (1961) An approach to cybernetics. Hutchinson, London.
Pask G. (1963) The use of analogy and parable in cybernetics, with emphasis upon analogies for learning and creativity. Dialectica 17(2/3): 167–202.
Pask G. (1970) The meaning of cybernetics in the behavioural sciences. In: Rose J. (ed.) Progress of cybernetics. Volume 1. Gordon and Breach, London: 15–44. http://cepa.info/1847
Pask G. (1975a) The cybernetics of human learning and performance. Hutchinson, London.
Pask G. (1975b) Conversation, cognition and learning. Elsevier, Amsterdam.
Pask G. (1976) Conversation theory: Applications in education and epistemology. Elsevier, Amsterdam.
Pask G. (1979) Against conferences: The poverty of reduction in sop-science and pop-systems. In: Proceedings of the silver anniversary international meeting of the Society for General Systems Research. SGSR, Washington: xiii–xxv.
Pask G. (1980) The limits of togetherness. In: Lavington S. H. (ed.) Information processing '80. North-Holland, Amsterdam: 999–1012.
Pask G. (1981) Organisational closure of potentially conscious systems. In: Zelany M., (ed.) Autopoiesis. North Holland Elsevier, New York: 265–307. http://cepa.info/2703
Pask G. (1987) Conversation and support. Research Programme Ondersteuning Overleving & Cultuur (OOC). Universiteit Amsterdam, Amsterdam: 5–43.
Pask G. (1996) Heinz von Foerster's self organization, the progenitor of conversation and interaction theories. Systems Research 13(3): 349–362. http://cepa.info/2706
Pask G., Kallikourdis D. & Scott B. (1975) The representation of knowables. International Journal of Man-Machine Studies 7: 15–134.
Pask G., Scott B. & Kallikourdis D. (1973) A theory of conversations and individuals (exemplified by the learning process in CASTE). International Journal of Man-Machine Studies 5: 443–566.
Piaget J. (1974) To understand is to invent: The future of education. Viking, New York.
Piaget J. (1977) Psychology and epistemology: Towards a theory of knowledge. Penguin, Harmondsworth.
Pickering A. (2010) The cybernetic brain: Sketches of another future. University of Chicago Press, Chicago IL.
Popper K. (1963) Conjecture and refutations. Routledge and Kegan Paul, London.
Rogers C. R. (1980) A way of being. Houghton Mifflin, New York.
Rosenblueth A., Wiener N. & Bigelow J. (1943) Behavior, purpose and teleology. Philosophy of Science 10(1): 18–24. http://cepa.info/2691
Ross L., Lepper M. & Ward A. (2010) History of social psychology: Insights, challenges, and contributions to theory and application. In: Fiske S. T., Gilbert D. T. & Lindzey G. (eds.) Handbook of social psychology. 5th Edition. Wiley, Hoboken.
Sanford F. H. (1966) Psychology: A scientific study of man. Second edition. Wadsworth, Belmont CA.
Schiepek G., Ludwig-Becker F., Helde A., Jagfeld F., Petzold E. R. & Kröger F. (2005a) Synergetics for practice: Therapy as encouraging self-organized processes. In: Bohak J. & Možina M. (eds.) Contemporary flows in psychotherapy. Slovenian Umbrella Association for Psychotherapy, Rogla: 25–33.

Schiepek G., Picht A., Spreckelsen C., Altmeyer S. & Weihrauch S. (2005b) Computer-based process diagnostics of dynamic systems. In: Bohak J. & Možina M. (eds.) Contemporary flows in psychotherapy. Slovenian Umbrella Association for Psychotherapy, Rogla: 34–51.

Schiepek G. K., Tominschek I. & Heinzel S. (2014) Self-organization in psychotherapy: Testing the model of change processes. Frontiers in Psychology 5:1089. http://journal.frontiersin.org/article/10.3389/fpsyg.2014.01089

Schlötter P. (2005) Vertraute Sprache und ihre Entdeckung. Systemaufstellungen sind kein Zufallsprodukt – der empirische Nachweis. Carl-Auer Verlag, Heidelberg.

Scott B. (1987) Human systems, communication and educational psychology. Educational Psychology in Practice 3(2): 4–15. http://cepa.info/1811

Scott B. (1993) Working with Gordon: Developing and applying Conversation Theory (1968–1978) Systems Research 10(3): 167–182. http://cepa.info/295

Scott B. (1996) Second-order cybernetics as cognitive methodology. Systems Research 13(3): 393–406. http://cepa.info/1810

Scott B. (1997) Inadvertent pathologies of communication in human systems. Kybernetes 26(6/7): 824–836. http://cepa.info/1809

Scott B. (2000) Cybernetic explanation and development. Kybernetes 29(7/8): 966–994. http://cepa.info/1807

Scott B. (2001a) Conversation theory: A dialogic, constructivist approach to educational technology. Cybernetics & Human Knowing 8(4): 25–46. http://cepa.info/1803

Scott B. (2001b) Cybernetics and the social sciences. Systems Research 18: 411–420. http://cepa.info/1804

Scott B. (2001c) Gordon Pask's conversation theory: A domain independent constructivist model of human knowing. Foundations of Science 6(4): 343–360. http://cepa.info/1806

Scott B. (2001d) Gordon Pask's contributions to psychology. Kybernetes 30(7/8): 891–901. http://cepa.info/1805

Scott B. (2002) Cybernetics and the integration of knowledge. In: Encyclopaedia of life support systems. EoLSS Publishers, Oxford. Web publication. http://cepa.info/1801

Scott B. (2004) Second order cybernetics: An historical introduction. Kybernetes 33(9/10): 1365–1378. http://cepa.info/1798

Scott B. (2006) Reflexivity revisited: The sociocybernetics of belief, meaning, truth and power. Kybernetes 35 (3–4): 308–316. http://cepa.info/1797

Scott B. (2007) The co-emergence of parts and wholes in psychological individuation. Constructivist Foundations 2(2-3): 65–71. http://constructivist.info/2/2-3/065

Scott B. (2009) Conversation, individuals and concepts: Some key concepts in Gordon Pask's interaction of actors and conversation theories Constructivist Foundations 4(3): 151–158. http://constructivist.info/4/3/151

Scott B. (2011a) Explorations in second-order cybernetics. Reflections on cybernetics, psychology and education. Edition echoraum, Vienna.

Scott B. (2011b) Heinz von Foerster: Contributions to psychology. Cybernetics & Human Knowing 18(3/4): 163–169. http://cepa.info/1789

Scott B. (2011c) Toward a cybernetic psychology. Kybernetes 40(9/10): 1247–1257. http://cepa.info/1790

Scott B. (2012) Ranulph Glanville's Objekte. In: Glanville R., The black b∞x. Volume 1: Cybernetic circles. Edition Echoraum, Vienna: 63–76. Originally published in 2005. http://cepa.info/1786

Scott B. (2014) Education for cybernetic enlightenment. Cybernetics and Human Knowing 21: 1–2: 199–205. http://cepa.info/1286

Scott B. (in press) Reflections on the sociocybernetics of social networks. In: Lisboa M. (ed.) Complexity and social actions: Interaction and multiple systems. Cambridge Scholars Publishing, London.

Scott B. & Bansal A. (2014) Learning about learning: A cybernetic model of skill acquisition. Kybernetes 43: 9/10: 1399–1411. http://cepa.info/1283
Scott B. & Shurville S. (2011) What is a symbol? Kybernetes 48(1/2): 12–22. http://cepa.info/1791
Shannon C. E. & Weaver W. (1949) The mathematical theory of communication. University of Illinois Press, Urbana IL.
Simmons J. P., Nelson L. D. & Simonsohn U. (2011) False-positive psychology: Undisclosed flexibility in data collection and analysis allows presenting anything as significant. Psychological Science 22(11): 1359–1366.
Sloan T. (2000) Critical psychology. Palgrave Macmillan, New York.
Soldz S. (1990) The therapeutic interaction: Research perspectives. In: Wells R. A. & Giannetti M. J. (eds.) Handbook of the brief psychotherapies. Plenum Press, New York NY: 27–53.
Sommerfeld P., Hollenstein L., Calzaferri R. & Schiepek G. (2005) Real-time monitoring: New method for evidence-based social work. In: Sommerfeld P. (ed.) Evidence-based social work: Towards a new professionalism? Peter Lang, Bern: 199–232.
Stapel D. (2014) Faking science: A true story of academic fraud. Translated by Nicholas Brown. http://nick.brown.free.fr/stapel/FakingScience-20141214.pdf
Stevens S. S. (1936) Psychology: The propaedeutic science. Philosophy of Science 3(1): 90–103.
Šugman Bohinc L. (2016) Social work: The science, profession and art of complex dealing with complexity. In: Mešl N. & Kodele T. (eds.) Co-creating processes of help: Collaboration with families in community. Faculty of Social Work, University of Ljubljana, Ljubljana: 41–63.
Tavris C. (2004) The widening scientist-practitioner gap. In: Lilienfeld S. O., Lynn S. J. & Lohr J. M. (eds.) Science & pseudoscience in clinical psychology. Guilford Press, London: ix–xviii.
Thumala-Dockendorff D. (2010) Proyecciones del Concepto de Sistema Psíquico de Luhmann y su vinculación con la Psicología [English Translation]. Cinta de Moebio 39: 186–191.
Tyler I. (2013) Revolting subjects: Social abjection and resistance in neoliberal Britain. Zed Books, London.
Umpleby S. (2003) Heinz von Foerster and the Mansfield Amendment. Cybernetics & Human Knowing 10(3–4): 87–190. http://cepa.info/1876
Varela F. J., Maturana H. R. & Uribe R. (1974) Autopoiesis: The organization of living systems, its characterization and a model. Biosystems 5(4): 187–196. http://cepa.info/546
Varga M. v. K. & Sparrer I. (2016) Ganz im Gegenteil. Tetralemmaarbeit und andere Grundformen Systemischer Strukturaufstellungen – für Querdenker, und solche die es werden wollen. Carl Auer Verlag, Heidelberg.
Vorhemus U. (2015) Systemische Strukturaufstellungen – Systemisch – Konstruktivistisch – Phänomenologisch. Systmedia Verlag, Aachen.
Vygotsky L. (1962) Thought and language. MIT Press, Boston MA.
Werbert A. (1989) Psychotherapy research between process and effect: The need of new methodological approaches. Acta Psychiatrica Scandinavia 79(6): 511–522.
White M. (2011) Narrative practice: Continuing the conversations. W. W. Norton, New York.
White M. & Epston D. (1990) Narrative means to therapeutic ends. W. W. Norton, New York.
Wiener N. (1948) Cybernetics: Or control and communication in the animal and the machine. MIT Press, Cambridge MA. The second edition was published in 1985.
Wittgenstein L. (1953) Philosophical investigations. Basil Blackwell, Oxford.
Yovits M. & Cameron S. (eds.) (1960) Self-organizing systems. Pergamon Press, London.

Consciousness as Self-Description in Differences

Diana Gasparyan

Introduction

1. In different areas of knowledge, consciousness or, more specifically, its explanation, is currently one of the most central and frequently discussed topics. This subject is invariably addressed by the cognitive sciences: philosophy of consciousness, neurosciences, cognitive psychology, etc. As for philosophy, since the mid-twentieth century consciousness has become one of its favored themes (Papineau 2002). The main purpose of these studies is to find the answer to the question "Can we explain consciousness?" (Chalmers 1997). Despite lengthy discussions, the question remains unanswered.

2. The problem of consciousness forces us to delve into the capabilities of the observation apparatus we have available, to think over what we are able to observe and what we can describe, acting as beings involved in observation of the objects we are going to describe.

3. Second-order cybernetics (SOC) can help to shed light on this topic. The fact is that no other area of knowledge is characterized by this situation, whereby a total disregard of the epistemological principles underlying SOC results in so many inconsistencies. This can be observed in consciousness-related studies, including one of the most popular and fast-growing areas of knowledge – modern analytical philosophy of consciousness. Philosophy of consciousness is a perfect choice for SOC application, and the lingering hesitation to apply it is unreasonable and perhaps simply an omission. One objective of this research is to rectify this situation.

4. The research is formed of two parts. In the *critique section* (Part 1) I will show that consciousness should be studied by giving the level of the observer priority attention, for its neglect will lead any theory of consciousness into paradoxes and finally to absurdity. Then, I will address criticism of the subject–object dualism from the SOC perspective. I will also detail the unproductive epistemological assumptions that may result from application of the subject–object model in studying consciousness. I will show that theories tending to explain consciousness from the subject–object dualism perspective invariably come to an epistemological deadlock. When dealing with theories attempting to explain consciousness, cognitive sciences indirectly introduce the idea of consciousness as a subject–object (dualistic) model where reflection can exist only in the form of metaknowledge. I will also demonstrate that in speaking about the self-description (reflection) of consciousness it would be inefficient to speak about a metalinguistic structure: in trying

to explain reflection in the subject–object paradigm (with object consciousness and standalone subject consciousness), we lay down a paradox. In the *positive section* (Part 2), relying on SOC principles, I will explain that consciousness should not be seen as an essence, and I will show how consciousness could be approached using the apparatus of differences.

Part 1: How consciousness does not work

Discussing consciousness using the language of consciousness: From the consciousness that is being observed to the observing consciousness

5. Although consciousness is an extremely popular topic that has attracted the attention of a great number of researchers, studies are frequently aimed at the "elimination of consciousness" or, in other words, at explaining it away (Chalmers 2003). Since the second half of the 20th century, most manipulations of consciousness have been performed to narrow it down either to physiological properties (Demircioglu 2013) or to some other objective forms (for example, social or linguistic relationships; Pinker 2007). Such "studies of consciousness" all have a similar purpose – to diffuse consciousness in objective processes. If we take a closer look at the general vector of consciousness-focused studies, which began in the second half of the 20th century, we clearly see a prevailing "preferred vocabulary," namely the vocabulary of natural sciences, which sets the criteria for passable solutions of consciousness problems (Van Gulick 1985).

6. The important point is that most modern theories of consciousness are based on principles typical of natural-science methods compliant with classical modern European (Newtonian) ideals. In accordance with Newtonian physics, the observer is essentially excluded from the view (Foerster 1995). This means that consciousness is seen as a certain essence or a phenomenon that can be studied *objectively* – disregarding the researcher-observer, who is perceived as a transparent medium. Consequently, it can be stated that most modern theories of consciousness as well as the entire analytical philosophy of consciousness are still based on first-order cybernetics (FOC). In the case of SOC, first-order cybernetics may be seen as a limited case where the link back from observed to observer is sufficiently weakened. Under such circumstances, we assume the observer simply and neutrally observes what is going on (Glanville 2002).

7. This approach is clearly manifested in different programs of reductionism, where the task is to reduce consciousness to various types of objective essences. This task is mainly implemented in the search for a causal relationship between the body and the mind, i.e., in understanding how the physical produces the non-physical – the so-called mind–body problem, which has been called the hard problem of consciousness by David Chalmers (1997). This also manifests itself in the search for other external reasons for consciousness, such as language, culture, and society. However, the application of FOC or, in other words, the conversion of consciousness into an object – while the observer is disregarded – generally results in the elimination of consciousness. Any attempts to speak "about" consciousness using FOC terminology are senseless, for we have to use this very consciousness to do so, having no other cognitive tools. *The object of consciousness is consciousness as such.*

8. For example, the most common type of reductionism – physicalism – argues that consciousness is completely reducible to neurophysiological processes and brain functions (Hellman & Thompson 1975; Dennett 1992). For example, the identity theory states that consciousness is identical to neuronal activity in the brain (Place 1956). According to this theory, any mental condition is identical to a certain condition of the brain, i.e., the mental condition and the respective neural condition mean the same. Therefore, advocates of the identity theory believe that while mental conditions can theoretically exist separately from the material systems that generate them and, as such, could exist on their own, in actual fact they coincide with them (Bennett et al. 2007; Tononi 2012; Churchland 2013; Bickle 2012).

9. In addition to physiological reductionism, there are other methods of eliminating consciousness. For example, some theories identify the functional capabilities of consciousness with consciousness. Such theories construe mental conditions as remaining in a certain functional condition. The core thesis of functionalism promotes the idea of carrying over consciousness from one medium (the human brain) to other possible media. In other words, some functional conditions can be "launched" in fundamentally heterogeneous physical systems; first of all, in computers (Funkhouser 2007). In fact, such theories also reduce consciousness – specifically to functional operations of consciousness (McCullagh 2000; Piccinini 2004; Shagrir 2005).

10. By now, reductionism – and its attempts to reduce consciousness and, in fact, to abandon it – has been severely and repeatedly criticized (Van Gulick 1985; Kriegel 2009). I will not dwell on the points of criticism; suffice it to say that most critics agree on the inadmissibility of reducing consciousness (mental facts) to material things and states (physical facts) (Levine 1983, 2001; Jackson 1986, 1982; Nagel 1986).

11. In the meantime, very few scholars have paid attention to the initial *epistemological problems* associated with studying consciousness or, more specifically, to the fact that understanding consciousness in the way expected by most modern theories assumes doing what cannot be done: moving beyond the limits of consciousness (this problem has been addressed by scholars such as Colin McGinn 1989, 1991 and Roger Penrose 1989, 1994). It is obvious that in being observers asking the questions "What is consciousness?" and "How is it connected with the body?" we ask these questions using the same consciousness that we are trying to convert into an object. But here SOC can offer the most appropriate and adequate explanation, namely that the observer is essentially included in the view through her frame of reference and her motion relative to the objects and events under consideration.

12. In fact, philosophy of consciousness, which claims consciousness of consciousness, should, similarly to cybernetics of cybernetics, take the observer into consideration, as in this case she is not only important as an interpreter of reality, but also related to the object of study (Ashby 1956). In this case, the object of study coincides with the method, since we study consciousness with the help of this very consciousness, without which we, as researchers, are bereft.

13. In SOC, we take into account the relationship between the observer (observing) and the observed, particularly when this relationship is understood to be circular. Second-order cybernetics presents a (new) paradigm, in which the observer is circularly (and intimately) involved with and connected to the observed. The observer is no longer neutral and detached, and what is considered is not the observed (as in the classical paradigm), but the observing system. Either the aim of attaining traditional objectivity is abandoned or what

objectivity is and how we might obtain (and value) it is reconsidered (Glanville 2002). The main strategy for using second-order cybernetics lies in the development of the philosophical idea that the concept of perception as reflecting the world in an objective manner is no more than illusion. Much of the work of embedding this idea into modern epistemology and science was undertaken by Heinz von Foerster (1981), Ernst von Glasersfeld (1987), Humberto Maturana, and Francisco Varela (1987). With a view to understanding human knowledge, cyberneticists studied the nervous system and came to the conclusion that observations independent from the observer are virtually impracticable. Underlying this conclusion is the philosophical reasoning according to which it is impossible to separate the cognizable from the instruments of cognition even logically, while it is impossible to prove the reality of the world beyond a certain system of perception (the reality of such a world might be introduced on the basis of a belief, a conviction, or a postulate, but not on proof of the procedure that was carried out). For example, as I am writing this article, I am using my own system of perception and cognition formed as a result of different interactions specific to my personality: cultural background, language, academic qualifications, personal interests, etc. If, according to the classical concept of science, scientific theories intrinsic as the observer's traits are not taken into account whilst observing, then by studying the nervous system, cyberneticists reveal a contradiction compared to the mainstream presentation of the philosophy of science.

14. Due to the fact that observations not dependent on the observer are impractical, the conclusion that all knowledge is judgmental suggests itself. This idea requires correct understanding: what it involves is that "objectivity" becomes *dispersed subjectivity*. This assertion in SOC means that people do not seek confirmation of their views in the "objective world." We do not have access to the world; however, we have access to other interpretations of the world, which one might aspire to bring into accord. Various interpretations become different prospects for the perception of one another even in cases where understanding of the world is not taken into account. This assertion in SOC will be used to support my own views in the last part of this article.

15. Therefore, if science relies on the objective nature of its objects, consciousness does not have this external criterion in its self-descriptions. When cognizing itself, consciousness discovers the meaning of its judgments only *in itself*. This means that consciousness is always studied by the consciousness, and any theories of consciousness constitute parts of self-describing consciousness and serve as tools for self-description. For this reason, consciousness can be seen as an example of the autopoietic system, knowledge of which is generated by the same system. Such a system is a unity, "defined as a network of production of components, which recursively, via their interactions, generate and implement the network which produces them" (Maturana & Varela 1980: 137). In turn, SOC is the most adequate model for description of this system, which is of "a circular nature: the person is learning to see himself as the part of the world he observes" (Foerster 1981: 239). Thus, we can assume that the "epistemology of the observer" should be seen as an adequate philosophy of consciousness (Foerster 1992), which focuses on internal description (Rockmore 2005).

16. In the development of both philosophy and science, SOC will be useful in the following ways:

a. SOC can help revise such types of reductive studies, in which we assume that consciousness is a certain object of study and we have a certain mysterious meta-language that has nothing to do with consciousness that we can use to discuss "object-consciousness." In particular, it might help to clarify the boundaries of physical reductionism itself: for example, when we suppose that we have consciousness and that there is a language of neural correlates, which generates it, or that there is consciousness and there are C-fibers that produce the "effect" of consciousness. Both these methods illustrate reductive techniques, and by using them we may speak about consciousness with the help of the language of physical processes and events.

b. By applying SOC, it is possible to bypass the strategy of a causative search for the factors generating consciousness – be it physical processes or groups of factors such as language, culture, or society (as the most frequently mentioned). In this case, SOC allows consciousness to be treated as something that is originally a causally closed system, with all reasons within itself.
c. SOC offers such a descriptive strategy, which would relieve us from the search for descriptive languages exterior to consciousness, which in turn merely lead to metalinguistic paradoxes since the descriptive languages are part of consciousness itself. By using SOC we can bypass the strategy of reductionism and replace it with a strategy whereby *consciousness will be talked about in the language of consciousness.* This means that consciousness should not be seen as an object observed from outside; understanding (description) of consciousness from inside makes better sense.
d. SOC helps eliminate subject–object dualism, the application of which regarding some significant manifestations of consciousness, in particular reflection, leads to paradoxes and unproductive theories.
e. SOC allows us to take a fresh look at the concepts of truth and objectivity within the framework of sciences studying consciousness. Under the conditions of totality of consciousness, the truth is understood not as remote objectivity but as "distributed subjectivity" – the confirmation of ideas and theories is not sought in the "objective world" but in the provisions of other theories or, more broadly, one system of knowledge within others.

17. When we speak about the difficulties we encounter when we try to reproduce someone else's individual consciousness with the help of our own consciousness, most people agree that the task is next to impossible. However, when we design different *theories of consciousness*, we frequently tend to ignore such difficulties. In the meantime, these difficulties arise every time we turn to traditional scientific theories of consciousness to explain consciousness with the help of *non-consciousness*. For example, physicalist theories of consciousness, which reduce consciousness to operations of the brain, seem to take a view overriding any mental experience and to offer a physical explanation of consciousness. This involves that the language of physics should be used to explain consciousness. However, this is impossible: the physicist acting as an observer still has consciousness, and her physical picture or theory has meaning only inside the conscious observation that understands this meaning. Therefore, if the task is to speak about consciousness *not using the language of the consciousness,* we encounter a paradox (Gennaro 2012). In turn, SOC entails a constructivist epistemology (theory of knowledge) that starts from the assumption that, "the thinking subject has no alternative but to construct what he or she knows on the basis of his or her own experience" (Glasersfeld 1995: 1).

18. Let us take a closer look at psychophysiological conditioning when *brain* activity defines the experience of consciousness. The weakness of this concept is obvious: we must say that even if the brain induces consciousness, the entire system exists in this consciousness (of the researchers, philosophers, and other observers who describe this system). In this system, the observer is essentially included in the view through his frame of reference and his motion relative to the objects and events under consideration (Foerster 1979). There is also a further type of conditioning: let us assume that *social mechanisms or any other cultural patterns* cause the generation of consciousness. The same logic as above can be applied here: conscious experience is *required* for putting social and cultural mechanisms into operation. Finally, we can say that consciousness is a consequence or product of *language* – however, we again encounter problems, as using language implies understanding the meanings inherent within it, and meaningfulness implies consciousness.

19. The three key grounds mentioned above – *language, culture, society (social relations)* – allow us to identify the paradox encountered in revealing generating reasons, which is even more obvious than in cases of reduction to the brain. Both language and culture, as well as society, *per se*, are just names given to different manifestations of consciousness. This means that, at first, we should assume the existence of consciousness in simply making what "causes" it possible. In SOC terms, this means that when describing consciousness, we skip one level of observation no matter what level of observation we assume as the initial level (Mead 1968). The conclusion is that consciousness *always precedes what is suggested as its cause*.

20. This then means that we can analyze consciousness only through the experience of consciousness, rather than through the experience of *brain, linguistic, or social activity*. The dependency of consciousness on the body can be displayed in the form of a conscious statement implying conscious understanding. But the meaning of the statement arguing that consciousness is induced by causes (for example, the physical activity of the brain) is not physical. By definition, the meaning is part of conscious experience.

21. Thus, programs that aim at studying consciousness by trying to narrow it down to physical processes or anything else (in fact, by abandoning consciousness) initially seem controversial. The explanation for this is that they try to convert consciousness into an object. Almost all theories in which consciousness is caused by something (that is, is seen as the product of non-consciousness) make the same epistemological mistake. As a rule, the reductionist puts consciousness into a black box and observes the processes accompanying the mysterious object's remaining in the box. However, such research methods do not offer anything that would help us understand consciousness. After sophisticated and extensive reductive manipulations, consciousness remains as mystical as it was before the research; however, the reductionist believes that he has been able to *explain* consciousness, though he was only able to describe something *associated* with consciousness (for example, neural processes in the brain).

The subject–object uncertainty

22. The main reason for this cognitive volatility of consciousness is the *subject–object uncertainty* with which it is associated. Should sciences (cognitive psychology, neurosciences, or philosophy of consciousness) enter their own object of study when seeing it as a phenomenon and, consequently, seeing themselves as its component? Strictly speaking, any theory of consciousness is self-description (of consciousness) or "consciousness of

consciousness," rather than description of the object (consciousness). Although this theory is distinct in having an observer (*the subject*) whose qualities, according to all scientific rules, cannot affect the target *object*, let alone penetrate this object, it (this theory) will invariably be part of the "object" when dealing with consciousness.

23. The above paradox implies that the concepts of *the subject* and *the object* turn out to be semantically inadequate for their application in the theory of consciousness. When dealing with consciousness, we have to reject the "old European tradition" that non-reflexively operated this type of "self-description" and wrongly assumed that it had some *substantial* content.

24. For theories focused on the subject–object model of cognition and associated FOC, consciousness is *cognitively inaccessible.* There is no external supra-conscious entity, such as the division into consciousness and non-consciousness or into what has and does not have consciousness, that works within consciousness. Consciousness is total in the same way as society is total in the sociology of Niklas Luhmann (1990, 2000). Therefore, the fact that *consciousness cannot be an object for us* implies that any attempt at describing it entails all the means and conditions, the origin of which must be identified.

25. As such, the theory of consciousness is the theory of the description of consciousness rather than the theory of its explanation. "The theory" of the description of consciousness confronts any other normal theory. To a certain extent, this theory of consciousness is a non-normal theory. With it we do not try to speak about consciousness by using languages other than the language of consciousness (for example, languages of neural correlates or computer programs), rather we analyze consciousness within the limits of consciousness. Here, we deal with a certain autopoietic system, in which there is nothing except consciousness. We can refer to it as an autopoietic system, following several interrelated features of consciousness, which we will continue to use in the present article. This is, first of all, the thesis of self-containment of consciousness – in its existence it is determined primarily by inward conditions. Second, it is self-construction, self-reproduction, and in the case of consciousness, an appeal to ourselves, as consciousness is impossible without self-consciousness. Third, it is organized without division into the producer and product – and in the case of consciousness into the conscious and something that is conceived. This system is seen as self-determining from within (Maturana 1980). It constitutes a fragment of reality, which is relatively isolated from the environment by its causal structure. The external environment is not able to determine the autonomous system from outside and to break its causal impermeability. The existence of the external environment can be taken out of context. If this environment has no effect on consciousness (it does not determine, create, or change it), it can be largely disregarded.

26. This "non-normal" theory can help us to explain the *special characteristics of consciousness*, which cannot be captured by a conventional theory. Special characteristics should be understood in context as when *the object under consideration is identical to its interpretation.* The "method of observation" and the "observed" turn out to be intrinsically indistinguishable from each other. Such effects generally escape the attention of conventional theories, which clearly differentiate between the "object" and the "method of observation"; therefore, they should be studied from the perspective of a special theory. This theory must provide for an approach that allows disregarding the difference between the interpretation and its object.

27. When applying this approach, we will have a description that is identical to what is described; in other words, "what" also means "how." This property (the identity of "what"

and "how") is the most significant characteristic of consciousness. When we encounter the identity between the object and the method of its observation, we come across the experience of consciousness. In other words, the experience of consciousness is *performative* in principle.

28. This fits quite well with Gregory Bateson's approach to the mind:

> [I]n no system which shows mental characteristics can any part have unilateral control over the whole. In other words, the mental characteristics of the system are immanent, not in some part, but in the system as a whole. (Bateson 1972: 338)

29. Let us go over the specifics of *performative utterances* – we say "I swear" or "I promise" and we perform an action through these utterances: we assert our oath or promise (Searle 1979). The form of such utterances is identical with their content – there is no need to expand these utterances so that they will perform their function (Austin 1962). Furthermore, if we make an attempt to perform such an expansion, we will see them turn into meaningless tautologies – we will have to say "I promise that I promise" or "I swear that I swear." If we keep expanding performative utterances, we will run into an infinite regress – utterances such as "I swear that I swear that I swear, etc." and "I promise that I promise that I promise, etc." (Searle 1979).

30. These specific characteristics of performatives give an excellent illustration of what is typical of the experience of consciousness. The acts of consciousness have a performative nature – the "what" of my thought is given to me having this thought as the "how." Strictly speaking, "what" I think and "how" I think are identical, as in the thought that my "how" is also my "what" (Gergen 1997; Harre 1989). Certainly, I can specify and clarify my thought or see some latent meaning from it; however, then it will be a different thought, in which "what" and "how" will be again identical (Gennaro 2004). When we encounter the above characteristic (*the characteristic of the performative*), we encounter what can be conventionally described as consciousness. The SOC principles should be applied to the phenomena of this type.

31. Thus, we know in the very least what should constitute the initial stage of our studies of consciousness: observation of the observation and description from inside. Adequate studies of consciousness imply that consciousness should be discussed in the language of consciousness without using the most commonly used reductionisms – applicable to the body (brain), language, and cultural experience. If we cannot study consciousness directly, we should try to study it through description. The theory of description, which allows for the principles of second-order cybernetics, is a specific "quasi-theory," the application of which makes us aware that consciousness is something we always and permanently have and that cannot be objectified or eliminated.

32. Then, *self-description* can be seen as the proper theory of consciousness. Following Luhmann's *The Society of Society* (Luhmann 1997), which tries to tell society about society on behalf of society, every observation is autobiographical. Therefore, SOC must primarily be considered in the first person and with active verbs: the observer's inevitable presence should be acknowledged, and written about in the first person, not the third, giving us an insight into who these observers are (Foerster 2003).

Reflection is always self-reflection

33. The subject and object concepts as well as the subject–object methodology are not only unsuitable for developing theories of consciousness, but also are quite inefficient for understanding more specific states of consciousness, for example, its most important characteristic – *reflection*. The subject–object dualism has inherent insoluble contradictions, which make it impossible to come up with an adequate idea about reflection. Any attempts made to resolve these contradictions tend to be counterproductive – it makes much more sense when we stop talking about consciousness in terms of a subject–object dualism, which implies insurmountable difficulties regarding the description of consciousness. I will try to prove that this is the case.

34. In most areas of modern analytical philosophy of consciousness (based on the classical epistemology of Descartes's type), consciousness is seen as a certain object or as what inheres in a person (a subject) (Gennaro 1995). In other words, the person is asked to comprehend his own consciousness in FOC terms – as something that is observable. The possession of consciousness is seen as the most reliable reference point (according to Descartes's well-known principle of the *cogito*, anything can be called into doubt except one's own consciousness) (Descartes 1966; Lähteenmäki 2007; Van Gulick 2000).

35. Besides, according to classical epistemology, consciousness can be defined through *reflection* or, in other words, through *knowledge about knowledge* (Merleau-Ponty 1962). *Consciousness* takes place when there is duplication of knowledge – it is not enough just to know; it is important *to know that you know*. Furthermore, reflection is seen in classical epistemology as a unique quality of *indifferent mirroring* (Wrathall & Kelly 1996). Both words are significant. "Mirroring" means the ability of consciousness to witness its own operation, while remaining *the same consciousness*. "Indifference" means that consciousness is given to itself as it is, "as true," without any distortions – the privileged access of consciousness to itself implies that knowledge about the knowledge is *always true*.

36. On the face of it, everything looks quite convincing; however, if we accept this theory, very soon we come across a number of paradoxes. The general principle "consciousness is knowledge about knowledge" brings up problems. The main objection is that subject–object dualism is applied to consciousness when it is seen as "knowledge about knowledge," which, in its turn, brings about a number of logical difficulties.

37. These difficulties can be demonstrated in the following way. If we intend to see consciousness as an object, then there must be something that perceives the consciousness that is objectified during the research. If we accept the "perceiving–perceived" tandem, we will need a third term to refer to the perceiving as perceived. From there, we have the following alternatives. We can opt for some element, and the entire series will move into the unperceived because of the optional stoppage. Or, if we accept infinite regress, it will bring us nowhere. The conclusion is as follows: the subject–object model cannot be applied to consciousness and, second, the consciousness that perceives is identical to the consciousness that is perceived. In other words, the consciousness that perceives constitutes a whole with the consciousness that is perceived. The latter circumstance makes us assume that the best approach is when consciousness is perceived only in a certain phenomenological circle (in the terms of phenomenology) (Merleau-Ponty 1962) or circularly, where there is no division into the producer and product (in the terms of SOC), i.e., autopoietically.

38. Therefore, we should avoid incorporating analysis that delimits consciousness (as the subject) and the perceived (as the object) into the philosophical model of reflection,

since the consciously perceived pair requires a third component so that the perceiving, in its turn, can become the perceived. Consciousness is not able to turn itself into an object, to act as the observer with regard to itself. In short, this means that the principle of the pair should not be applied to consciousness, since perceiving oneself is not a pair. If we want to avoid regress into infinity, we should make sure that consciousness is a direct relation of itself to itself. This relation of "itself to itself" is the pure immanence of the experience of consciousness, knowledge about oneself, or its autopoiesis.

39. After we have revised the dualistic models of description of reflection in favor of the autopoietic, we can offer a concept that meets the SOC principles, namely, the "*system of consciousness,*" which is offered as a solution to the paradoxes of the classical theory of consciousness. As there is inevitably a fusion of the observation with the observed on which the observation is focused, we have to speak about the primacy of the "system." This "system" offers a more universal level of description of consciousness compared to the subject–object level. The "system of consciousness" is not a subject that is seen as the universal foundation of observation within a framework of the reflexive procedure. Rather it includes qualities of both the object and the subject.

40. The aforesaid brings forth another implication, namely, revision of the classical idea about *reflection as mirroring.* This refers to doubts regarding the ability of consciousness to preserve indifference toward itself. Classical epistemology sees this wonderful attribute as the ability of consciousness to observe its own operation, which, in fact, means the ability to operate in a self-reporting mode (Husserl 1931). The key point in such self-observation is the passivity of the reflective function – it adds nothing and it lessens nothing; it only records the operation of consciousness as it is. This statement is consistent with what served as a significant assumption of classical (dating back to Descartes) philosophy of consciousness, according to which the most accurate reference point of any phenomenon of consciousness is the givenness of consciousness to itself (Humphreys 1992). This thesis assumes that consciousness has immediate experience of itself, which means that consciousness has such states, to which it can refer when claiming that it knows its rationale. This classical thesis, which can be found in many contexts of modern philosophy of consciousness, can be questioned when seen from the SOC perspective.

41. We can assume that any conscious act should involve such acts, which *took place outside and beyond any reflection,* being inaccessible to the latter. This means that there is a certain "non-objectified remainder" ("blind spot") in thinking, which, in fact, *is thinking per se*.

42. The difficulties encountered by methods of description of consciousness that rely on classical epistemology are caused by the fact that consciousness always comes across a certain "blind spot" that is inaccessible to reflection. These difficulties can be resolved if we give up our intention to achieve total understanding, which is typical of classical epistemology. Despite the postulates of this form of epistemology, there is no reflection primacy and consciousness has no miraculous ability to reflect upon itself. On the contrary: it is pre-reflexive consciousness that makes reflection possible (Husserl 1960). In other words, acts that have already taken place without any reflection and remain unreflected in consciousness constitute a condition of any conscious act. If von Foerster wrote about the non-transparency of areas of non-knowledge ("we do not see that we do not see" – Foerster 1979), in this case we speak about non-transparency of some functional areas of the innermost activity of consciousness.

43. These operations of consciousness result in its productivity, though they cannot be perceived (Nelkin 1989). If we understand something, the laws that govern our understanding cannot be understood and they cannot be included in the experience of understanding. Understanding cannot be grasped by understanding: it allows seeing itself as the *result* rather than the ongoing process; understanding always arrives with a slight delay – when the operation of the consciousness is completed – and it is given to us as an effect (or a result), having which we cannot deduce how it was obtained.

44. Thus, if a certain "non-objectified remainder" (pre-reflexive thinking) is the main active element of consciousness, the operation of consciousness takes effect due to the existence of a certain indefinite cognitive non-understanding, a cognitive blindness with respect to its own operations. However, we can also assume that this systemic drawback of consciousness is required for the successful functioning of the entire system of consciousness and its reproduction as an autopoietic system (Luhmann 1995).

Part 2: How consciousness works

Consciousness is "a difference that makes difference"

45. Therefore, if consciousness is an autopoietic system, we should try to understand *how* self-description works within it. This self-description can be analyzed at two levels:

- a the *local* level of consciousness of a carrier; and
- at the *global* level of consciousness, where the latter is seen as an aggregate of meanings as such and tries to form knowledge about itself.

46. In both cases, self-description of consciousness can be seen as a bundle of internal differences, where some parts, being differentiated, allow the existence of others (whereby I am assuming from the assertion that the description in terms of differences might be applied to any autopoietic system). *Self-descriptions make it possible to refer to themselves through differentiation* of some states from others (Luhmann 1990, 1995). In the first case, we speak about the functioning of *local consciousness*, which is represented by differentiation of its states such as memory, imagination, figure, background, etc. In the second case, self-description is represented by *global consciousness* – the science of consciousness (say, a certain theory of consciousness, whether dualism or physicalism, as its opposite), which only serves as a means of organization of areas of meaning, which are immanent to consciousness. However, in both cases, self-descriptions will constitute semantic effects, with the help of which consciousness describes itself. Self-descriptions are organized only with the help of internal relationships; they are not supported by any substance that could be objectified as an object. Below I will look into both levels – *local and global*. In general, we speak about the principle, based on which the objective thing does not exist as a substance (essence) but is perceived as the result of correlative processes in the internal system of relationships. The main problem here is how to understand consciousness if it is not represented by something substantial and objective. How can we understand what is not an object or a property of an object and what, as can be seen from the aforesaid, is difficult to define as *something*?

47. However, even if it is difficult to identify consciousness, we can still attempt a closer look at it, and we will see that it contains a certain criterion that makes it possible to differentiate consciousness from non-consciousness. At the same time, this criterion cannot be defined as something essential. It does not mean a positive attribute, for example, memory or imagination. We cannot set the criterion of differentiation between consciousness and non-consciousness through a positive attribute, for to understand what memory or imagination is we must already have consciousness. Summarizing the aforesaid, we can conclude that to understand what makes consciousness different from non-consciousness we must have consciousness.

48. This means that the experience of consciousness, which is basically *the experience of differentiation*, is a criterion of differentiation between the experience of consciousness and non-consciousness or non-experience. Here we encounter the same logic that allows us to agree with Bateson when we say that the *differentiation of information from non-information is, in its turn, information.* Keeping in mind its totality, when we actually have to say that everything is information, we must also add that information is nothing other than a universe of differences (Bateson 1972). Therefore, consciousness is the experience of *differentiation.*

49. In a more general sense, I understand the difference to be the *minimum unit of meaning*. One can say that the difference is everywhere that it makes sense, in other words: the meaning appears when there is a difference. This can be shown with the help of a logical apparatus. Judgments have a minimum meaning (concepts, from which judgments are made, still have no meaning, whereas conclusions are complex structures made of judgments). Judgments have the simplest subject-predicate form – "x equals y," which is formed with the help of a difference. In this respect, the judgment itself is in fact the difference. Thus, I understand the difference to be the basic element of logic (cogitation). Therefore, in the sense that consciousness takes into account such logic, it might be said that it is formed by differences. As far as the world of objects is concerned, in this respect they are products of our cognition (and according to the basics of epistemological constructivism are inseparable from it), therefore difference would also be fundamental for them. If we bear in mind the phenomenological rule "the order of ideas corresponds to the order of things," the difference will become the fundamental principle of organization, both for the arena of things and the arena of consciousness.

50. To some extent, the definition "experience of differences" can be applied to any system that focuses only on internal relationships and is total, autonomous, and self-sufficient (orientated at self-reproduction) – in fact, *autopoietic*. The array of difference will constitute the "essence" of this system (taking into consideration that it has no essence). To this extent, this system does not have any substances-essences and is not a substance by itself. This system has nothing but differences.

51. We should say that consciousness, though it is not a thing and does not constitute a unique set of properties and qualities, can still differentiate things, their properties, and qualities. We can say that the experience of differences is the simplest and "closest" experience available to anyone. If we take a closer look at the process of awareness, we will see that it contains nothing except differences. For example, we differentiate light from dark, warm from cold, hard from soft, one item from another item, contours of the item from its core, the cause from the effect, the assumption from the conclusion, etc. We fundamentally outline differences to form objects and, to this extent, we deal with the entirety of reality (Miller 1966). This is why we can assume that the world as such has no differences; they

exist only for the conscious observer. In this case, the basic difference will be the differentiation between the figure and the background or between the essential and the non-essential.

52. Thus, the experience of differences will be essential for consciousness as compared to any other autopoietic system. It is not just the way consciousness operates; it is consciousness itself. Indeed, if everything is given as one undifferentiated flow, we can understand nothing. Once we decide to apply conscious experience, we must perform a minimum differentiation or separation of one from another (Gasparyan 2015). Something must be hidden and something must be revealed. When I understand something (perceive something), I understand that there is *this* and *not-this*. This is a minimum basic level of perception: it is an ability to single out, to differentiate one from another. As a rule, all of us have this intuition for perception-differentiation. I am going to give an example from developmental psychology. When we show toys to a child and he just looks at them, we do not feel that he has any conscious response. He can even touch them; however, until a choice is made (all the toys are given approximately equal attention), his behavior does not look conscious to us. Once the child has chosen one toy out of many toys (for example, he stretches his arm and takes the toy or looks at it for a longer time compared with the other toys), we start interpreting his behavior as conscious at minimum (Piaget 1954). An illustration of this idea can be also found in more recent research interpreting consciousness as an implemented action (Noë 2004; O'Regan 2010). According to the sensorimotor theory of consciousness, sensorimotor contingencies constitute the core of phenomenal states of consciousness, and originate on the grounds of actively performed differentiations, allowing the action to be implemented. In turn, perception and action should be interpreted as equality, i.e., in a phenomenological sense (including a neuro-phenomenological sense, according to the meaning of Varela's theory), as one thing cannot precede another – one should realize before acting but act so as to realize. The difference in such an interpretation will be operative in two meanings:

a. in a methodological sense, allowing us, in spite of the equality, to talk about identity of two states: action and perception; and
b. content-wise: in order to act one should distinguish, and likewise in order to be conscious one should differentiate things (O'Regan, Myin & Noë 2005).

53. To explain the above-mentioned ideas, an example from the field of phenomenology can be given. Phenomenology or cognitive psychology tells us that in any state of consciousness we come across a step-by-step process of experience: differentiation–synthesis–recognition (Tye & Wright 2011). I want to point out that differentiation comes first as primary experience, due to which comparison and perception also become experience. Comparison cannot take place without previously differentiated correlative conditions (information about the color, shape, sound, smell, etc.).

54. We can raise an objection, saying that we do differentiate *something (identical to itself),* that we do differentiate one from another. As this "something" must exist, can it be that at first we recognize something (the object identical to itself) and only then differentiate it from another recognized something (the identical object)? If this is correct, can the experience of identity be more primary than the experience of differentiation?

55. The answer is as follows: synthesis and recognition are certainly required elements of the completed experience of consciousness, of the experience in which the "outlines" of the object are identified and synthesized, after which the object is configured and recognized. By so doing, we differentiate one object from another. If we take a closer look,

we will see that this full cycle of experience is also an experience of differentiation, for recognition results in differentiation. The complete structure of the perception experience will be as follows: differentiation–synthesis–recognition–differentiation. This means that synthesis and recognition are built into differentiation, and not vice versa. Differentiation turns out to be broader and it is differentiation that organizes both synthesis and recognition, which take place as temporary states since they quite quickly transit or are ready to transit into new differences (Ricoeur 1965).

56. Thus, the ability to differentiate characterizes psychic life in general, whereby human consciousness has a unique ability to spot differences and differentiate between types and hierarchies of meanings (self-reference, self-reflection). In turn, the difference between differentiation and identification in the traditional epistemology of the classical (Cartesian) theory of reflection is thought of as the difference between subject and object, or I and not-I, whereby it is not taken into account that these modes are parts of one consciousness in its different stages and manifestations. Reflection is no more than something external related to consciousness; however, a certain level of differentiation of differences is needed.

57. Thus, any "something" appears as an object of perception primarily due to differentiation. Distinguishing is performed not from the substantial material, but actually from other differences or, in this case, from other perceptions. Something is perceived because it is separated from another something, which, in turn, is also perceived at the same moment. If we speak about the simplest level of consciousness – perception, differentiation will constitute its essence, starting with the basic forms. Seeing, listening, feeling, smelling, touching, and experiencing taste sensations mean, first of all, differentiation. It is impossible to perceive something without differentiating (Bateson 1979). To put perception into effect, we must single out some information and take aside other information; we should see one and should not see another (Gasparyan 2015).

58. Thus, any difference envisages the difference between foreground and background; their principal "asymmetry" characterizes such an experience of consciousness as preference. In turn, the steady preference of a certain foreground and "ignoring" of the background implements the objectified function of consciousness. It manifests itself in postponing further contextual differentiations, which in this way allows the setting of the boundaries of the subject. Thus, the meaning of the objectivity of the subject is achieved by pausing in differentiating. The objectified function is the ground for recognizing the subject that might be interpreted as "shaped" and identified from the set of meanings, in which connection is included. The ability to differentiate defines the ability to direct attention, i.e., to separate and give a steady preference to this or that differentiated object, as well as to anticipate, forecast, and predict.

59. For instance, whilst differentiating between two colors, we immediately spot (differentiate) the context in which we conduct this differentiation: red and green might be the signals of traffic lights, symbols of social movements, degree of ripeness of certain fruits and vegetables, etc. Each of these contexts is at a certain level in the contextual hierarchy (embedded into another contextual differentiation): driver/pedestrian, elected person/voter, seller/buyer, etc. Differentiation is not an image, not a symbol, not an object, but a *source* of an image, symbol, or object. In turn, meaning is not a mental atom capable of amalgamating with other atoms, but the relation of contextual levels. In the case of traffic lights, the meaning of the light for us is the necessity of differentiating the movement of traffic flows or movement of transport and pedestrians. The meaning as differentiation defines a pos-

sible set of signs – the carriers of this meaning (a signal with the help of a color, a gesture, a traffic-controller).

60. In fact, this is the way in which perception turns into information. The scheme I offer matches the concept of information offered by Bateson. His range of differences is total – the differences that are not perceived are referred to by him as "potential differences," whereas "perceived differences" fall into the category of "effective differences." Our usual perception might be considered as consisting of a million potential differences, but very few of them become effective differences (i.e., units of information) in the mental process of a larger system of *observation. Therefore, information consists of non-indifferent differences* (Bateson 1972).

61. Taking into account the above-mentioned facts, I am assuming that we cannot see the differentiation – it is objectless. However, we can see the results of differentiation: we can see different colors, we can hear different sounds, etc. So the given is always a result or a consequence of some differences that themselves do not belong to this reality, and cannot be found within it. That through which reality is created is not a part of it. At the same time, it is important to note that these differences are for the most part strictly functional and operational; they can be operated, but cannot be recognized subjectively (objectified), which (if it were possible) would allow us to speak about them from the third-person point of view or make them universally observable.

62. In this sense, consciousness is the variety of differentiations (primary experience), as well as preferences (eliciting elements differentiated as the foreground) and identifications of the differentiated object. Therefore, differentiation is always primary in relation to the experience of synthesis and the experience of identification. If we try to prove the opposite and attempt to argue that, in enumerating differences, we identify them, we should keep in mind that the identification of a difference is nothing other than an experience of differentiation and its actualization. Such identification of differences, their enumeration and classification, is based on the base system of differences. In other words, any experience of givenness is based on the initial system of differences, which constitutes a special autopoietic system of consciousness.

63. Examples of basic differences constituting the primary experience of consciousness can be as follows. Differences in time: past, present and future; differences in spatial positions: top, bottom, right, left; differences in perception of color, sound, smell, taste, and touch; differences in "raw sensations" (qualia): light and dark, hot and cold, heavy and light, dry and wet; differences in basic modes of consciousness: perception, memory, imagination, doubt, assumption, etc.; differences in emotional states; transition from one state of consciousness to another, "mood changes"; difference between your own experience and someone else's experience; difference between waking consciousness and dreaming; difference between your own experience of consciousness and other types and kinds of experience, etc.

64. The aforesaid is in line with Bateson's interpretation of the mind, where differences are seen as an objective characteristic of the surrounding world (Bateson 1972). Bateson argues that perception is based only on differentiation. Any information is obtained through information about differences. The operation of the mechanism is explained by Bateson in the following way: the difference becomes non-indifferent for a certain perceiving system in terms of the initiation of further mental events within it, provided that this difference surmounts the differentiation threshold of this system, taking into consideration that any perception of difference is limited by the threshold value. When differences are

too weak or slow, they cannot be perceived, and they do not turn into food for perception. Bateson (1979: 134–136) outlines the following marginal characteristics of the system, which can be seen as characteristics of the mind:

- the system must operate with differences and be based on differences; and
- the system must consist of closed loops or networks of pathways, along which differences and transformed differences shall be transmitted (what is transmitted by a neuron is not an impulse, it is news of difference).

65. By seeing consciousness as experience of differences, we can, first of all, "retain" consciousness without reducing it to something physical; second, we can avoid substantivization of consciousness. Difference is a uniquely non-objective experience, which, in principle, cannot be substantivized. It cannot be found in the world as a thing; however, thanks to it we can find all other things – different and fragmented. These differences are primarily actualized non-reflexively (as the pre-reflexive thinking I have discussed earlier), but it is in them that the possibility of reflection as self-description is rooted. Therefore, the experience of differences is more primary than intentionality (the focus of consciousness on the object). The experience of differences allows for singling out such structures of consciousness as grasping and identification of the object. Consequently, reflection on this experience is a self-description of various states of the differentiated, which organize the operation of consciousness.

66. The principal proof of the fact that the experience of differences underlies all other types of experience stems from understanding the primacy of the experience of differences by everyone who attempts to reproduce it. Argumentation here, first of all, appeals to the immediate experience of consciousness. Differentiation might be directly "perceived" in the experiences of presentation and judgment, but also in a phantasy, reminiscence, evaluation, doubts, etc. (within the framework of the logical model "this is that" or "truth-false"); ethics and will (within the framework of the model "good-evil"); space and time ("up-down," "right-left," "past-present-future"); aesthetic sensibility ("beautiful-ugly"). Last, we have access to reflexive perception of stated differences only "on the basis" of differences themselves. In this sense, differentiation is the self-referential experience that does not need any justification.

Differentiation – local and global level

67. If we look at the *local level of consciousness*, self-description will be represented by reflection, which does not dominate the object similar to the subject, and does not double consciousness, as is typical of the dual pattern; in fact, it is the experience of differentiation, which is aware of itself.

68. As for the *global level of consciousness*, macro-mental formations such as culture, society, religion, economy, science, means of communication, and many others owe their existence to certain areas of consciousness. The organization of these macro-dimensions is also actualized through the experience of differences. Consciousness forms a number of movable semantic and value hierarchies that define the content of individual and intersubjective experience. The sequence of such hierarchies allows for talking about the history of human consciousness, avoiding nominalisation. This way one can eliminate periods – so-called primitive thinking – with prevailing specific differences and intuitive identifications,

and follow-up periods, which make up conceptual- and value-frames of certain epochs and cultures, where abstract differences and descriptive identifications prevail.

69. In this case, *self-description* is represented by the ability of these macro-mental formations to refer to themselves through the differentiation of some of their areas from other areas. An example of such a scheme can be found in Luhmann (1997). Luhmann talks about such an organization of society, which is kept together, for example, by differences between "external," "primitive," "barbarian" systems and "internal," "cultured," "civilized" systems. According to Luhmann, social systems, in contrast to physical-chemical and biological systems, operate on the basis of *meaning*, which he sees as the mechanism for processing differences. For Luhmann, "society" actually serves as a definition for the self-description of communications, and if during such self-description, any of these communications develop into a societal theory, they only assert that "society consists of communications." This means that description of society has a circular basis or, in other words, is self-description (Luhmann 1995, 1997).

70. Appealing to the example of Luhman's society theory, we see that this model, which applies to the form of consciousness that is known as society, is typical of the entire domain of consciousness. For example, in science, or rather in sciences addressing consciousness, consciousness reflects itself. Different related theories of consciousness (physicalism, dualism, functionalism, panpsychism, etc.) can be also seen as different communications – ways through which consciousness can speak about itself. These communications can exist only differentially – as theories different from one another. They are opposed to one another and acquire their "positive" meaning through this opposition. For example, physicalism is opposed to dualism and vice versa. The approach that eventually discusses the existence of these differentiated theories can be seen as the self-description of consciousness.

71. It should be noted that none of these communications (theories) relies on any "objectively" fixed *object* (consciousness), which means the impossibility of a criterion of truth in relation to the problem of consciousness under discussion, as well as the impossibility of the only reliable conditions of its observation. On the contrary, descriptive differentiation implies that consciousness sees itself through the "eyes" of its theories, which means that it has to describe itself by using different and sometimes mutually exclusive theories. In this respect, consciousness as the *object* of study of different theories in philosophy of consciousness is thus some sort of figment.

72. In turn, *the subject (the isolated observer)* is also a figment of self-description. It also constitutes the state of the system, which emerges during continuous correlating operations: communications in the system of consciousness. These operations exist only as alternating and are induced by one another. Then the subject will be represented by the system providing continuation of communications and experiences, and this operation by itself is a contribution to communication.

73. "Objectivity" in this case can be interpreted as *distributed subjectivity*. This assertion means that cognition is a map divided up into semantic regions from inside, where verification of one group of truths will be undertaken in contact with the other groups. Such a procedure is different from the correspondence theory of truth – in it, the theories and research seek confirmation of their views not in the "objective world" but in the provisions and results of other theories and research. We do not have access to the world; however, we have access to other interpretations of the world, which one might aspire to bring into accord with one's own. Different interpretations are described in relation to (1) different

prospects of perception regarding one another, and (2) those forms where consciousness exists, and these interpretations do not necessarily result in any contradiction.

74. By applying such a scheme, we might eliminate many errors and contradictions of reductionism, which simplifies the essence of things, as well as eliminating subject–object dualism. For this purpose, in the case of consciousness we have to think more metaphysically (due to its comprehensive nature and totality), to which we are accustomed within the framework of the modern scientist paradigm. But such an interpretation is justified by the very nature of consciousness, which is perceived by us through the perception of the facts that we and "it" are one and the same thing.

Conclusion

75. The focus on subject–object dualism based on the classical epistemology governing FOC, which is popular in modern analytical philosophy of consciousness, promotes the generation of numerous theories initially characterized by epistemic contradictions. As a result, analytical philosophy of consciousness has not moved far in its answers, while offering a great deal of varied theories and approaches. All reductive (and FOC) approaches to cognition share the same fate: the external observer fallacy (Bishop & Nasuto 2005). On the other hand, if we change the approach and replace FOC with SOC, at the very least we will have an adequate epistemology and avoid wasting our effort in trying to create contradictory or non-working theories.

76. We may ask what SOC can offer as a more successful theory. This will be the theory of the self-description of consciousness, which is different from traditional (subject–object) theories, which has no division into the subject and object, and which addresses self-understanding consciousness organized as a total and immanent domain of meanings-differences (Scott 1996). For example, this research is not a meta-linguistic explanation of consciousness, where consciousness is the object and the author is the privileged observer. The author of this research is part of the observation or, in fact, the self-observation of one large autopoietic system of consciousness. In turn, local consciousnesses similar to the existing theories of consciousness will, in fact, be individual self-descriptions of the system, where values will differ from one another by their larger or smaller scale and clarity.

77. Therefore, the strategy addressing consciousness in terms of differences, relationships, and correlative processes will be more efficient. The fact that consciousness is a bundle of significant differences can be demonstrated by the example of local consciousness, i.e., our own consciousness, which can be offered as the most illustrative means of justification. We only need to observe how the consciousness of each one of us operates. The same principle can be extrapolated to the total domain of consciousness. Thus, to a certain extent, consciousness is structured holographically – the general principle typical of the entire system can be found in each of its smaller local parts. Thus the system of consciousness is a system of differences composed of differences. Self-observation in this system is conducted with the help of observations of observations.

78. Turning back to the question regarding what SOC can offer to modern philosophy of consciousness, we should say that the task of the explanation of consciousness can, in a way, be redirected from the external question "How can consciousness be explained?" to the internal question "How does consciousness describe itself?" Even if we assume that the answer to the question "What is consciousness?" will remain a blind spot (the systemic

gap that is required for the operation of the entire system) for consciousness, internal self-description through different forms can be sufficiently transparent. Even if consciousness never finds out what it is, it is able to *be aware of* itself. Apparently, there are questions the system cannot answer. The most that can be done regarding these questions is to be aware of them. A certain number of such unanswerable questions can be singled out and perceived as system-inherent points of non-transparency, which are required for the operation of the system. Despite such areas of non-transparency, consciousness can become aware of itself through the differentiation of its own forms. Such awareness can be obtained without a contradictory "external observer." SOC will act as epistemology, adequately describing the principles underlying this system. Probably the best conceptual framework for avoiding the external observer fallacy and encompassing the fundamental characteristics of cognitive systems is the general framework of SOC.

Acknowledgments

The article was prepared within the framework of the Academic Fund Program at the National Research University Higher School of Economics (HSE) in 2016–2017 (grant no. 16-01-0032) and supported within the framework of a subsidy granted to the HSE by the Government of the Russian Federation for the implementation of the Global Competitiveness Program.

Open Peer Commentary: **On the Too Often Overlooked Complexity of the Tension between Subject and Object**

Yochai Ataria

1. Diana Gasparyan's target article aims to offer a new model for consciousness. However, it also raises some philosophical and methodological concerns, which will be the subject of my commentary.

Naturalistic dualism

2. David Chalmers offers different kinds of possible solutions to the hard problem in the study of consciousness. One of them is to create a nonreductive theory of experience by creating "new principles to the furniture of the basic laws of nature" (Chalmers 1995: 210). In turn, he says, "these basic principles will ultimately carry the explanatory burden in a theory of consciousness" (ibid). To make a long story short, Chalmers (1995) defines this kind of theory as *naturalistic dualism*. In a way, Gasparyan's solution goes in that direction because by saying that "differentiation is always primary in relation to the experience of synthesis and the experience of identification" (§62), she adds a new kind of law, and hence creates a nonreductive theory. However, by so doing, she embraces a dualistic approach. Indeed, from the following citation it becomes very clear that this new principle is in fact a new fundamental law:

> by seeing consciousness as experience of differences, we can, first of all, retain consciousness *without reducing it to something physical*; second, we can avoid substantivization of consciousness. Difference is a uniquely non-objective experience, which, in principle, cannot be substantivized [...] the experience of differences is *more primary than intentionality* (the focus of consciousness on the object). The experience of differences allows for singling out such structures of consciousness as grasping and identification of the object. (§65, my emphasis)

3. Thus to argue that the *experience of differences is more primary than intentionality* is to say that there is some basic experience that we do not have access to – as Gasparyan put it:

> I am assuming that *we cannot see the differentiation* – it is objectless. However, *we can see the results of differentiation*: we can see different colours, we can hear different sounds, etc. So the given is always a result or a consequence of some differences (§61, my emphasis)

In turn, these unobservable and inaccessible aspects are responsible for the structure of consciousness. This suggestion, so it seems, presents a new law.

4. By going in that direction, Gasparyan is giving up any attempt to explain consciousness rigorously and thus instead embraces the notion according to which "[c]ertain features of the world need to be taken as fundamental by any scientific theory" (Chalmers 1995: 210).

The language of consciousness

5. Gasparyan argues that "by using SOC we can bypass the strategy of reductionism and replace it with a strategy whereby consciousness will be talked about in the language of consciousness" (§16). However, it remains unclear what Gasparyan means by "language of consciousness." This is not a technical issue. It seems that in a way if we were able to define the language of consciousness we would, in fact, have already solved the hard problem in the study of consciousness. Thus, by arguing that one has the ability to *speak* in the language of consciousness, we assume that one knows all there is to know about phenomenal consciousness – to use Ned Block's (1990) terminology – for, unless one has this kind of knowledge, how can one have the ability to talk in the language of consciousness and to know that this so-called "language" is indeed the language of consciousness.

6. Furthermore, Gasparyan argues that "if the task is to speak about consciousness not using the language of the consciousness, we encounter a paradox" (§17). With this in mind, Gasparyan needs to confront the following problem: if consciousness has a language of its own that is unique merely for consciousness (and if I understand what Gasparyan is trying to say, this so-called *language of consciousness* describes the process of differentiations, which is, as we just saw, a new kind of a fundamental law), it is arguable that what Gasparyan is actually offering here is a *new* science of consciousness that has nothing to do with the current scientific project. If this is indeed the case, we do not have the means to examine Gasparyan's new model of consciousness from the very start with the tools we have today, given to us by traditional science.

7. The situation in the science of consciousness is not easy and as a result it might be tempting to call for a new paradigm for studying consciousness. Having said that, we must remember that any new paradigm needs to be established upon some new empirical findings or at least to be able to explain better some older findings – it goes without saying that

Gasparyan's model does not rely upon some new evidence and does not help us to solve old paradoxes or contradictions in the study of consciousness. Furthermore, Gasparyan's model does not predict anything in terms of scientific experiments; moreover, it does not even allow us to imagine new experiments that would support her model. So at this point, from the perspective of philosophy of science, the target article seems to offer nothing for the study of consciousness as a scientific project. Gasparyan writes,

> [consciousness] constitutes a fragment of reality, which is relatively *isolated from the environment by its causal structure*. The external environment is not able to determine the autonomous system from outside and to break its causal impermeability. The existence of the external environment can be taken out of context. If this environment has no effect on consciousness (it does not determine, create, or change it), it can be largely disregarded. (§25, my emphasis)

8. It seems that according to Gasparyan, the concept of how consciousness needs to be explained is fundamentally different compared to how the rest of the physical world needs to be explained. If my analysis is accurate, then Gasparyan's model is simply a version of dualism.

Science or a mystery

9. Gasparyan argues that

> Taking into account the above mentioned facts, I am assuming that we cannot see the differentiation – it is objectless. However, we can see the results of differentiation: we can see different colors, we can hear different sounds, etc. So the given is always a result or a consequence of some differences. (§61)

10. This appears to be a rather problematic suggestion. In his reply to Michael Kirchhoff and Daniel Hutto's "Never Mind the Gap" (2016), Michel Bitbol & Elena Antonova argue that Kirchhoff and Hutto "systematically misconstrues the original approach to the 'hard problem' of consciousness advocated by Francisco Varela under the name neurophenomenology" (Bitbol & Antonova 2016: §1). According to Varela, "the lived experience is where we start from and where all must link back to, like a guiding thread" (Varela 1996: 334). Yet with this in mind, the aim of Varela's neurophenomenology research program (NRP) was to "present a model that can account for both the phenomenology and neurobiology of consciousness in an integrated and coherent way" (Thompson, Lutz & Cosmelli 2005: 87). Varela argued that (and this stands at the core of the NRP working hypothesis) the "opposition of first-person vs. third-person accounts is misleading" (Varela 1996: 340). Notice that the NRP does not dismiss third-person accounts nor objective tools. Instead, the NRP tries to construct a robust and irreducible dialogue between first-person and third-person data, namely between neurobiological aspects of consciousness (third-person) and phenomenological features of consciousness (first-person). In contrast to NRP, Gasparyan's theory simply ignores the objective world, suggesting that "objectivity in this case can be interpreted as *distributed subjectivity*" (§73).

11. This limitation makes Gasparyan's theory appear to be a doctrine embedded within mystery rather than a scientific theory. For instance, it is difficult to understand how Gasparyan could have reached the following conclusion and how it can be examined: "Therefore, differentiation *is always primary* in relation to the experience of synthesis and the experience of identification" (§62, my emphasis). Furthermore, even if this notion is

correct, and differentiation is indeed a primary element, scientifically speaking, what are the implications of this statement? Does this notion allow us to reinterpret results in the field in the study of consciousness?

Chiasm

12. In his book *The Visible and the Invisible*, Maurice Merleau-Ponty presents the notion of chiasm. This concept is rooted within the "paradoxical fact that though we are of the world, we are nevertheless not the world" (Merleau-Ponty 1968: 127). He argues that we are both objects among objects like any other object in the world, yet at the same time we are subjects: "my body is at once phenomenal body and objective body" (ibid: 136). This structure of subject/object and touching/being-touched reveal the twofold dimensions of "reversible flesh." Merleau-Ponty uses the notion of flesh to describe our being-in-the-world in a less polarized manner (subject versus object), that is, as an integral part of the world. We are interwoven with the world, thus, the boundary between the body and world, according to Merleau-Ponty, is vague and hazy.

13. Gasparyan tries to develop a theory of consciousness that breaks the subject versus object structure – "SOC helps eliminate subject-object dualism" (§16d) – but fails to acknowledge the complexity of this structure, i.e., that it is impossible to describe the world merely subjectively or objectively. Although the Cartesian dualistic subject-object approach is indeed problematic, methodologically speaking, eliminating the subject-object structure is almost impossible (Ataria, Dor-Ziderman & Berkovich-Ohana 2015). Furthermore, phenomenologically speaking, it would be a fundamental mistake, for we are both a subject and an object, so by giving up one of these dimensions you will lose the very essence of being human. The tension is necessary: "A human body is present when, between the see-er and the visible, between touching and touched, between one eye and the other, between hand and hand a kind of crossover occurs" (Merleau-Ponty 1964: 163). Fundamentally then, if this sensing-(being)sensed structure collapses, the very essence of being a human being will lose its meaning and thus there "would not be a human body" (ibid: 164). Moreover, when this subject-object structure collapses, when one can no longer touch and being touched, it is not merely the end of the human body but, in fact, it is the end of humanity: "such a body would not reflect itself; it would be an almost adamantine body, *not really flesh, not really the body of a human being. There would be no humanity*" (ibid, my emphasis). Unfortunately, Gasparyan's subjective model ignores this complexity of the human being as both a subject and an object at the same time.

14. We need to understand this complexity; not to eliminate it. Thus, by eliminating the subject-object structure (tension), Gasparyan makes the same mistake as those who eliminate consciousness or the subjective experience. It is impossible to construct a science of consciousness while ignoring the subjective dimension, yet it would be meaningless and even pointless to ignore the fact that as humans we are part of the world, we are both a subject and an object.

Beyond the limits of consciousness

15. Gasparyan rightly notes that "understanding consciousness in the way expected by most modern theories assumes doing what cannot be done: moving beyond the limits of consciousness" (§11). One must wonder, however, does this observation not apply to

her article as well? The only way to answer this question is to assume something about consciousness. Indeed this might the target article's most serious problem: it begins from the very end – as if Gasparyan already knew something about consciousness and about the phenomenal subjective experience. It might be an intuitively appealing presentation; however, this intuition needs to be put to scientific examination. Yet instead of doing so, she offers a theory that, in essence, cannot be verified.

Open Peer Commentary: **Where Is Consciousness?**

Urban Kordeš

1. One of the highlights of Diana Gasparyan's target article – the idea to put distinction in the centre of the study of consciousness – is most intriguing, but, at the same time, it is also not quite new. The idea was put forward in the 1960s by polymath George Spencer Brown (1969), whose book *The Laws of Form* has inspired, among others, the authors of the autopoietic theory.[1] Spencer Brown grounds his work on epistemic logic on the imperative: "Draw a distinction!" He sees drawing a distinction as a condition and the fundamental act of cognition. In the absence of distinctions, one would be floating in an endless, shapeless void.[2] In general, it seems that Gasparyan agrees with Spencer Brown in concluding that without distinctions, the world would not be possible because "if everything is given as one undifferentiated flow, we can understand nothing" (§52).

2. The target article uses the term "difference" but I do not see a good reason for departing from Spencer Brown's term "distinction." Gasparyan uses the expressions "experience of differences" (§50) and "experience of differentiation" (§48), where the former sounds somewhat problematic; it suggests that the (experienced) differences are "out there." Spencer Brown with his "draw a distinction," on the other hand, meaningfully implies that the distinction lies in the hands of the beholder and is not a simple cognitive response to the contours of the "real" world. This view is closer to Gasparyan's "experience of differentiation."

3. Perhaps it would be suitable to follow Francisco Varela, Evan Thompson and Eleanor Rosch (1991: 172; see also Kordeš 2016: 383) in choosing the middle way by introducing the term "experience of enactment of distinction"? This, admittedly clumsy, denomination better emphasises the blurred line between perception and action; between representing and inventing – a quality that the authors of the term "enaction" wanted to affirm and that is also noticed by Gasparyan in §52.

Hiding in plain sight

4. The constructive role of the observer in the formation of distinction is perhaps best elucidated by the mind-body dichotomy, or better – the distinction between the experiential and the physical. It is interesting to notice how hard it is to pinpoint the former part of this

1. Varela (1979) saw Spencer Brown's kind of formal logic as a perfect analytical instrument to be applied in the then developing autopoietic theory.
2. Mystics might interject that letting go of distinctions is the very first step towards enlightenment, but let us leave this discussion for another time.

dichotomy (i.e., the experiential), despite it being our most intimate feature. Perhaps the best definition of the experiential is that about which we can ask ourselves with Thomas Nagel "What is it like to be?" The question "What is it like to be the reader of this text?" is answerable. The answers can vary substantially; nevertheless, each one of them will describe at least some kind of experience. If we, however, ask ourselves "What is it like to be this computer screen?", it becomes very hard to imagine the answer. The question therefore makes sense (is answerable) only when we are dealing with consciousness. One might therefore be tempted to describe experience as the answer to the question: "What is it like to be conscious?"

5. It seems that we have a blind spot for the fact that experience is the most basic and unavoidable medium of our being. Not only do we normally not notice how all our beliefs about ourselves and the world constitute experience; we do not notice that we do not notice. When we say "here is the screen," it is entirely natural to omit the part: "I experience/see/think that here is the screen." It is very hard to make ourselves notice that "here is the screen" necessarily presupposes experience, consciousness. This blind spot is related to what Edmund Husserl (1982) very aptly describes as the natural attitude: we organise, interpret and make sense of our experience with the constant help of a notion that all experience is the experience of something. This process of organising, interpreting and sense-making is so efficient and swift that it is not hard to overlook the medium into which it is inevitably submerged – consciousness.

6. The blind spot of the natural attitude prevents us from noticing that the dichotomy experiential/physical is not a genuine dichotomy. A genuine dichotomy has to split the content space into two, if at all possible, non-overlapping parts. The pair experiential/physical, however, muddles two content levels: the level of experience, which is the primary medium of being, and the level of the physical world, which is a way of organising the experiential content.

7. Let me point out that this is not a debate on the ontological existence of the physical world. Accepting the simple fact that we cannot perceive anything outside of our experience does not mean we have chosen a type of idealism, asserting that experience is primary and the physical world is but a frivolous play of mind. All we did (following Rene Descartes, William James, Husserl and other epistemologists) is notice the fact that experience is a medium into which we are immersed and from which we cannot escape.

8. Enactment of distinction as the fundamental modality of experience can be compared to the development of a scientific theory. If the enactment of distinction is to lead to a viable image of the world, it has to acknowledge constraints. From the constructivist viewpoint, we are, in both cases, constructing a functional theory – a theory adhering to the available data in a tightest possible way. Our beliefs about the world are, much like a scientific theory, a map, not the territory – they help us navigate. They are a way of organising experience in a meaningful and continuous way. Philosopher Paul Natorp (1912; see also Bitbol & Petitmengin 2013) has shown how the aforementioned blind spot leads to subsequent dichotomies, such as subject/object and "outer"/"inner." According to Natorp, we select the "parts" of our experiential field that are invariable in relation to (inter)personal, chronological and spatial situations. He calls this process, which leads to the feeling of a stable objective ("outer") world, "objectification." What is left is subjective, "inner" experience. Interestingly, Natorp notices that the boundary between the two changes throughout life (usually the subjective gives way to the objectification). Natorp sees the physical world

as a subset of the experiential,[3] which seems exactly opposite to our everyday attitude, which makes us see experience as a subset of the physical world. It is important to notice that these two seemingly opposite views are not symmetrical. In the first case (the notion of a physical world arises as a way of organising the experiential landscape), we are not talking about the actual physical world. Rather, we need to talk about our belief about the existence of such a world. Belief is, of course, a type of experience.

9. In our everyday "natural" intuitions, we overlook the experiential medium, which is the source of every possible perception, and accept the physical world as the foundation. If we overlook experience, what remains is a world filled with things. Some of these things exhibit behaviour that might hint at experience hiding behind it, but nowhere can we measure or clearly see this elusive entity. From this viewpoint, it is clear why, for a long time, experience did not belong to the scientific discourse. The rise of cognitive science forced researchers uneasily to accept the existence of this suspicious substance and to start looking for where and how it is hiding in the physical world. Experience chose a cleverer hiding spot than most cognitive scientists suspected: in plain sight. Everything is immersed in experience. Gasparyan, together with phenomenologists and the founders of second-order cybernetics, notices this immersion.

Can there be a theory of consciousness?

10. Up to this point, I have used the terms "consciousness" and "experience" almost interchangeably. From the view of the naturalistic cognitive science, we could define consciousness as that feature of organisms that enables experience. More precisely, that enables phenomenal reports or some other kind of behaviour that might hint at the presence of a phenomenal world. At this level, we can only talk about behaviour – as there is no experience to be found in the physical world.

11. There are at least three perspectives from which contemporary theories of consciousness try to approach their subject:[4]

Functional: What are its adaptive advantages and how has it (evolutionarily) come to be?

Explanatory: How does consciousness work or how does it emerge from the physical foundation?

Descriptive: What are the first-person features of consciousness?

12. Explanatory theories attempting to naturalise consciousness have to deal with the major problem, namely the fact that the universe they are operating in does not contain the entity they attempt to explain. So, as Gasparyan notices, all that remains are attempts to explain consciousness away (§5; Chalmers 2003). While searching for adaptational (and other) functional explanations of the emergence of consciousness might be interesting, it will probably not help much in understanding its essence.[5] What is left, then, is the descriptive approach, which can only be done from the first-person perspective. The target article promotes this option and I wholeheartedly agree.

---

3. In contemporary cognitive science, a similar view is held by Max Velmans (1990).
4. Here I partially follow Robert Van Gulick (2014). Güven Güzeldere (1995) enumerates even more levels of investigation that should be included in a theory of consciousness.
5. The scope of this commentary does not allow for a more thorough overview of the very interesting and broad field of physicalistic theories of consciousness. For a very broad overview of views and theories, see David Chalmers's list at http://consc.net/online/8

13. Gasparyan does not clearly articulate the relationship between consciousness and experience. If "consciousness is the experience of differentiation" (§48), I wonder, would it not be more appropriate to skip the "differentiation" part and start with "consciousness is experience"? Would such a position not mark the most fundamental level? As mentioned earlier, I am suggesting the relationship as: experience is what it is like to be conscious.

14. In any case, it is the experiential realm that the target article sees as primary for the discussions on consciousness, and I fully agree. Accepting such a(n) (immersed) perspective of description, "theory of consciousness" (§25) does not seem to be the proper term. A theory usually tries to describe a phenomenon with categories that are broader than the described phenomenon. If we agree that consciousness is a medium into which we are unavoidably immersed, then designing a theory of consciousness would be as if physicists tried to design a theory of the universe. One attempts to describe the features of the universe (as seen from within), but that is not the same as a theory of the universe. For such a theory, one should be able to step outside the defined phenomenon and describe how it came into being (from something else) and how it relates to other entities (outside it). If we are discussing the all-encompassing medium, then such an endeavour is meaningless. I believe that Gasparyan is stretching the term "theory" a bit too far when she writes: "As such, the theory of consciousness is the theory of the description of consciousness rather than the theory of its explanation" (§25). Would it not be better to simply state that we are aiming for the description of the phenomenal realm? Such a description could, of course, contain categories, description of various first-person modalities, and perhaps theories of those entities.

Attending the unattendable

15. The study of consciousness should be the study of consciousness as it presents itself, i.e., the study of experience. I especially agree with Gasparyan that second-order cybernetics would be the most appropriate epistemological foundation for such research. In the article "Going Beyond Theory: Constructivism and Empirical Phenomenology" (Kordeš 2016), I tried to point out the benefits of cooperation between second-order cybernetics (as an epistemological model) and the empirical study of experience.

16. The self-referential nature of studying experience is, in my opinion, one of the features where second-order cybernetics could offer an adequate epistemological framework. Nevertheless, the introduction of such a framework can by no means solve all the challenges posed by such research. Gasparyan's article addresses an exceptionally important one: how do we deal with experiential modalities, which are not (fully) explicable? I find this to be one of the most important novel insights delivered by her article.

17. Researchers in the field of so-called empirical phenomenology (Kordeš 2016) seem to have arrived at a (mostly unarticulated) consensus that it is possible – using appropriate techniques – to bring any kind of experience from the unattended fringe of consciousness to the focus of attention (Vermersch 2009). Considerations that some experiential phenomena cannot be fully explicated, however, are rare. The experience of enactment of distinction, discussed in the target article (under the term "experience of difference") might very well be one of those elusive experiential modalities. Following the target article's insight, I suggest serious consideration and further of the experiences that are intrinsically on the fringe of awareness. I refer to the phenomena that are, without a doubt, part of the experiential realm, but are not observable in the focus of attention. If the

purpose of many empirical phenomenological techniques is explication,[6] we are left with the question: How to study those parts of the experiential landscape, the principal quality of which is precisely that they reside on the periphery of attention? A deliberation on possible approaches to studying such phenomena would be a very important step towards understanding consciousness "from the inside," and Gasparyan's article seems to be one bold attempt in this direction.

Open Peer Commentary: **Theorizing Agents: Their Games, Hermeneutical Tools and Epistemic Resources**

Konstantin Pavlov-Pinus

1. In her target article, Diana Gasparyan continues her search for a general theoretical framework relevant to modern consciousness studies (Gasparyan 2015). This time, she concentrates not only on the ontological status of the theorizing agent, but also on methodological aspects that should be attributed to theorizing agents involved in consciousness research. In my opinion, Gasparyan's investigations have led her to the area where analytical philosophy meets phenomenology and hermeneutics for she ends up with self-description as a central epistemic concept; besides that, she finds the differentiation processes lying at the heart of the conscious life. These are all phenomenological themes. It seems to be right that the language of "second-order cybernetics" may play a bridging role between phenomenological and analytical styles of research. This gives us a chance to speak of a genuinely fundamental theory of consciousness (FTC), a brief outline of which is presented below. I will combine my critical notes with my personal constructive comments on this matter.

2. Generally speaking, any theory *Th* depends on five basic (meta)theoretical constitutive elements *Th* = *Th(language, problem/goal, criteria, method, theorizing agent)*. Here *language* stands for a certain discourse that we choose to deal with; the next three constitutive elements reflect the final *goal* to which we are oriented, the (set of) *criteria* for the theory's success and the relevant *methods* that we are allowed to follow. The last constitutive element requires explicit articulation of the ontological status of the *theorizing agent* and his epistemic resources. It recently became clear for all areas of theoretical research, from mathematics to cosmology and economics (Pavlov-Pinus 2015), that the theoretical status of the agent (i.e., our a priori assumptions about his computing, observing, comprehension and other abilities) influences significantly the ultimate architecture of the theory itself. Therefore Gasparyan is right to state that the theoretical status of the agent has to be taken into account seriously (Gasparyan 2015). It matters whether a theory is written as if on behalf of a godlike creature or from a viewpoint of a human with limited epistemic resources as this would significantly affect the corresponding theoretical approaches.

3. It is also widely known that, say, the predictive powers of a theory could be in conflict with the descriptive powers, the objectivity of the results could be incompatible with human comprehension abilities, and so forth. The architecture of a theory depends on our final goals and the target problems because whether we seek, say, deeper understanding

6. Most prominent is Pierre Vermersch's elicitation interview technique, formerly known as the explicitation interview (Vermersch 2016).

or more and more detailed "objective" descriptions is extremely important. From this, it could be predicted that any fundamental theory (within any fundamental area of research) will have the form of a *network of specific theories* in the same way as category theory in mathematics could be considered *both as* a general grounding theory for mathematics *and as* a network of specific mathematical theories at the same time. Of course, FTC will also have a network form. However, this is not the end of the story, for consciousness is a very specific subject of theoretical research.

4. Two more features will largely affect the architecture and the dynamics of FTC. First of all, this theory will depend heavily on hermeneutical definitions, rather than classical explicit definitions. Hermeneutic definitions, and hermeneutic procedures in general, are tightly related to the concept of self-description advocated by Gasparyan. Secondly, the performative nature of the subject we are aiming at, and its "nowhere" way of appearance, make it credible that FTC will have the form of multi-agent theoretical game, obeying its own game-theoretical laws and its own underlying logic. Let us take a closer look at both features.

5. The difference between explicit (classical) definitions and implicit (hermeneutical) definitions derives from the difference between explicit and implicit functions in mathematics. An explicit function has a form $f(z) = F(x, y, \ldots)$, with no *z*s appearing on the right-hand side. Implicit functions have a circular structure: $g(a, b, \ldots) = G(a, b, c, h(a), \ldots)$, with a appearing on both the right- and the left-hand sides. In the same way, we can say that classical definitions tend to have an explicit form $Def(z) = F(x_1, x_2, \ldots)$, i.e., the form of definition of z strictly in terms of $x_1, x_2, \ldots$ For example, a point is an entity with no parts. This definition has the form $Def(point) = F(entity, parts, negation)$, with F standing for a certain grammatical construction. However, far from all definitions in mathematics have such an explicit form. For example, "a vector is an element of a space of vectors" has a recursive structure. There is no vicious circle here at all. It is a primitive one-step recursion with an initial level, which has the form of an explicit definition. Most recursive definitions in mathematics have either a certain finite or even infinite number of steps. In general, however, we may talk of highly-complex ramified, implicit, recursive definitions without any starting point of recursion (in the same way as we can study implicit functions without any specific limitations). For example, in $Def_n(A) = \mathrm{R} = F_n(A, D_1(n), D_2(n), D_3(\mathrm{n}) \ldots)$, where n is a branch index, $D_1(n), D_2(n), D_3(n) \ldots$ are entities defining A in terms of relations of A with $D_1(n), D_2(n), D_3(n)$, while $= \mathrm{R} =$ stands for the switching rule between branches. These "ramified recursive definitions without an explicitly defined starting point" could be considered *as a formalization of hermeneutic definitions*. For the sake of simplicity could be called hermeneutic definitions themselves. In fact, implicit functions could be considered as historically the first example of an implicit definition of a certain variable in terms of its own functional relations with other variables.

6. A few illustrations. Note first that from an epistemic point of view, we must take care to ensure not only the formal correctness of our definitions but also their comprehensibility: "definitions" are epistemic procedures whose central theoretical role is to increase (or deepen) our *understanding*. If, for example, we take a closer look at the internal *epistemic* structure of such concepts as "order," "sequence," "word," and "sentence," we will see something that would be left out by any explicit definition. Namely, one can see that these phenomena require *unavoidable preliminary epistemic acquaintance* with them. Any definition of "order" will implicitly use ordering as a part of the definition, and without comprehending "ordering" as (the actual process of) ordering, one will never be able *to*

comprehend the definition of "order." A definition of "word" will necessarily consist of words that must be recognized as meaningful words, otherwise this definition will not be informative at all. Definitions of gods, or hobbits, or points do not require any preliminary acquaintance with these entities. The presence of these circular structures points to something much more general: certain aspects of human experience cannot be generated from scratch; they can only be *modifications* of (pre)existing forms of experience. Another illustration: any definition of "a human being" will be incomplete if we rule out the fact that "mankind" is the only generating source of any new human individual. It looks like a paradox: the whole society of humans should precede the existence of any particular human being. And yet these are the initial conditions of anthropological research: recursive structures of the type ((mankind → new humans) … → new humans) are the most informative source of knowledge about humans. Note that the recursion here has temporal ramifications: it moves step by step into the re-constructing past as well as into the expected model of the future. All other theoretical systems are less informative than, say, those that tend to explain human matters out of non-human matters, for example (dinosaurs → Shakespeare's *Hamlet*). While I do not have sound proofs so far, I think that the most informative theories of human matters are those that are able to trace the trajectory of human *experience modifications* both into the indefinite *human* past and into the indefinite *human* future. So my point is that (epistemically) complete definitions of such "circularly closed" entities should be able to grasp these interesting epistemic features. And not only that. Any definition is subject to revision. The reason for revision may come from many different contexts, and this is exactly why a multi-contextual ramification may take place.

7. Why is this relevant to consciousness studies? Since the *ultimate goal* of the philosophy of consciousness is to answer the question "*What happens in the world whenever an act of understanding occurs in there?*", there is no doubt that the most complete definition of consciousness has precisely this hermeneutical form. Gasparyan's intuitions could be formalized in terms of such recursive definitions, for consciousness could be defined only in terms of networks and systems that already include "consciousness" as a constitutive entity. For example, one branch of consciousness definition should reflect the circular structure of its social roots: $Def_1(consciousness) = F_1(social\ networks,\ consciousness)$. Another branch of this ramified definition should explicate the neurophysiological way of coding and decoding acts of understanding: $Def_2(consciousness) = F_2(neural\ networks,\ consciousness)$. There are also linguistic ways of coding and decoding the self-understanding of consciousness, and a few more; and each of them mutually affects the rest of them. It must be stressed that the possibility to encode and to decode certain properties of the self-understanding of consciousness in neurophysiological, or social, or linguistic terms does not imply the possibility of its ontological reduction to the corresponding phenomena. This is because to encode the description A in terms of D_1, D_2, … is not the same thing as to reduce A to the ontology described by D_1, D_2, …

8. There is one more thing on which I fully agree with Gasparyan: the central epistemic concept of modern analytical philosophy of consciousness so far has been "explanation" (§4); however, it is a priori not clear that all possible theoretical questions with which we may address consciousness could be exhausted in this way of asking. I think she is right in stating metatheoretical guidelines such as (1) Cartesian dualism, and (2) explanation as a method of construction of answers that are highly correlated with each other. It seems also that a guideline such as (3) usage of explicit forms of definitions must be added to this list. These theoretical orientations, in fact, show that the metaphysics that stands behind

modern analytical philosophy of consciousness significantly narrows down the horizon of possible questions with which we may address consciousness. This horizon should be enriched by phenomenological ways of inquiry which appears to be "translatable" to the language of second-order cybernetics.

9. Now let us take a look at dynamic features of the future of FTC. I see some sort of "abnormality" of this theory (as opposed to Gasparyan's point of view, see §25, where she calls her theory "non-normal"). This is not in that it deals with recursive structures and self-describing procedures but in that it should be able to interpret its own counterexamples, just because "consciousness" is a source of all possible counterexamples that human imagination may produce. In order to have internal resources for being able to deal with its own counterexamples, FTC has to have a dynamical form by itself. In other words, it must be not just a theory or a network of theories but should rather be *a multi-agent theoretical game*. Consciousness is not only a socially, biologically (etc.) conditioned family of specific phenomena but also is a performative source of theoretical novelty, FTC-architecture changes, and social reconfigurations. In particular, this means that the history of FTC must be an active part of FTC itself, at any moment. The only adequate way of theoretical reproduction of such a phenomenon is to reserve some space for theoretical dynamics in a game-theoretical form. Again, it could be compared, to a certain extent, with an application of category theory to the cosmological dynamics as described by Bob Coecke (2008) and Louis Crane (2007).

10. By the way, one more philosophical application of Gasparyan's ideas could be mentioned here, i.e., in discussing free will, which is quite popular in analytical philosophy. Most of these discussions appear to be meaningless in the context of modern physics because the assumed *theorizing agents* are purposely designed to be either "outside-of-the-world observers," or some sort of supernatural fictional creatures. As cosmologist Lee Smolin pointed out:

> The concept of an observer outside of the world is based on an elementary contradiction, for then there is a second world, larger than the first, that encompasses both what we called the universe and its fictional observer. In a truly fundamental theory that aspired to describe the whole universe, it should not be possible to make such logical error [...] To avoid this, I believe that we should ask more of a cosmological quantum theory than that it simply allow the possibility of an interpretation in terms of observers inside the world. We should require that the theory logically forbid the possibility of an interpretation of cosmological theory in terms of an observer outside of the world. (Smolin 1997: 269)

This means that free will discussions in a framework of such theories are in fact disputes about the free will of the assumed *theorizing agent(s)*. Therefore, if an assumed *theorizing agent* is a god or a godlike creature (like Laplace's demon), then within a given theoretical framework we are talking about the free will of the corresponding *theorizing agent* but not the free will of humans. It would be a categorical mistake to think otherwise. This argument could be extended to biological and other theories.

11. Finally, I have three more critical considerations.

a. I see no good reason to exclude traditional "objectivist" studies of consciousness (as Gasparyan suggests, say, in §23), for these results affect not only the sum of our "objective" knowledge about things, but also have a certain influence on our *self*-understanding. FTC as a network may easily include objectivistic theories inside of its evolving net of theoretical inter-relations.

b. The borderline between self-descriptions and other descriptions appears to be very vague and not constructive. For example, our mutual descriptions via social interactions seem to be very productive for understanding the nature of consciousness, but one cannot immediately conclude that these types of descriptions have a form of a self-description. Rather, we have here something like "descriptions-of-selves" within the semantic space of social interactions. Gasparyan comes closer to her own theoretical intentions when she speaks of "distributed subjectivity" (§16e), but unfortunately she does not pay much attention to this concept.
c. I cannot agree with Gasparyan's interpretation of phenomenological "intentions" (§65). The latter could be no more "objectivized" than Gasparyan's "differentiations," and it is misleading to treat them as "intentions to the *object*." Intentions are the structural part of *Dasein*'s way of being-in-the-world, the recursive step-by-step realization of which explicates its in-the-world orientation, governed by *Dasein*'s expectations about the future and other existential factors.

Open Peer Commentary: **How Can Meaning be Grounded within a Closed Self-Referential System?**

Bryony Pierce

1. The target article defends a methodological claim about research into consciousness, arguing that the epistemology of second-order cybernetics can offer a more effective means of investigation than methodologies relying on a subject-object view of consciousness. Diana Gasparyan proposes that consciousness be viewed as an autopoietic system: self-organising and self-contained. The observer, she claims, is an integral part of this (closed) system.

2. Ezequiel Di Paolo comments on the fact that there are unanswered questions in the primary literature on autopoiesis. He says that "several essential issues that could serve as a bridge between life and mind (like a proper grounding of teleology and agency) are given scant or null treatment" (Di Paolo 2010: 46). It is the problem of the grounding of meaning that I wish to address here. My claim is that the notion of consciousness as an autopoietic system that is *relatively isolated from the environment and causally impermeable*, as Gasparyan claims (§25), introduces a potential problem when we seek to ground meaning and norms. I will also consider whether Gasparyan's approach might be used to develop an integrated view of grounding in its various senses. I take the relevant senses of grounding to be:

a. symbol grounding (semantic content);
b. biological grounding (purposiveness); and
c. the grounding of reasons for action to oneself (meaningfulness/affective content).

3. The *symbol grounding problem* is that of how symbols within a symbol system, such as a language, can acquire meaning, while avoiding an infinite regress of explanations in terms of other intrinsically meaningless symbols (Harnad 1990). This is illustrated in John Searle's (1980) well-known Chinese Room thought experiment, in which a monoglot

English-speaking person follows detailed instructions in English to manipulate Chinese symbols to produce correct responses to questions in Chinese. Searle concludes that this fails to show that the English-speaking person thereby understands the meanings of the Chinese characters, or, by analogy, that a computer understands the strings of symbols it manipulates, even if it can simulate understanding.

4. Wolfgang Tschacher and Christian Scheier note that the symbol grounding problem is particularly problematic in a constructivist theory, when "adaptive action in a real world is to be understood within the context of an agent's internally constructed (instead of represented) 'reality,' alone" (Tschacher & Scheier 2001: 561). One possible solution to the symbol grounding problem, proposed by Stevan Harnad, is to ground symbols in sensorimotor activity, "in the capacity to discriminate and identify the objects, events and states of affairs that they stand for, from their sensory projections" (Harnad 1993: §7.2). This account sits well, *prima facie*, with Gasparyan's account of the language of consciousness as involving perception, differentiation and identification, though only if the autopoietic system is embodied and able to interact with objects in its environment. Gasparyan says that, "the observer is essentially included in the view through her frame of reference and her motion relative to the objects and events under consideration" (§11). I would like clarification of where the boundaries of the system lie. The existence of an external world could arguably be inferred, even if we can have no knowledge about it, from the notion of motion relative to objects, but a methodological approach that explicitly denies that we should incorporate the external world into our explanation of consciousness – advocating self-description and self-reference only – seemingly cannot ground meaning in anything external to itself or involving interaction with the external world.

5. As the grounding relation is a dependence relation in which one entity is explained in terms of another more fundamental entity, a methodology that seeks to explain the language of consciousness from within, with no reference to anything that might be deemed more fundamental, is potentially problematic. It may be that Gasparyan considers grounding an unnecessary notion in the context of the proposed methodological approach; it is not universally accepted as a useful notion. If so, I feel that some alternative way of explaining how meaning can arise within the system is needed, although I realise this was probably beyond the scope of the target article.

6. Gasparyan says that, "it is impossible to prove the reality of the world beyond a certain system of perception" (§13). My worry is whether, without certain knowledge of the external world, the observer can construct anything more than a coherent set of symbols that are internal to the autopoietic system and thus ungrounded in world-involving sensorimotor activity. Gasparyan says, "When cognizing itself, consciousness discovers the meaning of its judgments only *in itself*" (§15), but without [at least a presumption of] access to or interaction with that which is external to the autopoietic system, which is described as a 'unity' (*ibid.*), it seems that there can be no grounding of semantic content. It is not obvious how meaning can be discovered from judgements. Gasparyan says that meaning arises "when there is a difference" and that difference is "the basic element of logic" (§49). The system has "nothing but differences" (§50) and is described as a "domain of meanings-differences" (§76), but difference alone is not sufficient for meaning.

7. An element common to many accounts of grounding, including Harnad's (1993) account, is that there should be a causally efficacious realiser – information processed must have some means of acting upon the external world. If the autopoietic system is isolated from the environment and causally closed, it would inevitably lack a causally efficacious

realiser. It seems that Gasparyan's account would have to dispense with this feature, unless it could somehow be internal to the system. I am sympathetic towards a sceptical view of our epistemological limitations. My concern is that a theory of consciousness and action in which meaning – and reasons for action – are grounded, in the relevant sense, must at least posit the existence of an external world. For meaning to be grounded, there needs to be a coupling with the environment, as described in Andreas Weber and Francisco Varela's account:

> There cannot be an individuality which is isolated and folded into itself. There can only be an individuality that copes, relates and couples with the surroundings, and inescapably provides its own world of sense. (Weber & Varela 2002: 117)

There must also be a sense in which one entity is more fundamental than another, in order for a grounding relation to obtain.

8. *Biological grounding* of reasons for actions, conferring purposiveness, also presupposes an external world. An external world, in which organisms evolve and with which they interact, is necessary to ground reasons in biological utility. Biological grounding relies on the premises that replication is a kind of success and that survival is good, independently of the evolution of consciousness and rational thought. But these premises presuppose values that we can question in our reasoning. Biological utility cannot account adequately for normative reasons to pursue courses of action; it is not sufficient to give us reason to act from a subjective perspective or even to explain how reasons for action can be compelling, from a third-person point of view, because it "fails to stop the infinite regress of whys with respect to rational justification" (Pierce 2012: 84). Biological grounding tells an evolutionary story about how we have come to behave the way we do, but can provide no justification for attributing value to outcomes. Reasons grounded in biological utility lack meaningfulness unless also grounded in some other way.

9. *Grounding reasons for action to oneself* requires justification of one's reasons, such that they can be understood, subjectively and unquestionably, to be compelling. Sensorimotor activity is not an end in itself: we need affective content, as well as semantic content, for what I will call "subjective meaningfulness." I propose that it is *the qualitative character of affective responses* that stops the infinite regress of *whys* that can otherwise arise when we examine our reasons for action. It is always possible to question, intellectually, why we should value outcomes, including our own survival, but we need eventually to be able to stop asking why we ought (prudentially or morally) to do something, because it is self-evident that a reason justifies something, non-inferentially, without further explanation. I claim that it is the affective valence of anticipated outcomes of actions that provides non-inferential justification for reasons for action, halting the regress of *whys*. We cannot fail to believe that certain affective states are preferable to others. Affective valence is also necessary for the construction of conceptual frameworks used in evaluation: concepts such as "good" and "bad," "welcoming" and "hostile," or "broken" and "intact" require experience of affective valence in order to be meaningful in the range of contexts in which they are applied.

10. Emotion, which has affective valence as part of its content, appears only as an item on a list of types of differences in §63, and is accorded no special status in relation to value or meaningfulness. In §66, Gasparyan touches upon evaluation, good and evil, and aesthetic sensibility. These are concepts I argue can only be meaningful if one has the capacity to experience affective responses. As with reasons for action, their meaning

is grounded in affective experience, but in the target article they appear to arise out of the purely cognitive capacity for differentiation and categorisation. No explanation is given of how the meaning of these terms comes to be grounded or understood; it is stipulated that the experiences of presentation and judgement need no justification. The kind of grounding I am discussing now is distinct from symbol grounding, which merely allows symbols to refer to something external to the symbol system. Subjective meaningfulness has affective as well as semantic content. We *just know* what we value and desire and can use that knowledge as a premise in reasoning, without needing any further justification. Affective valence performs a regress-stopping role in grounding reasons for action *subjectively*, a role that is lacking when we rely on the other types of grounding discussed above.

11. On a positive note, perhaps Gasparyan's methodological approach could be used to give an account of the grounding of meaning *independently of knowledge of the external world*, by appealing to the role of affective valence within a self-referential autopoietic system. I can know how I expect to respond to anticipated experiences regardless of whether I can know that there is an external world. Reasons for action (from a position of external-world agnosticism, if necessary) can then be judged subjectively to be compelling or not. To allow a grounding relation, there could be subsystems within the autopoietic system. The qualitative character of affective responses to experiences might be deemed more fundamental than subsystems producing perceptual and other experiences to which meaning might be attributed. Which subsystems were fundamental could even be interchangeable and context-relative.

12. In conclusion, the methodological approach espoused by Gasparyan faces a number of challenges if the existence of an external world cannot be incorporated into an explanation of how meaning is grounded within the autopoietic system described. Despite my initial worries with regard to symbol grounding and biological grounding, one avenue to explore is that sensory perception might act as a subsystem within the system that is consciousness-as-presented-in-the-target-article, providing information about how the unknowable environment seems in such a way that values can be attributed to anticipated experiential outcomes. An integrated account of the grounding of semantic content (constructed from experienced perceptions and sensations, together with affective responses) and reasons for action, from which purposiveness might follow, might then be possible. Grounding would be in the qualitative character of affective responses to actual and anticipated outcomes, as experienced in consciousness.

Open Peer Commentary: Self-Description Alone Will not Account for Qualia

John Pickering

1. Consciousness is an aspect of life. Although the last few centuries have seen great advances made in understanding how living systems arise, persist and function, this is to study the vehicle for consciousness, not consciousness itself. It is notable that in his landmark essay *What is Life?* Erwin Schrödinger (1944) makes virtually no reference to consciousness. Even in his later work *Mind and Matter*, which deals explicitly with mental life, he implicitly assumes that physical objectification is the only proper epistemological stance for scientists to take (Schrödinger 1958). This stance, which treats consciousness as just another aspect of a world that, with time, could be as fully and quantitatively known as, say, electromagnetic radiation, inherits from the nineteenth century the ethos of what Thomas Nagel (2012) has termed the "materialist neo-Darwinian conception of nature." It is notable that part of Nagel's argument against this conception has centrally to do with qualia. Qualia, being the characteristics of all and any conscious experience, are what we know with most certainty. Despite this being so, many philosophers claim that we do not know what consciousness is (e.g., Strawson 2016). Perhaps the epistemological dead ends pointed out by Gasparyan contribute to this somewhat odd situation.

2. Nagel and other critics aside, the conception of nature that has no place for qualia is still the one most commonly adopted, implicitly or explicitly, in the physical and biological sciences. The existence of qualia is not in fact disputed; how could it be? However, the assumption is that in time and with greater knowledge, especially of the nervous system, they will be shown to be "nothing but" the workings of nerve cells (e.g., Crick 1994). In the social sciences, this conception of nature is under constant challenge, since it is transparently unproductive and misleading. In psychology especially, it has been a controversial issue right from the inception of the discipline, which William James named "the science of mental life" (James 1890: 1). James noted that the central phenomenon with which this science had to deal was consciousness. This, he felt, was something that, while patent, was essentially beyond the grasp of reductive methodologies. As he put it:

> [N]o mechanical cause can explain this process, nor can any analysis reduce it to lower terms or make its nature seem other than an ultimate datum, which, whether we rebel or not at its mysteriousness, must simply be taken for granted if we are to psychologize at all. (James 1890: 2).

The phrase "…this process…" in this particular passage referred to an act of recall, but towards the end of the same work, James expands and clarifies what he means:

> [T]he only thing which psychology has the right to postulate at the outset is the fact of thinking itself. I use the word 'thinking' […] for every form of consciousness indiscriminately. (ibid: 224).

3. Despite explicit early warnings against reduction by James and others, in the intervening years the majority of those studying consciousness have adopted a reductionist stance of one sort or another, where the aim is to objectivise subjectivity. As Gasparyan makes clear in the first section of her target article, this epistemological gambit is bound

to fail. Treating consciousness as if it were an object to be studied is only productive when what is being studied is the vehicle for consciousness, namely the life processes associated with it. But if this approach is extended to consciousness itself, the gambit fails; the observer and the observed become one and objectification becomes impossible, defeating the aim of the exercise and leaving the central problem untouched.

4. In her second section, Gasparyan recommends a second-order approach to consciousness based on the idea that to differentiate one subjective state from another is to notice what differentiates them. This makes a plausible start on defining what Gasparyan calls the "minimum unit of meaning" (§49). However, it begs the question of how the capacity to notice arises in the first place. Nor does it touch on the status of conscious beings, such as the higher vertebrates, who are certainly conscious, but not aware of being so.

5. Nonetheless, the differentiation of one state of consciousness from another is clearly a productive way forward. Gasparyan elsewhere observes "[…] that to understand what makes consciousness different from non-consciousness we must have consciousness" (§47). As this comes dangerously close to circularity, more needs to be said about what "different" means in this context. This is to be found, in part, in a somewhat Kantian section where Gasparyan concludes "[…] that the world as such has no differences; they exist only for the conscious observer" (§51).

6. This opens the way to connecting Gasparyan's proposal of a change to a second-order point of view with other discussions of differences and their origins. A fundamental difference, so fundamental in fact that it can stand as an epistemological axiom, is that between self and other. In psychoanalytic traditions, this has been a central issue, from Sigmund Freud's (2002) discussion of the origins of religious experience being a regression to the "oceanic feeling" of the infant mind to more recent pronouncements such as that of James Hillman (1995: xvi), who claimed: "There is only one core issue for all Psychology. *Where is the me?* Where does the 'me' begin? Where does the 'me' stop? Where does the 'other' begin?" The differentiation of self from non-self is patently second-order, since it assumes there is a self that is differentiating. This is made more explicit in the work of James Gibson. His notion of affordance is an illustration of how the perception of differences on which an observer relies to act effectively, crucially depends on the nature of the observer; more specifically on what the observer is able to do (Gibson 1977).

7. The difference between self and other is not merely a theoretical construct. In neurological studies of how the sensations reflecting actions in the world are distinguished from those reflecting actions of the observer, the differentiation of self and other is given quite explicit definition (e.g., Blakemore, Wolpert & Frith 2002). Likewise, studies of the immune system often propose that it is in fact a system that "knows" the difference between self and not-self (e.g., Varela 1994).

8. Further examples could easily be found but are not needed since the claim being made here should be clear. Merely to propose a change from first- to second-order cybernetics as the basis for observing and hence understanding consciousness, while useful up to a point, is not enough in itself. The type of observer needs to be taken into account.

9. But this in turn raises a further, productive, question. Are there as many types of consciousness as there are types of observers? At one level, the answer is clearly "yes." In his influential paper "What is it like to be a bat?", Nagel (1974) points out that while it would be unreasonable to deny that animals like bats are conscious, that does not imply that we can know what that means in terms of qualia, what it "feels like" to have the qual-

ity of consciousness of a bat. As he puts it, "Consciousness is what makes the mind-body problem really intractable" (Nagel 1974: 435).

10. This pinpoints the difficulty with Gasparyan's proposal. While it is useful in avoiding some of the dead ends that are all too common in the literature on consciousness, it leaves the central issue of subjectivity and qualia untouched. This, often called the "hard problem" (Chalmers 1995), like many other important problems, does not have a solution. It serves to create an arena of discussion within which ideas, such as Gasparyan's can be dialectically developed.

11. To develop Gasparyan's proposal, what is being advanced here is the idea that in order to be effective, a change of viewpoint also requires a metaphysical shift. This shift is towards the radical panpsychism of Charles Sanders Peirce and Alfred North Whitehead.

12. Peirce's triadic semiotics is an ontological system with three levels. The first is bare existence, the second is relation and the third is, crucially, semiotic interpretation. The first two levels are necessary for signs to exist, the third, and most significant level for what is being suggested here, is concerned with distinguishing signs and acting upon them. Peirce calls this level the "interpretant." Somewhat problematically, Peirce's triadic system puts the interpretant in third position. However, it is being suggested here that it corresponds to Gasparyan's second-order proposition: Peirce gives the interpretant a fundamental ontological role as the primordial unifier of the mental and physical worlds (see the numerous references to this volume in collections of Peirce's work, e.g., Houser & Kloesel 1992). Moreover, the interpretant is the means by which physical systems are able to respond to themselves. It is, hence, a treatment of consciousness with significant resemblance to Gasparyan's second-order approach. Biosemiotics, which combines Peirce's radical panpsychism with the rational biology of Jakob von Uexküll, provides the conceptual and methodological resources to clarify the types of differences of which any conscious being can become aware (e.g., Hoffmeyer 2008).

13. Perhaps more radically still, Whitehead's thoroughgoing organicism claims that the ultimate ontological primitives are moments of subjective experience. In each moment a process Whitehead calls "concrescence," prehends prior conditions of the world to produce new and unique conditions. In this manner, the universe enacts what he calls a "creative advance into novelty" (Whitehead 1929: 349). Whitehead holds that the process of concrescence has no parts – it is an indivisible aspect of the universe where no distinction can be made between mental and physical realms. There is a non-trivial resemblance here to Gasparyan's notion of a "minimum unit of meaning" (§49). It is also worth noting that the term "meaning" locates subjectivity. A subject "means" to act, and acts on the basis of what the world "means" to the subject.

14. Finally, it might also be observed that most of the literature cited by Gasparyan comes, reasonably enough, from the discussion of consciousness by Western academics working in a Western context, with the implicit ontological and methodological assumptions that this entails. In many Eastern traditions, the study of consciousness has proceeded for millennia with profoundly different assumptions. Western philosophers who have engaged with Eastern traditions, especially those who are experienced meditators, approach consciousness in a characteristically different way that often avoids some of the difficulties Gasparyan points out (e.g., Metzinger 2003). A postmodern synthesis of Eastern traditions and Western panpsychism may help to avoid some of the dead ends to which Gasparyan draws our attention.

Open Peer Commentary: **Consciousness as Self-Description and the Inescapability of Reduction**

Sergei Levin

1. Diana Gasparyan proposes applying the principles of second-order cybernetics (SOC) to the philosophy of consciousness as an alternative to the traditional approach, which is supposedly based on first-order cybernetics (FOC). The new approach is very ambitious and, if successful, may radically shift the paradigm in the study of consciousness.

2. Interestingly, the application of SOC principles results in a theory of consciousness that resembles the classical phenomenology. Both theories aim to provide a holistic understanding of cognition without separating an observer from his experiences and environment. The proposed theory distinguishes itself from the classical phenomenology, declaring "the experience of differences is more primary than intentionality" (§65).

3. There are various other aspects of the application of SOC principles in the philosophy of consciousness. I concentrate only on one. My major concern is the promise to get rid of the reductionist agenda in the study of consciousness. Gasparyan's new theory promises to escape reductionism and to provide a coherent self-description of consciousness. I argue that it actually includes the problem of reduction.

4. The author says that the task of reductionism in the philosophy of consciousness "is to reduce consciousness to various types of objective essences" and it "manifests itself in the search for other external reasons for consciousness, such as language, culture, and society" (§7). If it is possible to fulfill the reductionist's task, then consciousness fully depends upon something else that is not consciousness. If we assume that consciousness depends upon a number of factors X, then, in a metaphysical sense, X are prior to consciousness and X are more basic than consciousness.

5. It is hard to overestimate the appeal of the idea of the reduction of consciousness. Reduction of consciousness could help us to explain consciousness in scientific terms, it unifies the ontology and there are many philosophical arguments for reductionism. On the other hand, reductionism raises the concern that reduction does not explain consciousness but eliminates it. In effect, it just substitutes the phenomenon of consciousness with a list of external factors. The concern is clearly expressed in the target article and it serves as justification for replacing reductionism "with a strategy whereby consciousness will be talked about in the language of consciousness" (§16).

6. There are objections to attempts to study consciousness within the domain of the individual mind. For the sake of the argument, I would like to avoid going into these debates and would like to embrace the proposed methodology. My aim is to show that even if we agree with the initial premises and methodology of SOC in the study of consciousness, the question of reduction would arise yet again. In other words, Gasparyan claims that the "non-normal" theory of consciousness (§§25f) would replace the reductionist program with a self-description. However, after that, the reduction just does not seem to go away.

7. The "non-normal" theory of consciousness proceeds from the premise that "consciousness can be seen as an example of the autopoietic system, knowledge of which is generated by the same system" (§15). Consciousness is the source, and at the same time the instrument, of acquiring knowledge of itself. The application of SOC principles means that we should not try to describe our consciousness from the "external observer" point of view.

8. The main instrument of the study of consciousness for the "non-normal" theory is self-description without postulating a subject–object dualism and the objective world. The self-description should be as rich as our own consciousness; otherwise, it is not adequate. The self-description has different modes, levels and content. The latter does not become salient; it exists as "bundle of internal differences, where some parts, being differentiated, allow the existence of others" (§46). According to the "non-normal" theory, a core feature of self-description of consciousness is various differentiations.

9. Gasparyan has presented an elaborate theoretical model of self-description in terms of differences, but there are not many actual examples of such descriptions in the article. In this regard let us see how self-description in terms of differences give rise to the problem of reduction in three easy and, I suppose, universal steps:

a. There are parts of consciousness (experiences) that differentiate themselves by claims that they have consciousness. I call these experiences persons or people.
b. All people I have ever encountered have a relationships with other experiences. These experiences are their individual characteristics and factors around them. Some of those characteristics are related to claims about consciousness and some are not. For example, hair color has nothing to do with consciousness, but the attachment of the head to the body does seem to be very relevant factor.
c. In the self-description of my consciousness, I can search for the list of characteristics and factors that are necessary and sufficient for claims about consciousness.

10. In the third step, the question of reduction arises anew. During these steps of self-description of consciousness, I have followed the methodology presented in the target article (§32). I described my consciousness in the first person with active verbs and I acknowledged my presence as observer. Nevertheless, in the third step, I faced the question of reduction. Even if I had constructed the problem of reduction and had never perceived it directly, it is still the same problem, arising this time from the self-description of consciousness.

11. The search for the list of characteristics and factors that are necessary and sufficient for claims about consciousness is the question of reduction. The search for the list is the "search for other external reasons for consciousness" (§7) and this is the task of the reduction. The characteristics and factors are internal parts of my consciousness, so I still do not postulate any objective world. However, the list is external to people, since the experience of people and the experience of the factors and characteristics in the list are different.

12. The list does not even have to be the complete list of necessary and sufficient conditions. It may well be the case that consciousness is such a vague state that it is impossible to provide its comprehensive self-description or set rigid boundaries. The list may be more like Wittgenstein's family resemblance concept. Yet the search for this list would be a question of reduction of consciousness because the phenomenon in question could have various reasons (causes).

13. In the target article, Gasparyan has compared the experience of consciousness to performative utterances (§§27–30). *Prima facie,* it may look like this move is antireductionist because performative utterances are self-affirming. In fact, it is also compatible with the question of reduction. To be successful, the performative utterance depends on certain external factors. For instance, to name a ship, one must have the appropriate authority and one can only do it at the right time and at the right place. The same applies

to consciousness: claims about consciousness – even autobiographical claims – depend on certain factors.

14. One possible criticism of the proposed steps to reduction is to say that claims about the existence of the consciousness of others are not autobiographical and hence are not the part of an autopoietic system, which consciousness essentially is. These claims about consciousness bring back into discussion the third-person perspective and the objective world, which must be avoided in the new approach. That is how he problem of reduction appeared in the first place.

15. On closer examination of the proposed steps, it is evident that I do not postulate the existence of people or anything else. The word "people," etc., just refers to various parts of my consciousness. I do not see how anyone could conceivably deny the existence of the experience of others in the self-description of consciousness.

16. Furthermore, Gasparyan distinguishes access to the world and access to the other interpretations of the world. We do not have the former but we have the latter (§73). The interpretations of the world are not equal, some of them are better, some of them worse. Claims by other people that they have consciousness are their interpretations of the world. These interpretations are very convincing. A curious mind should start to wonder why such interpretations happen at all and what are the reasons for them – and with that very inquiry the question of reduction arises anew.

17. If the arguments I have presented are sound, then the self-description of consciousness in terms of differences does not remove the question of reduction. Even if epistemological and ontological reductions are impossible, there is no contradiction between self-description and the search for other external reasons for consciousness. The theory of consciousness as self-description in terms of differences may be a useful tool for the study of mind. It may even grasp the essence of consciousness. Still, the question of why consciousness happens this way would remain, and this is the search for external reasons for consciousness – hence the question of reduction.

Acknowledgments

The commentary was prepared with the financial support of the Russian Foundation for Basic Research, within the framework of the project "Unity of Consciousness: the Phenomenal Field and the Binding Problem" no. 16-03-00834.

Open Peer Commentary: The Non-Relationality of Consciousness

Adriana Schetz

1. In one of the most intriguing books of the last decade of the 20th century in cognitive science, Francisco Varela, Evan Thompson, and Eleanor Rosch developed a new approach to the phenomenon of consciousness. They insisted on abandoning a purely scientific, third-person methodology while discussing the realm of consciousness. The title of this book, *The Embodied Mind: Cognitive Science and Human Experience*, explicitly indicates which direction one should take to reach an adequate theory of consciousness: the idea of the embodiment of mind. Diana Gasparyan's target article fits squarely into this trend, and is yet another effort to overcome a conspicuous crisis in attempts to resolve the hard problem of consciousness, as described by David Chalmers (1995).

2. As I see it, the key problem Gasparyan presents is that contemporary cognitive science treats consciousness as a relational phenomenon – something that supposedly takes place between subject and object (§4). In this relationship, reflection is possible merely as meta-knowledge (§4). This approach is entirely coherent and in line with the spirit of cognitive science. However, it does not allow us to explain the main feature of consciousness – the fact that it is reflective in its very nature, and that this reflectivity is transitive not only outwardly, but somehow inwardly as well. Let me explain what this exactly means.

3. The author remarks that an "experience of consciousness is *performative* in principle" (§27). That seems, as she cogently claims, to fit well with John Searle's observations about expressions such as: "I promise" (§29). In the case of such a declaration, when saying "I promise," there is no need for further activity to be successful in promising something (Searle 1979). Similarly, every time our own thoughts, feelings, or desires become the object of our consciousness, the performative nature of such acts becomes noticeable: "what" one is thinking about is identical with "how" this appears in one's consciousness (§30). For example, when one thinks about the beauty of a flower (say, *this flower is very beautiful*), one is, at the same time, mentally somehow in a mode of performing the following thought: *this flower is very beautiful*. Finally, we get something that I want to call the non-relationality of consciousness, and that Gasparyan portrays as understandable from the perspective of second-order cybernetics (§§6f).

4. By second-order cybernetics, she understands a rejection of first-order cybernetics, i.e., rejection of the idea according to which consciousness forms a kind of relation between the subject (who is conscious) and the object ("aboutness" of consciousness). For second-order cybernetics, consciousness seems to be no longer relational. To be conscious means to be in an internal state of mental differentiation – "the experience of differentiation" (§48) – which eventually leads to undermining the subject-object distinction.

5. Admittedly, Gasparyan makes use of the concept of relation while speaking about the reflectivity of consciousness, but one may hold that this is only derivative relationality. It is derivative because it obtains between inner states of the mental system (mind) as a whole; in addition it is conceived as an autopoietic operation of that system (§50). The autopoietic nature of consciousness allows one to arrive at the conclusion that:

> The array of difference will constitute the 'essence' of this system [...]. To this extent, this system does not have any substances-essences and it is not a substance by itself. This system has nothing but differences. (§50).

6. The idea of the autopoietic nature of consciousness seems very plausible to me, and is fully consistent with Varela, Thompson, and Rosch's view on experience and auto-reflection. What bothers me is Gasparyan's attachment to the idea of a relational nature of consciousness. Here are some details of my concern.

7. The embodied mind approach suggests that consciousness is basically non-relational:

> The phrase 'unity of consciousness' refers to the idea that one understands all of one's experiences as happening to a single self. As Jackendoff rightly notes, however, there is an equally obvious *disunity* in consciousness, for the forms in which we can be consciously aware depend considerably on the modalities of experience. (Varela, Thompson, Rosch 1991: 55)

Moreover, Thompson, while speaking about John Searle's model of the conscious mind, notes that:

> According to this model, the neural substrates of individual conscious states should not be considered sufficient for the occurrence of those states, for those states themselves presuppose the background consciousness of the subject. Any given conscious state is a modulation of a preexisting conscious field. An individual experience or conscious state (such as visual recognition of a face) is not a constituent of some aggregate conscious state, but rather a modification within the field of a basal or background consciousness. (Thompson 2007: 351)

8. Perceptual consciousness provides a good example of what I mean by non-relationality, but without leaving out transitivity and differentiation of conscious states. The difference between, say, seeing and feeling some object by touch results from the difference in phenomenal character of those two experiences. The conscious content of visual experience is somehow identical with the way it presents itself to the subject of experience. When one is conscious of one's visual experience, one has conscious experience of one's visual experience. However, being conscious of one's visual experience should not be seen as a different or a separate state from having conscious experience of one's visual experience. For if this were so, we would certainly talk about a higher-order theory of consciousness.

9. Such a higher-order experience theory of consciousness, also referred to as a higher-order perception theory by Peter Carruthers (2011), was developed by John Locke (1975), and more recently by David Armstrong (1993). Unfortunately, this approach faces a number of difficulties. One of the most important is that in an attempt to explain this kind of consciousness, one risks getting trapped in an infinite regress: the experience of perceptual experience is explained by invoking the experience of that experience, and this experience of experience is in turn explained by the presence of the experience of experience of perceptual experience, and so forth. In order to avoid the regress, the content of the latter (i.e., visual) experience becomes a *mode of presentation* of the former (i.e., conscious) experience: visual experience modifies consciousness. The relation between conscious experience and perceptual experience (in this case, visual experience) is merely apparent or illusory. In this point, I strongly agree with Gilbert Harman (1990), especially with what he says about the transparency of perceptual experience. According to Harman, there is no "mental paint" – i.e., the concept of sense data – when we perceive an object, which should

be seen as being perceived by a higher-order conscious state. All that we are aware of in the act of perceiving is an object being perceived. Therefore, perception modifies consciousness without forming any kind of sub-state in relation to consciousness.

10. Harman's conception of the transparency of perceptual experience fits his representationalism. But, as I see it, since the content and the mode of perceptual consciousness presumably form one unified act, there is still a place for direct realism concerning perceptual experience. A dually constituted act of perception may still have its external object, i.e., reality. And then, when one is talking about the relational nature of consciousness, one is talking – I take it – about the interaction between the sensory system and an external object. Thus, in order to describe consciousness, one should reject the idea of "relationism," and replace it with the idea of knowledge, or better, of experience by differentiation. The reason I want to avoid relationism is that it brings in the subject/object distinction, which – as I agree with Gasparyan – should be abandoned if the phenomenon of consciousness it is to be explained. As I tried to show above, the idea of relationism leads directly to an infinite regress.

11. Therefore it is not relationality but modality that sets differentiation in the realm of mind, and hence determines the differentiation of consciousness. Moreover, this kind of differentiation is something perceptual modalities share, in my opinion, with a whole range of mental states, including beliefs, desires, emotions, etc. Thus, the idea of consciousness conceived as self-description in difference, as Gasparyan elaborates it, makes sense not only for perceptual consciousness. Second-order cybernetics is now possible, since self-description is no longer seen as a relation between subject (self) and object (content of consciousness). This inner or inward transitivity of consciousness goes hand in hand with its autopoietic nature, which preserves the unity of the conscious mind as a whole.

12. To conclude, the aim of my commentary was not to argue with Gasparyan's attitude towards the problem of consciousness. In fact, I agree with most of her claims. Instead, I merely wanted to strengthen the idea of consciousness conceived as a unified, non-dualistic phenomenon – in the sense of the inner distinction between subject and object. As I have tried to show, the insistence on the non-relational character of conscious states is a promising possibility in this respect.

Author's Response:
Phenomenology of the System: Intentionality, Differences, Understanding, and the Unity of Consciousness

Diana Gasparyan

Consciousness as a phenomenological system: Intentionality and differences

1. In his rather critical commentary, **Yochai Ataria** points out that I over-emphasized the subject by ignoring the tension between subject-object relations and by putting the object aside. I agree that this would turn my thinking into a version of rather naïve solipsism. However, this is not what I was trying to show. Rather I distanced myself from the subject-object paradigm because it all too often results in an unproductive act of balance between the two extremes – naïve realism and epistemic dead-end solipsism. In the first case, there is a substantial emphasis on the priority of things and objects of the external world, and in the second, on the priority of the consciousness that holds objects and the external world strictly within the boundaries of experience. Philosophically, realism is mistaken, for we never start with the outer world separately but only with the world as phenomenon, as it is being given to consciousness. However, by starting with the fact of consciousness, we risk never leaving it to reach the outer world – we meet only our own states of consciousness everywhere. While such solipsism may be a philosophically correct argument, it misjudges our everyday epistemic possibilities and renders cognition helpless. In order to eliminate such a stalemate, phenomenology offers a new set of instruments in order to overcome these equally bad options. In phenomenology, neither the subject nor the object is the grounding relatum, but they are rather mutually dependent. This is the primary virtue of putting the *intentionality* of consciousness as the starting point of every phenomenological analysis. Intentionality equals the subject and the object, insofar as it relates them to each other forever: Any object (phenomenon) exists only by being perceived by a subject (consciousness); conversely, any subject (consciousness) exists only by being directed to any kind of object. Simply put, there is no act of thinking prior to the objective thought and there is no objective thought without any act of thinking.

2. I agree with **Urban Kordeš** (§4) that the term "experience" is even more relevant than consciousness, as it certainly points to the indissolubility and priority of intentionality. In other words, the boundary between subject and object is not only blurred – which **Ataria** himself points out in §§12f it actually ceases to be a boundary and becomes a bridge or a connection point, which is described by the notion of intentionality. Most phenomenologists agree with this thought – Franz Brentano, Edmund Husserl, and Maurice Merleau-Ponty (to whom **Ataria** not only refers in §12 but with whom he also seemingly agrees). So the subject does not precede the object, and nor does the object precede the subject, but their relation is primary. Such a phenomenological view opens a wide range of possibilities. In fact, it envisages that there comes into existence neither a subject nor an object *by itself*, but that when they appear, they appear altogether as a *system of mutual relation and differentiation*. However, this model does not exclude that we can resort to the subject-object pair in cases when necessary. For example, science in general seems to require the use of this pair. As **Ataria** (§13) rightly notes, it is very difficult for us to give

up its use. However, here we can apply a "switch" that refers to the context – *naturalist* or *transcendental.* If we focus, within the scope of what is intentionally given to us, on the nature of objects, we follow a kind of naturalist approach. But if it is about how these objects are given in relation to consciousness – we use the transcendental approach. Thus, in spite of the subject-object unity, the intentional system allows for different accentuation or switches, comparable to gestalt switches between figure and background. However, it is important to understand that the notions "subject" and "object" themselves appear *after* we "turned on the switch," and in this sense they are to be considered *artificial constructs*, which are applied for certain (practical) purposes.

3. This is why it would be incorrect to interpret my approach as a variety of dualism, as **Ataria** supposes (§2). Of course, if one were to interpret my approach without leaving the narrow frame of the dualistic paradigm (§13), then my abandoning of the objective dimension would be regarded as an enclosure in the subjective dimension (the choice of solipsism whilst refusing realism). However, I abandon realism not in order to choose solipsism, but to find *a middle ground – the path of phenomenological analysis of intentionality*, hereby suggesting a way out of the realistic-solipsism dilemma.

4. What I suggested can hardly be called a naturalistic dualism, in particular because phenomenology does not restrict itself to the limits of any strong naturalist setting (which assumes that everything is a pure object, i.e., can be objectivated without the need for any reference to its state of being given to some consciousness). **Ataria** points out that the *addition* of the principle of differentiation as a new fundamental law serves as a reason for referring to my approach as naturalistic dualism. Such an assumption seems controversial to me – not every discovery of a new law within the framework of a given theory allows for treating this theory as naturalistic dualism (§3). The reason will be the nature of the law itself. David Chalmers (1996) underlines that those fundamental psychophysical laws that he suggests have a principally naturalistic nature are part of nature and might be discovered empirically (by means of a procedure of explanation). On the contrary, I insist that the principle of differentiation is a *metaphysical* (i.e., fundamentally non-empirical) law and empirical (explanatory) methods of cognition in this respect will turn out to be inapplicable.

Science or mystery: The role of understanding

5. The *non*-naturalistic direction of my approach is important for clarifying the issue of its "scientific status." In particular, another suggestion from **Ataria** (§7) underlines the fact that, according my approach, one cannot "predict" new scientific facts – but this seems to be rather irrelevant to my approach. The theory of consciousness I am talking about does not, in principle, claim to be empirically fruitful. This means that it is not the task of scientific explanation (i.e., external causative modelling) or prediction that falls to us, but rather the task of phenomenological description, which means the understanding of fundamental differences within the overall intentional system of relations.

6. In my opinion, second-order cybernetics (SOC), as well as the project of phenomenology, view epistemology more broadly than science. This, in its turn, means that science, at least in its reductionist dimension, is not the only form of cognition. The method I am trying to apply (which has different names: phenomenological rather than naturalistic, second- rather than first-order cybernetics, understanding/description rather than explanation) retains its rational nature from the point of view of *adequate directionality*. My thought is that the primary subject matter of study must be consciousness, i.e., the intentional di-

mension of the subject, such that the rational (adequate) stance will be a non-reductionist approach. However, one should not understand "the natural scientific approach" as "scientific nature." In this respect I agree with **Kordeš** (§14) that it is quite possible that it is not worth talking about descriptive theories of consciousness, but rather about description of the phenomenal realm as a method in its own right.

7. While this point would deserve a deeper discussion, I cannot elaborate it here. I can only refer to the founders of the neo-Kantian approach, who are in many respects the ancestors of phenomenology – Wilhelm Dilthey (1977), Wilhelm Windelband (1915), Heinrich Rickert (1962) – who suggested that we differentiate all knowledge into two types of sciences – *science of spirit (consciousness)* and *science of nature*: we *explain* the life of nature and *understand* the life of consciousness. **Konstantin Pavlov-Pinus** (§3) talks about the same division when referring to the method of hermeneutics. The necessity of dividing methodology becomes evident because mental life (understood in the very broad sense, not only as the mentality of individuals – a point that is important for my discussion of **Sergei Levin** below – but as the consciousness of whole societies or the aggregate of senses developed by humankind) cannot be explained in terms of causative connections building on some general law but must be reconstructed on the basis of some fundamental laws of the comprehensive intentional system.

8. For example, the *explanation* of an event such as the falling of a body is totally exhausted by pointing to gravitation as the general law, the particular case of which would be the falling of a particular body. Such an event is regular and repetitive. Also, physical processes do not have a qualitatively heterogeneous inner nature. By contrast, it is difficult to point to a general law underlying, for example, some historical event such as the execution of Robespierre during the French revolution. This dramatic event, first of all, is singular and not repetitive, and second, to understand it, it is not enough to state the closest natural reason (physical cutting-off of the head as a reason for death). If we only apply external scientific explanations, we will only be able to construct the knowledge of the physical and physiological dimension of this process but will not understand anything about such the notional dimension of a political leader being eliminated, or of treason in terms of revolutionary ideas, of the downfall of the pathos of humanism, etc. If we want to understand factual events of such a type, we are interested in *meanings,* which are always founded by *values* and envisage *understanding*. For example, the discovery of an artefact will become meaningful for us, not when we perform a comprehensive chemical and physical analysis of this object (because it is useless to weigh or dissect a statuette in order to understand what it is like), but when we try to find out whether it is an object of a religious cult or simply a decoration. And for this, in its turn, it is necessary to understand what religious, aesthetic, and other values various peoples have cultivated, that is, it is necessary to go deeper into the sphere of meanings.

9. In this respect, I found the thoughts of **Bryony Pierce** in §2 on the notion of *norms* (the issue of values) particularly interesting. It seems that the sphere of values is, in fact, the dimension of qualia, which in principle retains its prospects in the first-person. Being a phenomenal aspect, the sphere of values cannot be explained in an external causal way. This impossibility poses a serious problem for philosophy, i.e., the separation between "what is" and "what ought to be," which was presented in the most articulate way by David Hume:

> In every system of morality, which I have hitherto met with, I have always remarked, that the author proceeds for some time in the ordinary way of reasoning, and establishes the being of a God, or makes observations concerning human affairs; when of a sudden I am surprised to find, that instead of the usual copulations of propositions, is, and is not, I meet with no proposition that is not connected with an ought, or an ought not. (Hume 2010: 213)

And further, "the distinction of vice and virtue is not founded merely on the relations of objects, nor is perceived by reason" (ibid: 214).

10. This also accords with the position of Ludwig Wittgenstein (1965), who denied that values (morals) were facts or revealed in the world of particular fact-events, which therefore cannot be formulated in terms of sentences. That is why with their help it is impossible to respond to the question "why?", regarding the cashier ("Why did you return the money to the cashier who was in the error?" – "I considered it correct," – "Why did you consider it correct?" – "I do not know"), as they do not continue the causal chain but rather interrupt it in a wilful way that remains unclear for the procedure of explanation.

11. I am inclined to interpret this topic in accordance with the idea stated in §7 and §11 above–regarding the impossibility of an external explanation of the inner measurements of a subject. So the point here is not about explaining but about *understanding* some events – in this case acts of moral action. I agree that emotions mean the end of grounding. I do not deny that the emotional sphere is an area inside the sphere of consciousness, as it might seem, according to **Pierce** (§10). The point is that emotions still need to be conceived. Thus *joy* cannot be phenomenologically separated from the *realization of joy*. Therefore, phenomenology speaks about intentionality – emotion must also refer to a subject of consciousness in order to become effective. This does not contradict, it seems to me, **Pierce**'s statement in §10 that emotional arguments do not behave like rational arguments – the latter lead to regress, whereas the former stop it.

12. Correspondingly, when I am talking about self-consciousness, I mean exactly this global sphere of meanings, the distinction of which would be in the fact that it always understands itself from within itself. It goes without saying that such a global sphere of meanings understands itself and envisages a serious metaphysical shift, which **John Pickering** (§11) points out, who uses "meaning" in Hegel's terms. In order to specify this abstraction, suppose that under this global sphere of meanings, the whole universe of the symbolic can be understood – any sign formations (semiosis) – languages, cultures, values, images, and ideas. This model well accords with what **Pickering** maintains, bearing in mind a certain model of radical panpsychism (§11). Obviously, inside this sphere, the procedures of reducing certain things to others are implemented (as **Levin** §9 points out), for example, from simple to complex. Bearing in mind that any understanding requires reducing one thing to another, I have to specify that when I speak about the inapplicability of reductionism to consciousness, I mean the *heterogeneous (not homogeneous) reductionism* such as the reduction of the mental to the physical, or subjective to objective. However, I do not deny any homogeneous procedures of reduction, but rather point to the irrelevance of heterogeneous reductionism. Therefore, I agree with **Levin** that inside self-consciousness there are parts that can be reduced (or as I would say, referred) to others (§9), but under no circumstances would I say that we are dealing with reduction (affine), which I oppose.

13. In contemporary analytic philosophy of consciousness, the above-mentioned division between "understanding" and "explanation" corresponds to the division of first-person access and third-person access, respectively. According to the first approach, we understand, and according to the second, we explain. The search for external reasons for

consciousness, which **Levin** speaks about in §17, is part of the second strategy. In general, this search remains irrelevant for the work on consciousness, for the simple reason that consciousness does not allow talking about itself in terms of the external. It is not possible to come closer to consciousness via non-consciousness. It is in this, and probably only in this sense that reductionism and its infeasibility is referred to in my target article (§§7–10).

14. Exactly this motive retains its importance in case the possible objectivity or observability from the third-person perspective (third person access) is considered the only criterion in the study of consciousness, as **Ataria** (§11) envisages. By contrast, I am orientated towards an understanding in philosophy (which also exists in the philosophy of consciousness, within the boundaries of anti-reduction programs) maintaining that there are things that would be impossible to define in terms of objectivity. Above all, consciousness refers to this. But also differentiation supposedly refers to it in the sense of the intentional system of relations described above. In itself, differentiation is not objective (for example, it is not a sign, if we talk about language– see below §§19–22 of this response), but it makes all other elements (i.e., signs of language) objects.

15. In the following respect, **Ataria** (§7) is certainly right: the prognostic functions of sciences studying consciousness and the totality of mental phenomena are by far not all the possibilities that the sciences possess in order to study natural facts. But it is important to understand that this would be connected not with the "defects" of the method of understanding but with the specific nature of the "object" itself – in this case consciousness (presumably of a non-determinant nature). Equally, this does not cancel out the fully-featured heuristicity of knowledge about consciousness, but points to the necessity of applying another more relevant methodology. Treating it as the only explanatory (reductionism) strategy in science, this explanation of understanding (explanation) in terms of a strategy of pairing science and mysticism crucially impoverishes the picture of human consciousness and definitely does not leave us with any chance of solving the puzzle of consciousness.

How closed is the system? Differences and the unity of consciousness

16. The model of a closed system, which, as **Pickering** (§11) rightly notes, has actually some features of panpsychism, suffers from a series of problems. The most obvious of these, as **Pierce** (§10) notes, are the two problems of

a. establishing meanings:, and
b. justifying meanings

(It seems to me that by talking about the problem of justifying decisions **Pierce** also means the problem of establishing meanings in general). **Pierce** writes that it is not quite clear how meanings are established inside the system itself (§6).

17. To solve the first problem (establishing meanings in general), I would like to look at the way in which language is organized (to the extent that language is an example of an autopoietic system, the principle of internal differences can also be applied to it.) This problem is described in structural linguistics as follows: "language is a form rather than substance," "the difference of the sign from others is everything that constitutes it," "language has nothing except for differences," etc. (de Saussure 1983: 116). In general terms, the linguistic model offers a view of the essence of the thing (its meaning) when it is determined not by its positive content, but *differentially* – through the relation to essences of other things (their meanings). In this respect, there is no substance if we understand it as

some fixed and autonomous content as such, but there are meanings – values differentiated relatively to each other.

18. According to Ferdinand de Saussure, this model is justified thanks to the fact that the linguistic sign is not the connection of the physical object and its name but the unity of concept (signified) and the acoustic image (signifying). The concept is *the image of the object* in our consciousness, and the acoustic image is *the image of the sound.* These two sides of the sign have a psychic origin, i.e., are ideal and exist only in our consciousness. This phenomenon is also interpreted in phenomenology – this is not the object outside the boundaries of consciousness but the image of the object (always processed and constructed to a certain completeness). It is very difficult to falsify such a definition, as it is clear that the physical object cannot be part of the sign or phenomenon.

19. This definition will also help to shed light on the second problem mentioned by **Pierce** (§10) – i.e., how meanings can be justified inside the system itself. As **Pierce** rightly notes, common sense tells us that beyond signs there are objects (§4). However, if we take into account that the sign does not possess any substantial characteristics, i.e., does not have any sovereign meaning outside the boundaries of the system in general, it does not follow that signs correlate, first of all, with each other. Thus the meaning of any chess piece is formed not as a result of its objective characteristics (material of manufacture, shape, etc.), but by what place (position) it takes on the chess-board. Within the framework of language, "everything relates in correlation," i.e., no sign can have meaning on its own, but only in relation to other signs. Meanings in language are not defined positively in terms of their content, but negatively – in contraposition to other items of the language of the same system. Accordingly, any separate item of the language is characterised by what the other is not. Thus, in relation to the signs of writing, Saussure says that the "significance of letters is quite negative and differential – all we need to do is to distinguish one letter from the another" (Saussure 1983: 118). And further on, about the system of language as a whole: "The whole mechanism of the language [...] is based on oppositions of such a kind and on phonetic and conceptual differences accompanying them" (ibid: 119).

20. As the signifying and the signified also refer to each other, *first of all* the signified cannot be isolated (taken out of the structure of the language) and put outside, and *second*, it is impossible to define what precedes what – the signified, the signifying, or vice versa. If we accept the structure of the sign, we must agree to the fact that between them there are no relations of precedence – they exist *at the same time,* concerning and in relation to each other; that is always, when there is one (signifying), there is another (signified) (in terms of phénomenology: noema always refers to noesa, and vice versa). This assumption, naturally coming out of the definition of the sign, however, also seems counter-intuitive, as common sense tries to insist on the fact that the signified has priority in terms of primordiality, and exists *before* the signifying. However, here we need to ignore the data of everyday representation, as the signifying and the signified are retained by each other such that they cannot have any semantic or objective advance. Consequently, we cannot appeal to the primordiality of things outside the system of signs – we can suppose them (things), but cannot work with them. But then it is not worth waiting for meanings to be transformed by means of referring to "real" things.

21. In turn, the question of the close or open nature of the system should be interpreted so as to suggest that, in all probability, the system has its own external. But the key word is "own." Any system has an external that it *can* have. In spite of the presence of this external and active interaction with it (leading to transformation and change of the system), the sys-

tem never meets it but includes, processes, and totally changes (the whole of it). Therefore, the system will not let us come to the "real external." But this does not mean that it is closed to the external or that there is no external whatsoever, as **Ataria** suggests in relation to my approach (§10). It possibly exists; however, it always comes up as represented by different interpreters. As **Kordeš** says, we are immersed into our experience and cannot leave it (§7). Such an interpretation plays a fundamental role in Peirce's theory of significance, which **Pickering** speaks about (§12).

22. In this respect, I am grateful to **Pavlov-Pinus** for a number of significant remarks. I agree that the definitions of the systems supposing recursion are always opened towards the future (§5). Such a system is in principle an indefinite and actively transforming organism that absorbs its own description, which in itself is a new move in the game, and in this respect its description is already part of what is to be described – the rules of the game are also acted out. I agree with **Pavlov-Pinus** in that such a system cannot be considered complete in any of its states (§5) – its own attempts at thematisation lead to a state of imbalance time after time. In all probability, it is exactly in this state that all the most productive entireties exist that are known to us – consciousness, society, culture, and others. That is why even if the system is open towards its external, and the external is present for this system, the system operates proactively again – by being external it meets itself as renewed in connection with the processing of the external.

23. Following the curious remarks of **Adriana Schetz** (§9) about a higher-order experience theory of consciousness, I agree that the differences inside the system are certainly not arranged hierarchically. Even if we talk about relativity, the point is that all relations are *horizontal*, and what is more important, inter-referenced (relational). Such a model allows paradoxes of endless regress to be avoided. When talking about the system of differentiations, I mean a non-hierarchical and *not a vertical* system. As I write in the section on reflection in my target article (§§33–44), reflection is not correctly defined as knowledge on knowledge. This is connected with the fact that in this case, as **Schetz** rightly notes, there is a risk of endless regress (§9). On the contrary, in a situation where all references are mutually crossed and belong to one level, such a danger is avoided.

24. I agree with **Schetz** that the system must be understood as a unity (§11). The inner system of differences must not wash out this unity. On the contrary, the very possibility of implementing a system of inner differences is only possible because these differences beyond the system do not have any significance. As pointed out above, a good example is the system of chess – each element of the system is allotted meaning only due to the presence of other elements and the unity of the system on the whole. In all probability, speaking about the work of consciousness, it is necessary to admit this connectivity of inner elements.

25. At the end I would like to emphasize that the ideas of intentionality, differences, understanding, and the unity of consciousness make sense not only for phenomenology. Since self-description is no longer seen as a relation between subject (self) and object (content of consciousness), genuine second-order cybernetics is now possible. Intentionality removes the subject-object dualism by fixing that there is no subject before object or object before subject. If so, this system no longer requires an external or independent observer who could explain it to the outside. Therefore, not external explaining but only inner understanding is possible. This understanding takes place between mutually differentiated elements that require each other in order to have any significance. Such an inner or inward

self-maintenance of the system goes hand in hand with its autopoietic nature, which preserves the unity of the consciousness as a whole.

Acknowledgement

I am extremely grateful to my respected colleagues for reading my target article so attentively and suggesting very precise, profound, and original thoughts. Some of these comments were very critical, while others supported and developed my ideas. But without any doubt, all of them were very productively directed to the clarification of the target ideas.

I would also like to thank Peter Gaitsch for his critical and helpful feedback on this text.

Combined References

Armstrong D. M. (1993) A materialist theory of the mind. Revised edition. Routledge, London. Originally published in 1968.

Ashby W. R. (1956) An introduction to cybernetics. Chapman and Hall, London. https://archive.org/details/introductiontocy00ashb

Ataria Y., Dor-Ziderman Y. & Berkovich-Ohana A. (2015) How does it feel to lack a sense of boundaries? The case of mindfulness meditation. Consciousness and Cognition 37: 133–147.

Austin J. L (1962) How to do things with words. Harvard University Press, Cambridge.

Bateson G. (1972) Form, substance, and difference. In: Bateson G., Steps to an ecology of mind. University of Chicago Press, Chicago IL: 448–466.

Bateson G. (1979) Mind and nature: A necessary unity. Hampton Press, Cresskill NJ.

Bennett M., Dennett D., Hacker P. & Searle J. (2007) Neuroscience and philosophy: Brain, mind, and language. Columbia University Press, New York.

Bickle J. (2012) A brief history of neuroscience's actual influences on mind-brain reductionalism. In: Gozzano S. & Hill C. (eds.) New perspectives on type identity. Cambridge University Press, New York: 43–65.

Bishop J. M. & Nasuto J. S. (2005) Second-order cybernetics and enactive perception. Kybernetes 34(9/10): 1309–1320. http://cepa.info/835

Bitbol M. & Antonova E. (2016) On the too often overlooked radicality of neurophenomenology. Constructivist Foundations 11(2): 354–356. http://constructivist.info/11/2/354

Bitbol M. & Petitmengin C. (2013) On the possibility and reality of introspection. Kairos 6: 173–198. http://cepa.info/2298

Blakemore S. J., Wolpert D. M. & Frith C. D. (2002) Abnormalities in the awareness of action. Trends in Cognitive Sciences 6: 237–242.

Block N. (1990) Consciousness and accessibility. Behavioral and Brain Sciences 13(4): 596–598.

Carruthers P. (2011) Higher-order theories of consciousness. In: Zalta E. N. (eds.) The Stanford encyclopedia of philosophy. Fall 2011 Edition. http://plato.stanford.edu/archives/fall2011/entries/consciousness-higher

Chalmers D. (1995) Facing up to the problem of consciousness. Journal of Consciousness Studies 2(3): 200–219.

Chalmers D. (1996) The conscious mind: In search of a fundamental theory. Oxford University Press, New York.

Chalmers D. (1997) Facing up to the problem of consciousness. In: Shear J. (ed.) Explaining consciousness: The "hard problem." MIT Press, Cambridge MA: 9–30.

Chalmers D. (2003) Consciousness and its place in nature. In: Stich S. & Warfield F. (eds.) The Blackwell guide to philosophy of mind. Blackwell, Malden MA: 102–142.
Churchland P. S. (2013) Touching a nerve: The self as brain. W. W. Norton, New York.
Coecke B. (2008) Introducing categories to the practicing physicist. in: Sica G. (ed.) What is category theory? Polimetrica, Monza: 45–74.
Crane L. (2007) What is the mathematical structure of quantum spacetime? arXiv:0706.4452.
Crick F. (1994) The astonishing hypothesis: The scientific search for the soul. Charles Scribner's Sons, New York.
Demircioglu E. (2013) Physicalism and phenomenal concepts. Philosophical Studies 165(1): 257–277.
Dennett D. (1992) Consciousness explained. Back Bay Books, New York.
Descartes R. (1966) Meditations on first philosophy. Translated by John Cottingham. Cambridge University Press, Cambridge.
Di Paolo E. (2009) Overcoming autopoiesis. In: Magalhães R. & Sanchez R. (eds.) Autopoiesis in organization: Theory and practice. Emerald, Bingley: 43–68. http://cepa.info/2366
Dilthey W. (1977) Descriptive psychology and historical understanding. Translated by R. M. Zaner and K. L. Heiges with an introduction by R. A. Makkreel. Martinus Nijhof, The Hague. German original published in 1894.
Foerster H. von (1979) Cybernetics of Cybernetics. In: Krippendorff K. (ed.) Communication and Control in Society. Gordon and Breach, New York: 5–8. http://cepa.info/1707
Foerster H. von (1981) Notes on an epistemology for living things. In: Foerster H. von, Observing systems. Intersystems, Seaside CA: 258–265. Originally published in 1972. http://cepa.info/1655
Foerster H. von (1992) Ethics and second-order cybernetics. Cybernetics & Human Knowing 1(1): 9–25. http://cepa.info/1742
Foerster H. von (ed.) (1995) Cybernetics of cybernetics. Future Systems, Minneapolis MN. Originally published in 1974.
Foerster H. von (2003) On constructing a reality. In: Foerster H. von, Understanding understanding: Essays on cybernetics and cognition. Springer, New York: 211–227. Originally published in 1973. http://cepa.info/1278
Freud S. (2002) Civilization and its discontents. Penguin, London. German original published as: Freud S. (1930) Das Unbehagen in der Kultur. Internationaler Psychoanalytischer Verlag, Vienna.
Funkhouser E., (2007) Multiple realizability. Philosophy Compass 2(2): 303–315.
Gasparyan D. (2015) What can the global observer know? Constructivist Foundations 10(2): 227–237. http://constructivist.info/10/2/227
Gennaro R. (1995) Consciousness and self-consciousness: A defense of the higher-order thought theory of consciousness. John Benjamins, Amsterdam.
Gennaro R. (ed.) (2004) Higher-order theories of consciousness. John Benjamins, Amsterdam.
Gennaro R. (2012) The consciousness paradox. MIT Press, Cambridge MA.
Gergen K. J. (1997) Realities and relationships: Soundings in social construction. Harvard University Press, Cambridge MA.
Gibson J. J. (1977) The theory of affordances. In: Shaw R. E. & Bransford J. (eds.) Perceiving, acting, and knowing. Lawrence Erlbaum Associates, Hillsdale NJ.
Glanville R. (2002) Second order cybernetics. In: Encyclopaedia of life support systems. EoLSS Publishers, Oxford. (Web publication). http://cepa.info/2708
Glasersfeld E. von (1987) The construction of knowledge. InterSystems Publications, Salinas CA.
Glasersfeld E. von (1995) Radical constructivism: A way of knowing and learning. Falmer Press, London.
Güzeldere G. (1995) Problems of consciousness: A perspective on contemporary issues, current debates. Journal of Consciousness Studies 2(2): 112–143.

Harman G. (1990) The intrinsic quality of experience. Philosophical Perspectives (4): 31–52.
Harnad S. (1990) The symbol grounding problem. Physica D42: 335–346.
Harnad S. (1993) Grounding symbols in the analog world with neural nets. Think 2: 12–78.
Harre R. (1989) Metaphysics and methodology: Some prescriptions for social psychological research. European Journal of Social Psychology 19(5): 439–453.
Hellman G. & Thompson F. (1975) Physicalism: Ontology, determination and reduction. Journal of Philosophy 72: 551–564.
Hillman J. (1995) A psyche the size of the Earth. A psychological foreward. In: Roszak T., Gomes M. & Kanner A. (eds.) (1995) Ecopsychology. Sierra Books, San Francisco CA: Xvii-xxiii.
Hoffmeyer J. (2008) An examination into the signs of life and the life of signs. University of Scranton Press, Scranton PA. Originally published in 2005.
Houser N. & Kloesel C. (1992) The essential Peirce. Indiana University Press, Bloomington IN.
Hume D. (2010) A treatise on human nature. Edited by David Fate Norton & Mary J. Norton. Oxford University Press, Oxford.
Humphreys N. (1992) A history of the mind. Chatto and Windus, London.
Husserl E. (1931) Ideas: General introduction to pure phenomenology. Translated by W. Boyce Gibson. MacMillan, New York. German original published in 1913.
Husserl E. (1960) Cartesian meditations: An introduction to phenomenology. Translated by Dorian Cairns. Martinus Nijhoff, The Hague.
Husserl E. (1982) Ideas pertaining to a pure phenomenology and to a phenomenological philosophy: First book, general introduction to a pure phenomenology. Kluwer, Boston MA. German original published in 1913.
Jackson F. (1982) Epiphenomenal qualia. Philosophical Quarterly 32: 127–136.
Jackson F. (1986) What Mary didn't know. The Journal of Philosophy 83(5): 291–295.
James W. (1890) The principles of psychology. Henry Holt, New York.
Kirchhoff M. D. & Hutto D. D. (2016) Never mind the gap. Constructivist Foundations 11(2): 346–353. http://constructivist.info/11/2/346
Kordeš U. (2016) Going beyond theory. Constructivist Foundations 11(2): 375–385. http://constructivist.info/11/2/375
Kriegel U. (2009) Subjective consciousness. Oxford University Press, Oxford.
Levine J. (1983) Materialism and qualia: The explanatory gap. Pacific Philosophical Quarterly 64: 354–361.
Levine J. (2001) Purple haze: The puzzle of consciousness. Oxford University Press, Oxford.
Locke J. (1975) An essay concerning human understanding. Clarendon Press, Oxford. Originally published in 1690.
Luhmann N. (1990) Essays on self-reference. Columbia University Press, New York.
Luhmann N. (1995) The paradoxy of observing systems. Cultural Critique 31: 37–55. http://cepa.info/2707
Luhmann N. (1997) Die Gesellschaft der Gesellschaft. Suhrkamp, Frankfurt. English translation: Luhmann N. (2012/2013) Theory of society. Stanford University Press, Stanford.
Luhmann N. (2000) The reality of the mass media. Translated by Kathleen Cross. Stanford University Press, Stanford. German original published in 1996.
Lähteenmäki V. (2007) Orders of consciousness and forms of reflexivity in Descartes. In: Heinämaa S., Lähteenmäki V. & Remes P. (eds.) Consciousness: From perception to reflection in the history of philosophy. Springer, Dordrecht: 177–201.
Maturana H. R. (1980) Biology of cognition. In: Maturana H. R. & Varela F. J. (eds.) Autopoiesis and cognition. Reidel, Dordrecht: 5–58. Originally published in 1970. http://cepa.info/535
Maturana H. R. & Varela F. J. (1980) Autopoiesis: The organization of the living. In: Maturana H. R. & Varela F. J. (eds.) Autopoiesis and cognition. Reidel, Dordrecht: 73–134. Originally published in 1974. http://cepa.info/552
McCullagh M. (2000) Functionalism and self-consciousness. Mind and Language,15(5): 481–499.

McGinn C. (1989) Can we solve the mind–body problem? Mind 98: 349–366.
McGinn C. (1991) The problem of consciousness. Blackwell, Oxford.
Mead M. (1968) Cybernetics of cybernetics. In: Foerster H. P. H., White J. & Russell J. (eds.) Purposive systems. Spartan Books, New York: 1–11. http://cepa.info/2634
Merleau-Ponty M. (1962) Phenomenology of perception. Translated by Colin Smith. Routledge and Kegan Paul, London.
Merleau-Ponty M. (1968) The visible and the invisible. Edited by C. Lefort and translated by A. Lingis. Northwestern University Press, Evanston.
Metzinger T. (2003) Being no one: The self-model theory of subjectivity. MIT Press, Cmbridge MA.
Miller J.-A. (1966) La suture. Cahiers pour l'Analyse 1: 37–49.
Nagel T. (1974) What is it like to be a bat? Philosophical Review 83: 435–456. http://cepa.info/2399
Nagel T. (1986) The view from nowhere. Oxford University Press, New York.
Nagel T. (2012) Mind and cosmos: Why the materialist neo-Darwinian conception of nature is almost certainly false. Oxford University Press, Oxford.
Natorp P. (1912) Allgemeine Psychologie nach kritischer Methode. J. C. B. Mohr, Tübingen.
Nelkin N. (1989) Unconscious sensations. Philosophical Psychology 2: 129–141.
Noë A. (2004) Action in perception. MIT Press, New York.
O'Regan J. K. (2010) Explaining what people say about sensory qualia. In: Gangopadhyay N., Madary M. & Spicer F. (eds.) Perception, action and consciousness: Sensorimotor dynamics and two visual systems. Oxford University Press, Oxford: 31–50.
O'Regan J. K., Myin E. & Noë A. (2005) Skill, corporality and alerting capacity in an account of sensory consciousness. Progress in Brain Research 150: 55–68.
Papineau D. (2002) Thinking about consciousness. Oxford University Press, Oxford.
Pavlov-Pinus K. (2015) Human knowledge and "as-if" knowledge of ideal observers. Constructivist Foundations 10(2): 239–240. http://constructivist.info/10/2/239
Penrose R. (1989) The emperor's new mind: Computers, minds and the laws of physics. Oxford University Press, Oxford.
Penrose R. (1994) Shadows of the mind. Oxford University Press, Oxford.
Piaget J. (1954) The construction of reality in the child. Basic Books, New York. French original published in 1950.
Piccinini G. (2004) Functionalism, computationalism, and mental states. Studies in the History and Philosophy of Science 35: 811–833.
Pierce B. (2012) Is the function of consciousness to act as an interface? In: Paglieri F. (ed.) Consciousness in interaction. John Benjamins, Amsterdam: 73–88.
Pinker S. (2007) The stuff of thought: Language as a window into human nature. Penguin Books, London.
Place U. T. (1956) Is consciousness a brain process? British Journal of Psychology 47(1): 44–50.
Rickert H. (1962) Science and history: A critique of positivist epistemology. Van Nostrand, Princeton. German original published as: Rickert H. (1926) Kulturwissenschaft und Naturwissenschaft. 6th and 7th expanded editions, Mohr Siebeck, Tübingen.
Ricoeur P. (1965) De l'interpretation. Ed. du Seuil, Paris.
Rockmore T. (2005) On constructivist epistemology. Rowman & Littlefield, Oxford.
Saussure F. de (1983) Course in general linguistics. Translated by Roy Harris. Duckworth, London. French original published in 1916.
Schrödinger E. (1944) What is life? Cambridge University Press, Cambridge.
Schrödinger E. (1958) Mind and matter. Cambridge University Press, Cambridge.
Scott B. (1996) Second-order cybernetics as cognitive methodology. Systems Research 13(3): 393–406. http://cepa.info/1810
Searle J. (1979) Expression and meaning: Studies in the theory of speech acts. Cambridge University Press, Cambridge.

Searle J. (1980) Minds, brains and programs. Behavioral and Brain Sciences 3(3): 417–424.
Shagrir O. (2005) The rise and fall of computational functionalism. In: Ben-Menahem Y. (ed.) Hilary Putnam. Cambridge University Press, Cambridge: 220–250.
Smolin L. (1997) The life of the cosmos. Oxford University Press, Oxford.
Spencer Brown G. (1969) Laws of form. Allen & Unwin, London. http://cepa.info/2382
Strawson G. (2016) Consciousness isn't a mystery. It's matter. New York Times 16 May 2016. http://www.nytimes.com/2016/05/16/opinion/consciousness-isnt-a-mystery-its-matter.html
Thompson E. (2007) Mind in nature: Biology, phenomenology, and the science of mind. The Belknap Press of Harvard University Press, Cambridge MA.
Thompson E., Lutz A. & Cosmelli D. (2005) Neurophenomenology: An introduction for neurophilosophers. In: Brook A., Akins K. & Brook A. K. (eds.) Cognition and the brain: The philosophy and neuroscience movement. Cambridge University Press, New York: 40–97. http://cepa.info/2374
Tononi G. (2012) Phi: A voyage from the brain to the soul. Pantheon Books, New York.
Tschacher W. & Scheier C. (2001) Embodied cognitive science: Concepts, methods and implications for psychology. In: Matthies M., Malchow H. & Kriz J. (eds.) Integrative systems approaches to natural and social dynamics. Springer, Berlin: 551–567.
Tye M. & Wright B. (2011) Is there a phenomenology of thought? In: Bayne T. & Montague M. (eds.) Cognitive phenomenology. Oxford University Press, Oxford: 326–344.
Van Gulick R. (1985) Physicalism and the subjectivity of the mental. Philosophical Topics 13: 51–70.
Van Gulick R. (2000) Inward and upward: Reflection, introspection and self-awareness. Philosophical Topics 28: 275–305.
Van Gulick R. (2014) Consciousness. In: Zalta E. N. (ed.) The Stanford encyclopedia of philosophy (Spring 2014 Edition). http://plato.stanford.edu/archives/spr2014/entries/consciousness
Varela F. J. (1979) Principles of biological autonomy. Elsevier, Amsterdam.
Varela F. J. (1994) A cognitive view of the immune system. World Futures 42(1): 31–40.
Varela F. J. (1996) Neurophenomenology: A methodological remedy for the hard problem. Journal of Consciousness Studies 3(4): 330–349. http://cepa.info/1893
Varela F. J., Thompson E. & Rosch E. (1991) The embodied mind: Cognitive science and human experience. MIT Press, Cambridge MA.
Velmans M. (1990) Consciousness, brain and the physical world. Philosophical Psychology 3(1): 77–99.
Vermersch P. (2009) Describing the practice of introspection. Journal of Consciousness Studies 16 (10–12): 20–57. http://cepa.info/2416
Vermersch P. (2016) Notes on the coupling between the observer and the observed in psycho-phenomenology. Constructivist Foundations 11(2): 391–393. http://constructivist.info/11/2/391
Weber A. & Varela F. J. (2002) Life after Kant: Natural purposes and the autopoietic foundations of biological individuality. Phenomenology and the Cognitive Sciences 1: 97–125. http://cepa.info/2087
Whitehead A. N. (1929) Process and reality. An essay in cosmology. Cambridge University Press, Cambridge.
Windelband W. (1915) Geschichte und Naturwissenschaft. In: Windelband W., Präludien. Aufsätze und Reden zur Philosophie und ihrer Geschichte. Volume 2. Fifth expanded edition. J. C. M. Mohr (Paul Siebeck), Tübingen: 136–160. Originally published in 1894. http://digi.ub.uni-heidelberg.de/diglit/windelband1915ga
Wittgenstein L. (1965) A lecture on ethics. The Philosophical Review 74: 3–12.
Wrathall M. & Kelly S. (1996) Existential phenomenology and cognitive science. The Electronic Journal of Analytic Philosophy 4. http://ejap.louisiana.edu/EJAP/1996.spring/wrathall.kelly.1996.spring.html

Design Research as a Variety of Second-Order Cybernetic Practice

Ben Sweeting

Introduction

1. In recent years there has been a resurgence of interest in cybernetics amongst designers. This has been prompted in part by the increased availability and affordability of technologies with which to augment the environments we design, and those we design in, which has fuelled interest in ideas regarding interactivity. While this technological focus is an important aspect of what cybernetics offers design, the relations between the two fields run much deeper. These connections have been explored explicitly in the work of Ranulph Glanville (1999, 2006a, 2006b, 2007a, 2007b, 2007c, 2009a, 2011a, 2014b, 2014c), whose work I use as a point of departure in this article.[1]

2. Drawing on Gordon Pask's (1976) conversation theory and the common characterisation of design in terms of conversation (such as by Schön 1991), Glanville (2007c, 2009c) has suggested a close analogy between cybernetics and design, understanding both as "essentially constructivist" activities (Glanville 2006a: 63; 2013). The parallels Glanville draws are significant enough for him to claim that "cybernetics is the theory of design and design is the action of cybernetics" (Glanville 2007c: 1178).

3. While part of Glanville's motivation in developing the connection between cybernetics and design has been the insight that the former might bring to the latter, it is an important aspect of his position that the converse is also the case: that design can set an example to cybernetics in terms of practice and so inform it, not just vice versa. Thus the relationship between cybernetics and design is to be understood as one of mutual overlap and support and, as such, one that avoids the difficulties that can follow from the application to design of theories external to it (a problem that seems to recur in architecture in

1. Together with Neil Spiller, Glanville supervised my PhD research, and although this article has been developed after his passing, it is significantly influenced by my conversations with him. In addition to his work, on the relationship between cybernetics and design see also: Dubberly & Pangaro (2007, 2015); Fischer (2015); Fischer & Richards (2015); Furtado Cardoso Lopes (2008, 2009, 2010); Gage (2006, 2007a, 2007b); Goodbun (2011); Herr (2015b); Jonas (2007a, 2007b, 2012, 2014, 2015a, 2015b); Jones (2014); Krippendorff (2007); Krueger (2007); Lautenschlaeger & Pratschke (2011); Lobsinger (2000); Mathews (2005, 2006, 2007); Pratschke (2007); Ramsgard Thomsen (2007); Rawes (2007); Spiller (2002); Sweeting (2014, 2015c).

particular) and the more general shortcomings that can follow from our tendency to see the relation of theory and practice as predominantly the application of the former to the latter (Glanville 2004a, 2014a, 2015; see also Sweeting 2015c).

4. More specifically, Glanville's understanding of design as being the action of cybernetics is part of his characterisation of second-order cybernetics (SOC) as being concerned with how cybernetics is to be practiced rather than, as can tend to be the case, a theoretical reflection on this (Glanville 2011b; Sweeting 2015b). This concern was particularly evident during his time as President of the American Society for Cybernetics (ASC), during which he often referred to Margaret Mead's (1968) challenge, delivered in her address to the inaugural ASC conference, to practice cybernetics in line with its own ideas. While the principal legacy of Mead's remarks has been the epistemological concerns of SOC, as developed by Heinz von Foerster (1995, 2003a) and others, their original context is that of the practice of the society itself. It is this aspect to which the ASC returned during Glanville's presidency, in terms of both the form and content of its conferences, which explored cybernetics' relation to practice using conversational, cybernetic, formats (Baron et al. 2015; Glanville 2011b, 2012; Glanville, Griffiths & Baron 2014; Glanville & Sweeting 2011; van Ditmar & Glanville 2013).[2]

5. In contrast to this understanding of its relation to practice, Andrew Pickering (2010: 25f) has characterised SOC as a turn away from the more tangible modes of experimentation in earlier phases of cybernetics, and towards the linguistic. This view can be countered: SOC is a reflection on the performative involvement of observers within their observations, in contrast to the separation of observer and observed in conventional science. This is very much in line with Pickering's own emphasis, for example in his comments on R. D. Laing's psychiatry as taking seriously "the idea that we are all adaptive systems, psychiatrists and schizophrenics alike" (ibid: 8) or his reference to Pask's account of the "participant observer," who tries to maximise interaction with what he or she observes in order to explore it (ibid: 343f).

6. However, even its advocates must admit that SOC can run the risk of becoming overly introverted, especially given its central concern with self-reference. Recent thinking regarding von Foerster's development of SOC has addressed this concern by understanding it as the beginnings of a research programme rather than as primarily a form of worldview, and as prompting the "new course of action" suggested in this journal under the heading of "second-order science" (SOS) (Müller & Müller 2007; Müller 2008, 2011; Riegler & Müller 2014).[3] In this light I suggest that Glanville's understanding of design, and particularly his (1999, 2014c) account of the relation between design and science that I discuss below, allows us to view the currently expanding field of design research as a contemporary variety of SOC practice, whether SOC is explicitly invoked or not.[4] My purpose

2. See also the recent special issue of *Constructivist Foundations* on alternative conference formats, which was inspired in part by these ASC conferences (Hohl & Sweeting 2015; regarding the ASC conferences, see especially: Richards 2015; Sweeting & Hohl 2015).
3. See also http://www.secondorderscience.org
4. Given that cybernetics stresses the interdependency between acting and understanding, and so between theory and practice (see e.g., Glanville 2014a; Sweeting 2015c), I could equally refer to design research as a contemporary variety of second-order cybernetics as to one of second-order cybernetic practice. Nevertheless, I feel it is important to stress the practical here, given that SOC, and constructivism generally, currently risk being seen more as a worldview than an active research tradition.

in doing so here is not primarily to add to what SOC can bring to design research, which has been explored in depth elsewhere by many others. Rather, my focus is on what design can bring to cybernetics, in line with what I have understood as being part of Glanville's own motivations for developing this analogy, as noted above. Design research offers an example of how SOC can develop as a practice-based and outward looking enquiry, while also suggesting a way of integrating the legacy of tangible experimentation from earlier cybernetics with its contemporary concerns.

Method and practice in design and research

7. During the period of scientific and technological optimism that followed the Second World War, there was a tendency, as evident in what is usually referred to as the design methods movement, to see design as something that should be put on rational scientific foundations.[5] Since around 1980 this view has been countered by arguments that have seen design as a discipline in its own right and so as being of the same status as science rather than something to be corrected by it. Amongst these, the account that Glanville (2014c) presented at the 1980 *Design: Science: Method* conference, later expanded as the journal article *Researching Design and Designing Research* (1999), is particularly strongly framed, reversing what had been the more usual hierarchy. Rather than seeing design research as one specific form of scientific research, Glanville argues that, instead, we can see science as a specific form of design enquiry. This follows from the way that scientific research inevitably involves design activity, for instance in devising and setting up experiments, but not vice versa.[6] Design is, it follows, the more general case and, therefore, "it is inappropriate to require design to be "scientific": for scientific research is a subset (a restricted form) of design, and we do not generally require the set of a subset to act as the sub subset to that subset any more than we require [that] the basement of [a] building is its attic" (Glanville 1999: 87f).

8. This argument and others like it around that time, such as those put forward by Bruce Archer (1979), Nigel Cross (1982) and Donald Schön (1991), consolidate a shift during the 1970s from trying to base design on the scientific method to the idea that it has its own epistemological foundations, independent of science (for an overview of this shift, see Cross 2007b). The attempt to order design according to a linear version of the scientific method, understood as moving from analysing the problem at hand to testing and optimising solutions to it, failed for reasons that seem obvious in retrospect: because design involves the creation of new situations, design questions cannot be fully formulated in advance but shift and change as they are explored and as proposals are enacted. One of the most important accounts of these limitations is that developed by design theorist Horst Rittel, who, writing with the urban designer Melvin Webber, characterised the situations that designers encounter as "wicked problems," the complex interdependencies of which make them unsolvable using conventional linear problem solving (Rittel 1972; Rittel & Webber 1973, 1984).

5. See for instance: Alexander (1964); Broadbent & Ward (1969); Simon (1996). For a critical discussion of the design methods movement, see Gedenryd (1998).
6. This is not to say that designers do not make use of scientific research but that doing so is not essential to what design is, whereas design is a core aspect of research and so science.

9. On the face of it, Rittel and Webber's observations mark an incompatibility between design and science in terms of method. Indeed the exhaustion of the design methods movement by the 1970s – with leading figures such as Christopher Alexander (1984), John Christopher Jones (1984) and, indeed, Rittel distancing themselves from it – along with the unravelling of modernism more generally during that decade, marks something of a parting of the ways between design and science (architecture, for instance, would increasingly turn towards history and philosophy, rather than science, for theoretical support). However, given Glanville's SOC-inspired argument noted above, this separation between design and science is not what we might expect. If science is a limited form of design, then is it not the case that scientific approaches should be commensurable with design, even if not a basis for it? This apparent disjunction is only the case if we follow the changes in how design was thought about during this period without also following the comparable changes regarding science.

10. Design research and the philosophy of science broadly parallel each other over this period. Both move from a concern with method in the 1960s through a critique of this in the 1970s to new foundations from the 1980s onwards, focusing on what designers and scientists actually do in practice rather than on what seems ideal in theory. As noted above, this led to design being seen as a discipline in its own right (Archer 1979), with its own "designerly ways of knowing" (Cross 1982) and a refocusing from methodology to broader and more practice-based concerns, under the heading of design research (for an overview, see for instance: Grand & Jonas 2012; Michel 2007; Rogers & Yee 2015). In the context of science, there was a comparable turn during the 1970s and 1980s towards understanding it in terms of the social and material agency of research as practiced, with the growth of the fields of the sociology of scientific knowledge (SSK) and science and technology studies (STS), such as in the work of Karin Knorr Cetina, David Gooding, Bruno Latour and Pickering amongst others (for an overview see Pickering 1992). These accounts are suggestive of a more designerly paradigm in science, in line with Glanville's argument. Indeed, accounts of experimentation in SSK/STS can be read almost as if describing the activities of a design studio; see for instance: Gooding (1992), Pickering (1993, 1995) and Knorr Cetina (1992), who even uses a direct analogy with architecture.

11. In this light, what appears to be a rupture between design and science during the 1970s is instead a close parallel. Indeed, key critiques advanced in each area – that of Rittel in design, and that of Paul Feyerabend (1970, 1982, 1993) in science – have similar content. Rittel and Feyerabend were colleagues at UC Berkeley while they were developing their ideas. Both were influenced by thinking in cybernetics and systems at that time. Rittel worked with Ross Ashby at the Ulm School of Design (Fischer & Richards 2015), while Feyerabend (1982: 64) refers to "new developments in systems theory," which was flourishing at Berkeley (which was also home to C. West Churchman) and elsewhere in California at the time (where Gregory Bateson, amongst others, was based), and his (1982: 18) comments regarding participant observers reflect contemporaneous preoccupations of SOC.

12. Science, like design, involves creating new ideas and understanding; therefore, as in design, the criteria and methods that are appropriate will change as part of the process and cannot be defined in advance if science is to progress:

> …to ask how one will judge and choose in as yet unknown surroundings makes as much sense as to ask what measuring instruments one will use on an as yet unknown planet. Standards which are intellectual instruments often have to be *invented*, to make sense of

> new historical situations just as measuring instruments have constantly to be invented to make sense of new physical situations. (Feyerabend 1982: 29)

13. Feyerabend's (1970, 1993) *reductio ad absurdum* argument against the predefined methods that were characteristic of the philosophy of science at the time concludes by showing that the only criteria that can be given in advance, that will not inhibit scientific progress, is that "anything goes." This also appears in Rittel and Webber (1973: 164), while Rittel (1972: 393) has "everything goes": because designers inevitably encounter new and ambiguously defined situations (it being the purpose of design to create the new), they have no well-defined problems to solve or enumerable lists of options to pick from and "any new idea for a planning measure may become a serious candidate" (Rittel & Webber 1973: 164). This phrase is also anticipated by theatre director Joan Littlewood (1964: 432) in describing the Fun Palace project, on which cybernetician Pask was a key collaborator along with architect Cedric Price (see e.g., Lobsinger 2000; Mathews 2005, 2006, 2007; Spiller 2006: 48–50), and that is equally concerned with the in-principle unpredictable. Furthermore, Feyerabend's (1982: 202) comment that the proponents of scientific theory are out of touch with scientific practice echoes the situation in design, where design methods had become an academic game divorced from practice, as both Alexander (1984: 309) and Jones (1984: 26) point out.[7]

14. While Archer (1979) differentiated design as a third disciplinary pole with the same status as the traditional "two cultures" of the arts and sciences (Snow 1961), Glanville (2014c) argues against this separation and, instead, characterises all research as being a design-like activity. This designerly continuity across different fields is, however, obscured by popular misrepresentations of science as a logical and predictable activity, such as are perpetuated in the structure of traditional scientific papers, which Peter Medawar (1996) has critiqued as a fraudulent account of what scientists actually do in practice. Glanville (2014c: 111) calls for honesty about how research is practiced in all disciplines, and suggests that this will make similarities clear between apparently quite different fields. In this, Glanville reflects a willingness to transcend disciplinary boundaries that is characteristic of cybernetics' origins, which had cut across distinctions between research fields as well as those between objectivity and subjectivity, human and machine, and mind and body.

15. In stressing the continuities between design and other disciplines, Glanville (2014c) contrasts his account with that of Archer (1979), whose positioning of design in terms of its own disciplinary pole, separate to the arts and the sciences, risks isolating it from other research traditions. Glanville's understanding, however, still gives design research the special status of Archer's account: given the parallels he draws between design and research, Glanville recognises design research as a self-reflexive activity of *researching research* (Glanville 2014c: 116–119). That is, as design is a core part of research activ-

7. As one of the anonymous reviewers of this article suggests, the shift away from science in design can be thought about in terms of a search for forms of rigor that make sense in a design context, such as for example those described by Schön (1988). Feyerabend's (1982, 1993) argument, however, indicates that the scientific method as it had been promoted was unsatisfactory not just in making sense of design but also in accounting for scientific practice itself (Feyerabend demonstrates that examples commonly regarded as paradigmatic by the advocates of method violate the methodological principles they propose). That is, in this period, the need for an understanding of rigor that makes sense in the context of practice is a feature not just of design but also science. Thanks to the anonymous reviewer for prompting my thoughts on this point.

ity, to research design is to inquire into an aspect of research activity itself. In so doing, Glanville anticipates recent discussions in *Constructivist Foundations* regarding second-order science (SOS) as research activity focused on research itself (Müller & Riegler 2014a, 2014b; Riegler & Müller 2014). I return to this below.

Design research and second-order cybernetics

16. SOC was developed in the context of the shifts in understanding science and design that I have summarised above, and parallels these concerns. As such, SOC sits in a pivotal position within cybernetics' wider history. In consolidating its epistemology and, with it, an ability to address rigorously the issues of self-reference towards which a field concerned with circularity is inevitably drawn, it is with SOC that cybernetics reaches maturity as a discipline. That this happens simultaneously with the fragmentation of the field during the 1970s – under pressure from changes in the external funding climate and professional accreditation (Umpleby 2003; Umpleby & Dent 1999) – has consequences not just for the ideas of SOC but also for how we understand earlier, and other, aspects of cybernetics.

17. Firstly, as the earlier work occurred before the maturity of the field, it is inevitable that it contains inconsistencies in epistemology, approach and terminology. This is further complicated by the way that the fragmentation of the field is often associated with the emergence of critiques of science and technology during the 1970s, of which SOC is one instance. SOC has often, for this reason, been presented in contrast to first-order cybernetics (FOC), which tends to be associated with the earlier work. The "first" and "second" should not, however, be understood as implying a sequence or the surpassing of one by the other.[8] Rather, SOC is specifically the application of cybernetics to itself – "the cybernetics of cybernetics," as von Foerster (2003b: 302) titled Mead's (1968) paper.[9]

18. The terminology of "first" and "second" can obscure the continuity between SOC and earlier cybernetics. While Glanville has spent considerable effort in distinguishing the two (e.g., Glanville 1997, 2004c), he has also recognised that cybernetics always involves second-order considerations and did so even at its origins (Glanville 2013: 28) and that "it would be better, nowadays, to talk only of Cybernetics, without orders: thus bringing the different approaches into proximity" (Glanville 2002). This is certainly the case for Mead and Bateson, whose backgrounds in anthropology involved a consideration of the participation of observers in what they observe (see e.g., their discussion in Brand, Bateson & Mead 1976), and for Ashby in his (1991) understanding of the black box system as involving the observer as part of it. Even Norbert Wiener, according to Pask (as reported by Glanville 2002; 2013: 33), recognised that there were further steps to take in developing the

---

8. While Fischer and Richards (2015), rightly point out considerable overlaps between the development of SOC and Rittel's characterisation of first- and second-generation design methods, it should be remembered that "first" and "second" are used in different ways in each context.

9. This development was both necessary for the field to consolidate its own disciplinary foundations, and has also made possible innovatively reflexive research programmes that are of particular interest in exploring those questions regarding cognition, society, epistemology and ethics that inevitably involve self-reference. For a fuller discussion of SOC see e.g.: Glanville (1997, 2002, 2004c, 2011b, 2013); Müller & Müller (2007); Müller (2008, 2011); Scott (2003, 2004, 2011); Foerster (1995, 2003b); Foerster & Poerksen (2002).

subject. Indeed, Alvin Toffler's (1970) *Future Shock*, a book that is emblematic of the criticisms of science and technology that are often assumed to apply also to cybernetics (e.g., Lobsinger 2000: 134), is anticipated two decades earlier in the similar, cautionary account of technological change in Wiener's (1950) *Human Use of Human Beings*. In addition, while Pickering (2010) sees SOC as being in contrast to the tangible modes of exploration of the earlier cybernetics in which he is interested, the performative quality of the devices through which Pask, Ashby, Grey Walter and others explored their ideas is an example of the participation of observers in observation on which SOC reflects and places value.

19. Secondly, it is difficult to judge the consequences of SOC for practice, as the field within which these implications would have been explored had broken up by the time the possibility of doing so had emerged. The tendency of SOC to be largely theoretical in orientation – which leads Pickering (2010: 25f) to view it as a form of linguistic turn – needs to be understood in this context of a lack of opportunity for experimental work.

20. With the break up of cybernetics, many of its ideas were absorbed back into its constituent fields. Some research in other disciplines, such as for instance robotics or complexity, can be recognised as a continuation of its ideas and research programme, including its performative approach to experimentation (see, for instance, the discussion of Rodney Brooks, Stephen Wolfram and Stuart Kauffman in Pickering 2010: 60–64, 156–170). Given its continuities with cybernetics, as introduced in part above and discussed further below, the field of design research can be thought of, similarly, as one such successor field.[10]

21. There is a longstanding history of connection and influence between cybernetics and design, as has been summarised by Hugh Dubberly and Paul Pangaro (2015). In particular, Ashby and Pask both engaged directly with design.[11] Ashby lectured at the Ulm School of Design with Rittel (see Fischer & Richards 2015) and was also a significant influence on Alexander.[12] Pask, meanwhile, became increasingly involved in architecture from the 1960s onwards. He was a significant contributor to the prominent Fun Palace project with Price and Littlewood, and collaborated with Nicholas Negroponte at MIT, for whose *Soft Architecture Machines* (Negroponte 1975) he contributed a chapter. In addition he held a consultant position at the Architectural Association in London, wrote explicitly on architecture and design (Pask 1963, 1969) and influenced the development of interactive architecture through Negroponte and others such as John and Julia Frazer (Frazer 1993, 1995; Furtado Cardoso Lopes 2008, 2009; Spiller 2006: 204–210). More recently, figures such as Pangaro, Glanville and Klaus Krippendorff, influenced particularly by Ashby (Krippendorff) or Pask (Glanville, Pangaro), have made prominent contributions in both design research and cybernetics, while many others have worked in one field in a way informed by thought in the other.

---

10. By "successor field" I do not mean to imply any sense of superiority, but rather the inheritance of ideas.
11. Other figures could also be mentioned. Dubberly and Pangaro (2015) and Müller and Müller (2011) also stress the interest of Heinz von Foerster in design. He addressed design audiences (e.g., Foerster 1962) and was connected to figures such as architect Lebbeus Woods and Stuart Brand, who can be mentioned in his own terms as a cross-over figure. Fischer (2015) has suggested connections between Wiener and recent work in design, while the work of Bateson, who introduced Brand and von Foerster to each other, is a point of reference for contemporary discussions of architecture and ecology (see e.g., Goodbun 2011; Rawes 2013).
12. Although, as Upitis (2013: 504f) notes, Alexander's (1964) use of Ashby's ideas can be questioned.

22. As well as this continuity of people, there is a significant continuity of ideas and approach such that cybernetics can be thought of as design's "secret partner in research" (Glanville 1999: 90f). While this is not the place for a full discussion of these parallels – I defer here to the accounts of Glanville and the others who I have cited – key points include the following:

- There is a conversational, and so cybernetic, structure that is central to what is distinctive about the way designers work (see for instance Schön's (1991: 76) characterisation of design in terms of a "reflective conversation with the situation"). Glanville has developed this parallel to the extent that, as I have noted, he claims that "cybernetics is the theory of design and design is the action of cybernetics" (2007c: 1178) while it is also what lies behind his (1999, 2014c) characterisation of research in terms of design, as discussed above.
- Both design and cybernetics are concerned with the new, as supported by the tendency of conversation to involve invention at every turn. Both are "essentially constructivist" activities (Glanville 2006a: 63; see also: Glanville 2006b; 2013; Herr 2015b) that enable a form of "forward-looking search," as Pickering (2010: 18) has described cybernetics, developing new ideas and possibilities rather than looking to correspond to, or replicate, the real or the optimal.
- The way that designers use drawings and models *for* exploring ideas rather than as representations of them (Glanville 2009b) resonates closely with the performative nature of the work of Pask and others, who played out their ideas using physical, experimental devices in much the same way (as emphasised in Pickering's (2010) account). In contemporary practice-based design research, some work has strong continuity with the sorts of devices made in earlier cybernetics (e.g., that of Mette Ramsgard Thomsen (2007), Jennifer Kanary Nikolov(a)[13] or Ruairi Glynn[14]), but even the use of more analogue media (such as the sorts of pen drawings with which I work; see Sweeting 2014) has a similar attitude to modelling as part of thinking rather than as a representation of thought.
- Design research is often concerned with epistemological questions regarding the interrelations of designers, other stakeholders, working methods and the knowledge embedded in what is designed. This has often been articulated in terms of differences between research *about/into*, *through/by* and *for* design, following Frayling (1993) and others, and as reviewed and synthesised by Jonas (2012, 2015a, 2015b). These distinctions distinguish between that research which looks at design from the outside or which is applied to it, from that which is conducted as an integral part of it. This resonates strongly with SOC concerns regarding the participation of observers in their observations, and the active difference made by how this participation is configured. Jonas (2007b, 2012, 2015b) in particular has explicitly used the framework of cybernetics, drawing on Glanville (1997), to clarify these points. I return to this below.
- Design is a self-reflexive activity in much the same way as cybernetics, both involving circular reflective processes and being examples of disciplines that can be applied to themselves, in the design of design or the cybernetics of cybernetics.

---

13. http://www.labyrinthpsychotica.org/Labyrinth_Psychotica
14. http://www.ruairiglynn.co.uk

23. While design research and cybernetics mostly differ in their subject matter, the above parallels are significant. They share both ways of working – a conversational forward-looking search and an interactive, non-representational use of modelling – and also core concerns with observer positions and self-reflexivity in the constitution of their research processes. These parallels hold to the extent that, while design research continues to make reference *to* cybernetic ideas (for instance in exploring the possibilities of new technologies (e.g., Ramsgard Thomsen 2007; Spiller 2002), or in understanding the relationship between research and design (e.g., Jonas 2007b, 2012, 2014, 2015a, 2015b), I suggest we can also understand it as a contemporary variety *of* cybernetic research, whether the connections with cybernetics are made explicitly or not. Seeing design research as an example of SOC in this way suggests a continuity between the epistemological focus of SOC and the tangible experimentation of earlier cybernetics, a connection that can easily become obscured, as is evident in Pickering's (2010) account.

Second-order science

24. As well as helping integrate the more practice-oriented legacy of early cybernetics with SOC, design research can also provide an important point of reference for contemporary discussions of SOS, which have been a recent focus of *Constructivist Foundations* (and which have led to the present volume). Karl Müller and Alexander Riegler (2014a) proposed SOS as "a new course of action" in order to reinvigorate SOC – and constructivist approaches generally – as an active research field. They characterise SOS as a reflexive form of research, either in methodological terms through the inclusion of observers as participants (a direct continuation of von Foerster's (1995, 2003a) SOC as the "cybernetics of observing systems"), or through self-reflexive domains of research, in the sense of the science of science or, similarly, the cybernetics of cybernetics or the sociology of sociology, such as through meta-analyses of the products or practices of other scientific enquiry.

25. Müller and Riegler position SOS as a specific research agenda within the significant transformations currently underway in the landscape of science (Müller 2008, 2011; Müller & Riegler 2014b). These have partly been, as noted above, in terms of how science has come to be understood in terms of its practice by fields such as SSK and STS, but also through significant changes in this practice itself. This has included: a change of focus away from a mechanistic and reductionist paradigm (associated with Newton and Descartes) towards one based in complexity, adaptation and evolution, which Rogers Hollingsworth and Müller (2008) have labelled in terms of a transition from Science I to Science II; significant changes in the organisational structure of knowledge production, with an increased emphasis on its social robustness and the context of application, which has been labelled as a shift from Mode 1 to Mode 2 (see Nowotny, Scott & Gibbons 2006); and growing interest in transformative and transdisciplinary aspects of research (e.g., Nicolescu 2012; Schneidewind & Augenstein 2012).

26. These various changes in science have all had the effect of science moving towards a more designerly paradigm, in line with Glanville's (2014c) argument discussed above (as noted by Jonas 2014, 2015a). Given this convergence and the historical and conceptual connections that I reviewed above, there is reason to consider SOS as a potential point of interchange between design and science. This is especially so given that there is a considerable overlap between core interests of design research and the two "motivations"

for SOS that Müller and Riegler (2014a: 2f) have put forward: self-reflexivity, and the inclusion of observers.

27. Firstly, self-reflexivity is important in design research in various ways. In a general sense, designers often do this implicitly as they work, reflectively redesigning their design processes to suit the specifics of the situations they encounter. More explicitly, design is a field that, like cybernetics, can be applied to itself in the sense of the design of design. This includes such instances as: the design of particular design methods (e.g., Alexander 1964) or of technologies with which to design (e.g., Frazer 1995; Negroponte 1975; or contemporary developments such as building information modelling); the way that a design research conference is something that itself needs to be designed (Durrant et al. 2015; Sweeting & Hohl 2015); and the way that the products of design can allow for a continuation of the design process in them, such as in the architecture of Price (as Price 2003: 136 himself remarks).

28. Specific design projects can also explore aspects of design itself, as for instance in Peter Downton's (2004) practice-based reflections on epistemology, or the work of Peter Eisenman (Bédard 1994). Indeed, Eisenman's Cannereggio project, for instance, can be considered a meta-analysis in Müller and Riegler's (2014b) sense for the way it takes Le Corbusier's unbuilt Venice Hospital scheme for the same site as its starting point.

29. Most significantly for SOS, understanding design as a core part of research, as per Glanville's (2014c) account discussed above, positions design research as a field of researching research. This observation holds possibilities yet to be fully explored, offering design research a field of application in science rather than vice versa, as is more often the case.

30. Secondly, as noted above, the position of the observer has been a theme of particular importance in design research as part of the field's shift from its mostly professional origins to being seen in more academic terms. This has included careful delineations between ways in which designers and others observe and participate in design, and of the ways in which material artefacts operate variously as part of the research process, as the object of enquiry, as output or dissemination and sometimes as more than one of these depending on their context. As noted above, one important and widespread way in which these distinctions have been made is by distinguishing in terms of research *about/into*, *for* and *through/by* design. As Jonas (2012: 34) discusses, the value of this sort of categorisation is that it differentiates on the basis of the attitudes and intentions of designers, rather than in terms of subject matter (which would not make sense in design because of its tendency towards diverse and ambiguously delineated content). This has helped clarify where design is used actively as a research process to explore a topic (*through/by*), where separate research is applied in design, such as in research and development or market research (*for*) and where design is the object of separate study by another discipline, such as history or sociology (*about/into*). In elaborating on and clarifying these distinctions, which were initially rather ambiguous, Jonas has drawn on Glanville's (1997) description of different observer positions and orientations as a foundation, associating research *through* with the engaged SOC observer, and *for* and *about* with the detached observer of FOC. Jonas distinguishes a new category of research *as* design to correspond to where, in Glanville's scheme, the observer is inside the inquiring system and looking inwards, and interprets this in terms of "design

as the inaccessible medium of knowledge production" and the role of abductive reasoning (Jonas 2015b: 35).[15]

31. Categorisations of this sort are very much in the spirit of SOC and are highly relevant for SOS; and we can think of research *for*, *about/into*, *through/by* and *as* in this context in much the same way as in design. It is the observer-included modes of research *through/by* and *as* that are of most relevance (these being associated with SOC). Examples include Glanville's approach to conference design in terms of using cybernetic processes (so the content of the conference can be acted out in its form; Glanville 2011b; Sweeting & Hohl 2015) and the performative aspects of the devices of Pask and others, as stressed by Pickering (2010). The more detached modes of research *about/into* or *for* also have their counterparts, and would include historical and theoretical work, including this present article and also accounts such as that of Pickering and others to which I have referred.[16]

32. While Jonas has used the terminology of FOC and SOC to give a foundation to these designerly categories, in turn they offer complementary possibilities back to cybernetics. Whereas the phrasing of FOC and SOC invites a sharp distinction in terms of whether the observer is included or not, and can be confusingly interpreted in terms of a chronological sequence as discussed above, the categories of *for*, *about/into*, *through/by* and *as* distinguish something of the nature of an observer's involvement, not just the acknowledgement of it, enabling these different observer positions to be seen in productive combination. This latter point is important for SOS, especially where it is conceived in terms of reflexive operations such as meta-analyses, as it requires a close relationship to the more conventional first-order science on which it is to operate (Müller & Riegler 2014b).

33. Given these significant overlaps, design research is a productive point of comparison for SOS. In particular, it suggests a possible example for how SOS can be constituted as a research field that is practice based and outward looking, both aspects that are important in this "new course of action" (Müller & Riegler 2014a). This is partly through the connections between SOC and earlier, more tangible, forms of cybernetics that are suggested by design research, and also through examples of research through design, which is notable for the way that even some of its most abstracted and introverted moments retain rich potential for concrete connections with the world.

15. Given Glanville's (1997) enigmatic silence regarding this category, it makes sense to associate it with the role of tacit knowledge in design, especially when seen in the context of Jonas's (2015a, 2015b) presentation of these categories in terms of their relations with each other. Locating the tacit here can help clarify the relation between the research involved in any design act and research through design, which is in need of more explicit articulation, even if this could still be through various media or embedded in artefacts.

16. Note that to write *about* SOC is a first-order activity. This is why neither von Foerster (2003b: 301) nor Glanville (2002) see the need for any third or fourth orders of cybernetics; these would simply be instances of its first or second orders.

Conclusion

34. I have drawn on the continuities, both of concepts and participants, between SOC and the field of design research in order to position SOC in terms of practice rather than as a mainly theoretical perspective. I have drawn, in particular, on Glanville's (2014c) account of scientific research as a form of design activity, understanding this in the context of the shifting relationship between design and science during the formative period of both SOC and design research, and since.

35. I have suggested that design research is not just a field that is influenced by SOC but a contemporary variety of it, whether this connection is made explicitly or not, in a similar way that other fields can be regarded as continuing or reinventing cybernetic concerns. Understanding design research in this way suggests a continuity between the epistemological concerns of SOC and the material experimentations of earlier cybernetics, in contrast to the way that SOC is sometimes regarded as a turn away from these more tangible qualities.

36. These connections with cybernetics' past are also relevant to contemporary discussions of SOS. Given that design research shares the central concerns of SOS with both self-reflexivity and the inclusion of observers as active participants, it is suggestive of ways in which SOS may develop as a field of research.

Acknowledgements

This article was developed from a presentation given at the 2015 conference of the International Society for Systems Science in Berlin (Sweeting 2015a) in a session chaired by Peter Jones, whom I thank for his comments and encouragement. Thanks also to the American Society for Cybernetics, who funded my attendance at that conference as part of the 2014 Heinz von Foerster Award. The ideas presented here have been influenced by discussions with Nick Beech, Murray Fraser, Tim Ivison and Simon Sadler at the Canadian Centre for Architecture in Montreal as part of an ongoing collaborative research project funded by the Andrew W. Mellon Foundation. Thanks also to Tanya Southcott, Tilo Amhoff, Stuart Umpleby, the editors and the anonymous reviewers for their comments and assistance.

Open Peer Commentary: Design Cycles: Conversing with Lawrence Halprin

Tom Scholte

1. Ben Sweeting's target article provides an informative outline of the conceptual confluences of design research and second-order cybernetics, explicit and otherwise, as they have unfolded over the last several decades. A practitioner absent from Sweeting's summary (and from any other cybernetic overview of design practice of which I am aware) whose work might be fruitfully included in this analysis is environmental architect Lawrence Halprin. This OPC will endeavor to provide a brief sketch of the second-order cybernetic features of Halprin's RSVP cycles in the hopes that they may find their way into the ongoing discourse on the cybernetics of design that Sweeting has framed.

2. In the 1960s, while second-order cybernetics was incubating in Heinz von Foerster's Biological Computer Laboratory, architect and environmental designer, Halprin, in collaboration with his wife Ann, choreographer and artistic director of the San Francisco Dance Workshop, were engaged in their own inward examination of group creative processes in search of a theory outlining their main features. Similar to Ranluph Glanville, Halprin explicitly rejected "the attempt to make a science out of community design" claiming that...

> [h]uman community planning cannot ever be a science anymore than politics can rightly be called political science. Science implies codification of knowledge and a drive toward perfectibility none of which are possible or even desirable in human affairs. (Halprin 1969: 4)

What Halprin did desire was a "means to describe and evoke (creative) processes on *other* than simply a random basis" in the hopes that it "would have meaning not only for (the) field of environmental arts and dance-theatre, but also for all the other arts where the elements of time and activity (particularly of numbers of people) would have meaning and usefulness" (ibid: 1). It may be argued that, in his own way, Halprin may also have been looking for something one might call "rigour," but not as a means of justifying design's place in the academy on intellectual grounds. He simply wanted to help people work more efficiently on a purely pragmatic level and, at the same time, avoid the undesirable outcomes of a narrowly linear, dare call it "scientific," approach to the transcomputable complexities inherent in any and all design processes. He formalized his findings in the 1969 book *The RSVP Cycles: Creative Processes in the Human Environment*, describing a recursive schema of iteration and evaluation bearing striking resemblances to the conversational conception of second-order cybernetics.

3. Below are the four components of the RSVP cycles as defined in Halprin's book (ibid: 2):

> *R – Resources* which are what you have to work with. These include human and physical resources *and* their motivation and aims.
>
> *S – Scores* which describe the process leading to the performance.

> *V – Valuaction* which analyzes the results of action and possible selectivity and decisions. The term 'valuaction' is one coined to suggest the action-oriented as well as the decision-oriented aspects of V in the cycle.
>
> *P – Performance* which is the resultant of scores and is the 'style' of the process.

4. While the arrangement of the acronym RSVP (the request for a response) was chosen for its elegance in naming an essentially conversational process (ibid: 2), a typical iteration of the cycle would more accurately be expressed as RSPV: the articulation of an inventory of the resources available, and desirable, for inclusion in the project, the articulation of a score indicating what is to be done with/to the resources, the performance (implementation) of the score, and a period of valuaction during which the results of the performance are evaluated and re-enter the next iteration of the cycle as new resources, for which a new score will be articulated.

5. From a second-order cybernetic perspective, it is significant that the "motivations and aims" of all of the individuals involved in the project must also be articulated and taken into account in addition to the purely physical or financial resources at play. This is, in fact, the ethical foundation of the entire schema, as "its purpose is to make procedures and processes visible, to allow for constant communication and ultimately to insure the diversity and pluralism necessary for change and growth" (ibid: 5). This ethical foundation seems entirely commensurate with the "desirable ethics" of Glanville (2004b).

6. Halprin opens his book with a definition of scores:

> Scores are *symbolizations of processes* which extend over time. The most familiar kind of 'score' is a musical one, but I have extended this meaning to include 'scores' in all fields of human endeavor. Even a grocery list or a calendar, for example, are scores. (Halprin 1969: 1)
>
> The essential quality of a score is that it is a system of symbols which can convey, or guide, or control (as you wish), the interactions between elements such as space, time, rhythm, and sequences, people and their activities and the combinations which result from them. (ibid: 7)

7. Halprin goes on to expand his list of sample scores to include plans for buildings, mathematics, stage directions and dialogue for a play, Navajo sand paintings, the intricacies of urban street systems as well as plans for transportation systems and the configurations of regions, and much more. The most significant feature of any score is its position on a spectrum from "open" to "closed" in terms of the amount of control it exerts.

> The real nub of the issue [...] is what you control through the score and what you leave to chance; what the score determines and what it leaves indeterminate; how much is conveyed of the artist-planner's own intention of what is to happen and to what degree what actually happens and the quality of what actually happens is left to chance; the influences of the passage of time; the variables of unforeseen and unforeseeable events, and to the feedback process which initiates a new score. (ibid: 7)

8. As to the performance phase of the RSVP cycles, an analogy between scientific experimentation and the performing arts employed by philosopher Robert Crease might help further position Halprin's schema at the intersection of design research and second-order cybernetics described by Sweeting. Crease tells us that "the structure of performance is essentially the same in the theatre arts and experimental science" when we consider that "[p]

erformance involves the conceiving, producing, and witnessing of actions in order to try to get something that we cannot get by consulting what we already have." In both domains, "the representation (theory, language, script) used to program the performance does not completely determine the outcome (product, work), but only assists in the encounter with the new" (Crease & Lutterbie 2010: 165). Of course, the phenomena generated by both experimentation and performance might well differ significantly from the expected outcome. Larry Richards reminds us that is is the dynamics of performance that account for these potential suprises and, in the spirit of second-order cybernetics, open up new horizons of possibility to be explored in a subsequent iteration.

> Formal languages remove the dynamics absolutely; in fact, the value of formalism is that it removes the dynamics to leave a skeleton of constraints to guide action and performance (like a script or score) [...] A poem, a piece of music, a play, and their performance are ways to use a language to play with dynamics. They don't *cause* things to happen; they *trigger* a dynamics of interaction that can lead to new distinctions. Contradictions and paradoxes become desirable as avenues to new ideas, new alternatives, new choices. (Richards 2010: 16)

9. For Glanville, the second-order cybernetic conception of design is in direct opposition to the "slogan" in modern architecture, attributed to Louis Sullivan, that "form follows function" (Glanville 2007b: 88). The level of complexity in most design challenges calls for another approach entirely.

> Rather than try to specify every requirement and every relationship between these requirements, and then find an optimal solution, design starts more or less 'aimlessly' and gradually constructs an 'evolving' form that not only changes but, in doing so accommodates the required functions also, often in a novel and surprising manner, where normal relations between functions are enriched or even replaced by new ones that are unexpected, different, and often very good! (Glanville 2007c: 1196)

10. Glanville tells us that "the drawing, sketch or doodle" is "central to the process of design" and that "[t]hese are often made without much purpose" (ibid: 1179). Throughout his corpus, Glanville sings the praises of purposelessness and the "gifts" that it can bring; a position that might seem, to some, to be at odds with the goal-directed preoccupations of cybernetics. It is, however, yet another theoretical commitment shared by Halprin, who claimed that "becoming *goal oriented* is "one of the gravest dangers that we experience" through our tendency to pursue social goods, based on "incontrovertibly 'good ideas,'" by "the most direct means possible" resulting, through an "oversimplified approach [...] in the chaos of our cities and the confusion of our politics (or other politics – fascism and communism are clear statements of this approach)" (Halprin 1969: 4).

> When ekistitcians, for example, say that the 'search for the ideal is our greatest obligation' they are making the same basic error that all goal-oriented thinking does – a confusion between motivation and process. We can be scientific and precise about gathering data and inventorying resources, but in the multivariable and open scoring process necessary for human lifestyles and attitudes, creativity, inquantifiable attitudes and openness will always be required. (ibid.)

11. If, as Halprin suggests, the "confusion of our politics" is equally a result of a flawed design process that is too dependent on narrowly defined goals and insufficiently sensitive to feedback, then, perhaps, it is not going too far to expand Glanville's audacious claim that science is but a subset of design and make a similar claim regarding governance;

a term that is, after all, also commonly understood to be virtually synonymous with the term "cybernetics." A conception of governance informed by the kind of second-order cybernetic approach to design espoused by Glanville and encapsulated in Sweeting's article would have no option but to acknowledge openly the inevitability of error and eliminate the peddling of supposedly iron-clad, fool-proof "solutions" in which the politicians of every liberal democracy currently traffic. And where might that lead us? But that is a conversation for another time.

12. Sweeting's article does valuable work in consolidating Glanville's legacy of design cybernetic theorization as it evolved alongside a growing awareness within the design research community that first-order, non-reflexive "scientific" models are insufficient to deal with the emergent functional, aesthetic and ethical complexities of actual design practice. This provides a robust foundation from which a whole generation of cybernetic designers influenced by Glanville (Thomas Fischer, Candy Herr, Michael Hohl, Tim Jachna and others) can further develop and disseminate this rich body of theory and practice to the generations to come. As a theorist/practitioner who independently evolved a recursive, conversational approach to design so thoroughly embodying the ethical commitments of second-order cybernetics, an additional reflection upon the work of Halprin has much to offer this on-going endeavour.

Open Peer Commentary: **Understanding Design from a Second-Order Cybernetics Perspective: Is There a Place for Material Agency?**

David Griffiths

1. The main focus of Ben Sweeting's target article is to examine the terms "design" and "second-order cybernetics," together with the practice designated by them, and to discuss their relationship. This task is simply described, but leads inexorably into deep waters, in part because of the entangled relationship between the terms, and in part because both terms are contested. In the main, Sweeting navigates this complexity with skill, but inevitably there are loose ends in the argument, which are worth pulling on to see if they lead to further insight.

2. The argument is founded on Sweeting's analysis of Ranulph Glanville's ideas on design and second-order cybernetics (SOC), a task that he is particularly well-positioned to undertake, given his long relationship with Glanville as both a student and a collaborator. Sweeting cites Glanville as stating that "cybernetics is the theory of design and design is the action of cybernetics" (§2), and reports that "Glanville [...] characterises all research as being a design-like activity" (§14) and that he "recognises design research as a self-reflexive activity of researching research" (§15). On the basis of Glanville's work, exemplified by the above quotations, Sweeting makes the core proposal of the article, suggesting that

> Glanville's understanding of design, and particularly his [...] account of the relations between design and science [...], allows us to view the currently expanding field of design research as a contemporary variety of SOC practice. (§6)

This proposal is both well-founded and useful.

3. I also find Glanville's argument regarding the relationship between science and design, and Sweeting's discussion of it, to be convincing: "Design is, it follows, the more general case and, therefore, 'it is inappropriate to require design to be 'scientific': for scientific research is a subset (a restricted form) of design…" (§7). The argument is in line with the critique made by authors such as Stuart Umpleby (2014) and Karl Müller (2014), who have contributed greatly to second-order science (SOS), to which Sweeting dedicates a substantial section. This critique focuses on the important role of the scientist as an observer and active constructor of the scientific process, a role that is systematically erased from positivist accounts of scientific activity.

4. Sweeting thus establishes two alignments: between design research and SOC, and between design and SOS. The question that arises in the reading of the article is the degree to which it is possible to extrapolate from the alignment between these discourses in order to draw conclusions that are applicable to science as it is carried out beyond the cybernetic tradition and to design that is carried out without a reflexive turn.

5. When Glanville spoke about design, he did so not as an external observer surveying the field, but as a participant explaining his experience of the process of design (including his design of musical environments and performances). Indeed, given the view of cybernetics that he sustained and lived by, we should not expect anything less. Sweeting does not discuss Glanville's practice but implies that it was in line with Horst Rittel's argument that "'everything goes': because designers inevitably encounter new and ambiguously defined situations (it being the purpose of design to create the new), they have no well-defined problems to solve or enumerable lists of options to pick from" (§13), and that the problems encountered by designers are "wicked" (§8) because of their complex inter-dependencies. Much design practice is illuminated by an analysis conducted from this position, but many design problems are perceived by designers in much simpler terms, and are not seen as being wicked. The Chambers Dictionary definition of the verb "design" is "to develop or prepare a plan, drawing or model of something before it is built or made," and readers will be able to confirm that other dictionaries have similar definitions. This definition includes many contexts where designers are convinced that they are working with well-defined problems, and that enumerable lists are available, including much of the field of engineering. A reading of Sweeting's article with a focus on this volume is complicated by the fact that the logic of the argument leads to thematic sections that discuss both design research (which necessarily has a reflexive aspect) and design (which, in the view of many practitioners, does not necessarily involve a self-reflexive aspect).

6. The designers of scientific instruments such as the CERN particle collider have a well-defined goal, in this case to provide an apparatus capable of detecting the Higgs Boson. But even in design that does not involve engineering, well-defined problems can be identified. The builders of musical instruments provide a good example of designers who have well-defined problems with lists of options. Iris Bremaud describes the choice of woods for construction in the case of the designers of xylophones and slit-drums in Africa:

> Many species could be encountered in either xylophones designed for temporary use, or slit drums with strong aesthetical meaning, involving the ability of wood to be intricately carved […]. On the contrary, the more prominent the purely 'acoustic' function of instruments was, the higher the proportion of use of Pterocarpus […]. This choice is nearly exclusive in most elaborate xylophones and in slit-drums that were used for message transmission – up to more than 10 km distances. (Bremaud 2012: 812)

These designers are clearly making choices from a list of predefined options, and deploying their design expertise in making the trade-off between the contrasting benefits of different materials and the range of pre-defined purposes to which the instrument will be put.

7. In a rather different musical context, Brian Eno, often described as a sound designer, also explains the act of creating a musical composition in terms of selection:

> What the composer had was a kind of menu, a packet of seeds, you might say. And those musical seeds, once planted, turned into the piece. And they turned into a different version of that piece every time. (Eno 2011)

Eno relates this approach to the influence of Stafford Beer, and perhaps this cybernetic connection should not be surprising given the importance of selection in cybernetics since the early work of Claude Shannon (1948).

8. The purpose of this digression into music, a field that was one of Glanville's main areas of activity, is to argue that there exist design practices that are well-defined, involve selection from a list of pre-determined options, or both. I suggest, therefore, that Sweeting's characterization of design is best seen as an accurate description of a particular type of design. It may also be an argument and exhortation to other designers who do not share these ideas or practice to consider more deeply the recursion involved in their design activity, and I believe that this was the intention of much of Glanville's work. The question arises, however, how far (if at all) it is possible to make a convincing argument about design in general on the basis of this SOC analysis to those who do not share the epistemological position of the field, a challenge that is common to SOC as a whole. I see Sweeting's discussion of Andrew Pickering as being central to this question.

9. Sweeting cites Pickering extensively, and mostly with approval. However, he disagrees with Pickering's characterization of SOC as "a turn away from the more tangible modes of experimentation that characterized earlier phases of cybernetics, and towards the linguistic." Sweeting counters this argument by pointing out that "SOC is a reflection on the performative involvement of observers within their observations" (§5), but that the opportunity to carry out this function was limited because the field of cybernetics had "broken up" (§19) by the time that SOC emerged. I have some sympathy with this view, but nevertheless I believe that it is incumbent on those who feel there is value in the heritage of cybernetics to investigate Pickering's point more deeply. Specifically, we need to assess the degree to which the risk that Sweeting identifies that SOC can become "overly introverted" (§6) may have played an active part in the break up of the field. Sweeting's concern is not to conduct such an inquiry into the decline of cybernetics, but rather to explore how its legacy can be applied and revived in design research. Nevertheless, I believe that there is a key point at issue here, as I now discuss.

10. The examples that are given of Pickering's performative approach can indeed be situated within SOC (R. D. Laing's work on therapists, Pask and the participant observer). But there are many aspects of Pickering's thinking about the performative that are not easily situated in this way. Pickering describes his conception of the performative as an "…image of science, in which science is regarded as a field of powers, capacities and performances, situated in the machinic captures of material agency" (Pickering 1995: 7). In his book *The Mangle of Practice*, Pickering examines the history of the bubble chamber in physics research. He argues that we should see this as a "dance of human and material agency" (ibid: 51). Pickering goes on to describe how…

> [r]esistance (and accommodation) is at the heart of the struggle between the human and material realms in which each is interactively restructured with respect to the other – in which, as in our example, material agency, scientific knowledge, and human agency in its intentional structure and its social contours, are all reconfigured at once. (ibid: 67)

Here, I think, is the heart of the problem of the generalizability of insights from SOC. The idea that the object of investigation (or design) has material agency that pushes back at the scientist (or designer) is one that sits uncomfortably with an SOC view of constructivism, and certainly of the radical constructivist tradition within SOC as exemplified by Ernst van Glasersfeld (1995). To put it another way, the conception of the performative within design research as described by Sweeting, and perhaps within SOC as a whole, may be different from that which Pickering proposes.

11. In my view, SOC does not necessarily preclude the ascription of agency to the material world. For example, the reformulation of the scientific method undertaken by Humberto Maturana (1990: 18) implies constraints on our ability to engage with the agency of the material, but it does not preclude its existence, and is compatible with Pickering's "mangle of practice." The analysis proposed by Sweeting, however, does not encompass the agency of the material. He does mention "the ways in which material artefacts operate variously as part of the research process, as the object of enquiry, as output or dissemination and sometimes as more than one of these depending on their context" (§30), but there is nothing to suggest that the physical world "pushes back" at the designer, or even that such a thing might be possible. I do not see this as a problem for the analysis proposed by Sweeting per se, as the design practice described may indeed consist of a recursive interaction between the designer, the design and the people for whom it is intended. Moreover, from a radical constructivist perspective, it may be argued that the perception of material agency is no more than a perception, and that a methodology based on this is intellectually misleading and practically unreliable. It does, however, raise a problem for the claim that design is a category that subsumes science. Sweeting's argument that scientific activity is a kind of design holds for a broad definition of design, but the specifically SOC view of design put forward in this article does not map well onto mainstream conceptions of science. The same applies even to first-order cybernetics in the performative mode, for example for Grey Walter, whose robotic "tortoises" addressed a well-defined problem: "to model goal seeking and, later, learning. But he did so as economically as he could" (Boden 2006: 244). The problem of mapping from design to science can be resolved in one of two ways. One option is to broaden our understanding of design so that it includes material agency, in line with Pickering's mangle of practice. This would enable the insight from SOC into the role of the designer in a recursive process of construction to be generalized across the whole range of scientific and design activities. Alternatively, we can make it clear that we are adopting a critical view of science, engineering and craft. This would embrace the differences between different types of design and scientific practice, and challenge practitioners to question the externality of the material agency that they ascribe to the surrounding environment and independent of themselves. There is indeed a role for such a practical critique. Sweeting refers to "pre-defined methods that were characteristic of philosophy of science" in the 1970s, but a glance around the bodies funding research today would show that this preference for pre-defined methods is alive and kicking.

12. Divergent opinions on the performative may in turn account for Sweeting's disagreement with Pickering on the linguistic turn in SOC. Sweeting comments that "SOC is a reflection on the performative involvement of observers within their observations" (§5). However, material agency is at the core of Pickering's view of the performative but is not

represented in design seen from a SOC perspective, as represented in this article. Consequently, from Pickering's perspective SOC is lacking an account of material agency and its effects, whereas Sweeting does not discuss any such lack. It is the discrepancy on this lack, I suggest, that leads Pickering to identify a linguistic turn in SOC, and also leads Sweeting to disagree with him.

13. In conclusion, the important contribution of this article is to bring together and extend the thinking of Glanville, and to show how this can both inform design research and serve as "continuing or reinventing cybernetic concerns" (§35). In doing this, Sweeting offers a much-needed response to the lack of practical research being carried out within SOC, a concern that Glanville also shared. In doing this, the article also raises important issues, going beyond its main focus, about the nature of the relationship between second- and first-order cybernetics and the possible role of material agency as a point at issue in the understanding of the performative in these two aspects of cybernetics.

Open Peer Commentary:
What Can Cybernetics Learn from Design?

Christiane M. Herr

Differentiating externally motivated application and internally motivated practice

1. Ben Sweeting's focus on the relationship of cybernetics and design presents a valuable counterpoint to recent attempts at renewing interest in cybernetics by framing it primarily in reference to science (§24). Based on Ranulph Glanville's (2007c: 1178) characterization of design as the action of cybernetics, and cybernetics as the theory of design, Sweeting positions design research as a variety of second-order cybernetic (SOC) practice (§4). This central point of Sweeting's article deserves further strengthening, as practice is not to be understood in this context as the *application* of theory (§§6–10). As argued by Sweeting based on Glanville (2014a, 2015) (§3), SOC should not be conceived of as a theory preceding and determining subsequent action. When seen from the perspective (and experience) of design, theory is more appropriately understood as a framework for making explicit thoughts developed in and through action. While generated from action, such a theory can then also be used for abstract argument and analysis, but this should not be seen as its primary purpose. Design reasoning is typically implicit: a form of thinking immanent in, expressed, and developed through acting. This is illustrated in Donald Schön's (1991) well-known characterization of design processes as *reflection in action.* It is this recognition of the fundamental involvement of the observer in the process that sets design (and cybernetics, specifically SOC), apart from the sciences, as Sweeting shows (§14).

2. Cybernetic descriptions of processes frequently revolve around goals, in particular the pursuing of goals within circular processes (Ashby 1957). While such processes may be described from the perspective of an outside observer identifying purposeful actions, it makes much more sense to shift perspective to that of the involved inside observer. From the perspective of the involved observer, goals appear more flexible, as they are deliber-

ately selected, often temporary, and typically subject to change in response to various constraints encountered in the process of acting (Fischer & Richards 2015; Glanville 2007c). Consideration of this constructed and process-oriented nature of goals is essential when aiming to understand the actions of designers. With goals as well as ways to pursue them being the subject of choices, the resultant cybernetic process relies strongly on personal values and ethics. In the context of cybernetics, this observation has led Heinz von Foerster (1992) to distinguish what he termed *in principle undecidable questions* – questions that cannot be decided objectively or from an external perspective. This is well known by designers, who must rely on personal values for much of their decision making (Trimingham 2008), as any kind of design practice involves questions of an ethical nature. While designers rarely make this explicit, cultivating personal values forms part of what can be described as *design rigour*, in reference to conventional scientific research. This observation may lead to further examination of the role of personal ethics in cybernetic practice.

The pleasure of constructing the world

3. While design may be understood as describing a particular kind of process (§§12, 22), it may be argued that design is also, and perhaps most importantly, a way of thinking and perceiving the *other* (Glanville 2007c: 1197). This way of thinking and perceiving cultivates not only keen awareness of the *what is* but also of the *what could be*. To designers, the world is always a constructed world (Herr 2015b), where self and other are dynamically merged: this worldview may be described as much as an analytical one as an aesthetic-appreciative and constructive one. From this perspective, it becomes obvious why designers are typically flexible in their employing of a wide spectrum of tools and methods, ranging from science to art. Engaging with the other in this manner initiates conversational processes of exploration that may be started by premeditated goals, but are in essence driven by perceptions of potential and possibility. I would argue that cybernetic processes may be understood in a similar manner.

4. While designers construct their realities, they tend to pay little attention to the nature of the world in which they construct their realities (Glanville 2006a; Herr 2015b). When engaged in explorative processes of designing, designers typically cast away theoretical preconceptions in favour of what is found to be practically viable in a given particular situation. What matters most is the immediate response of the other generated from action, and the changes in thinking and perception this response in turn generates in the designer. Designers construct realities through processes of informed participation, which resonates with radical constructivist theory (Glanville 2006a) as well as Margaret Mead's call for cybernetically informed ways of acting (§4).

5. Although designers are usually comfortable acting, they are not necessarily comfortable or able to make explicit the nature and mechanics of the processes in which they engage. This generates challenges when communicating beyond specific instances of design processes and especially beyond disciplinary boundaries. In addition, the implicit nature of designing makes it difficult to discuss design processes in educational settings. It is here where I see great potential for integrating the formal rigour of the cybernetic body of thought with action-oriented design. As Sweeting has argued (§14), cybernetic vocabulary was specifically developed to transcend disciplinary boundaries and is well suited to supporting designers in describing, perhaps also in fine-tuning, their acting. In addition to transcending cross-disciplinary boundaries in this manner, I would argue that design-

cybernetic perspectives can also help in transcending cross-cultural boundaries, as I have previously discussed (Herr 2011).

Cybernetic machines for thinking and showing

6. Sweeting (§22) points out that material experimentation, as it happened in earlier cybernetics, could inform similar experimentation in contemporary SOC to allow it to be more outward looking (§33). In this respect, cybernetics could adopt techniques well honed in design, where models are employed not only for purposes of representation or prediction, but mainly to support the exploration of ideas (§22). For a design-based variety of cybernetic practice, the continuation of making automated models of an explorative and performative nature seems a particularly fruitful direction. Such machines can be understood to be similar to conceptual models in design and may be developed based on cybernetic themes. Besides the precedents in design research discussed by Sweeting (§22), there are further examples of devices constructed in this spirit and relating explicitly to cybernetics.

7. Over the past 15 years, Thomas Fischer and myself have built and documented various – often automated – devices for conceptual idea exploration in architecture based on cybernetic ideas. We have characterized these devices as *machines for showing* (Herr & Fischer 2013, 2004), intentionally sidestepping expectations for prediction or immediate applied utility. I have recently continued this line of thought with an analysis and discussion of cellular automata models as they are used in design, where I have emphasized the explorative and flexible nature of such models (Herr 2015a). Once designers work with tools, they tend to adapt them to their own purposes, which results in rules being interpreted in a flexible manner and tools being used against their intended purposes (Fischer & Herr 2007). In a similar manner, designers tend to adopt vocabulary and theory from various fields other than design (§3). What is often not reflected well is that such processes of adopting external theory and vocabulary should be understood as part of creative conversations, in which terms are interpreted flexibly and vocabulary as well as theory is typically transformed to fit a particular design situation.

8. One recent occasion where cybernetic devices were presented was an informal exhibition held as part of the 2014 annual conference of the American Society for Cybernetics, http://asc-cybernetics.org/2014. Most of the devices displayed were, however, of a representational nature, with only a few of the presented items intended to create questions or initiate new thoughts. With cybernetics' growing maturity and increasingly complex body of ideas, it seems contemporary cyberneticians are primarily concerned with explaining existing principles clearly rather than inventing new ones or playing with these ideas in an open-ended – and perhaps messy or incongruent – manner. It is this manner of acting and thinking through acting in a playful and explorative spirit that cybernetics can learn from design.

Open Peer Commentary: **Rigor in Research, Honesty and Values**

Michael Hohl

1. In §14 of his target article, Ben Sweeting examines Ranulph Glanville's concept of honesty. I think the concept of honesty, while not being central to the main argument of the target article, deserves some more reflection, especially in relation to the concept of rigor.

2. While "rigor" in research is often mentioned, I think its constituents are rarely thoroughly discussed. I would like to use this opportunity to discuss these further, as Sweeting's article allowed me to get a much deeper understanding of Glanville's concept of honesty, especially linking it to post-rationalisation, which I found very enlightening.

3. When I was conducting my PhD research at Sheffield Hallam University, between 2003 and 2007, we had regular debates about academic rigor and what constituted rigor in the research process of artists and designers. Adopted from research in the sciences, the significant terms associated with rigor, and associated with PhD research, were that the research had to be "*thorough, exhaustive, accurate, and systematic*." In art and design *critical* and *reflective* were often added as well. In our seminars, it emerged that "thorough" and "exhaustive" were related and could described as together forming a "T"-shape: the horizontal line of the "T" consisting of an exhaustive, broad and comprehensive overview of what is considered the context of research and related practice, while the focus area, the vertical element of the "T," consisting of going deep into it and being thorough in one's own contribution. I assume "objective" might have been included in earlier definitions of rigor in research in art, design and architecture, however in the research of artists, designers and architects, the requirement of the term might have been abandoned at some time. In artistic research, the individual creative process involves necessarily subjective, intuitive and explorative phases in which adhering to "objectivity" might be more of a hindrance and lead to post-rationalisation. More about this below.

4. When we examine the next term, "accurate," meaning "correct in all details" or "faithful representation," it is perhaps to this that Glanville's demand for honesty is most related. How may "accuracy" be possible from a constructivist perspective? Does the demand for accuracy refer to observations, measurements, models and analysis only? Then how might it include a playful exploration, intuitive insights, creative leaps of mind, random iterations or doodling conversations (Glanville 1999) that may lead to new understanding, insights, methods, techniques or discoveries? Glanville views such creative moments as "[…] pointless, undirected, seemingly purposeless, playful and dreamy activity that is at the heart of design" (Glanville 2006a: 105). When such designing is at the heart of research, then research has dreamy and purposeless aspects to it. How might these be documented and interpreted accurately?

5. In the spirit of honesty, I would like to reflect upon my own PhD research process. In retrospect, it had aspects of double-bookkeeping: presenting my methodology, plans and intentions to my supervisors (and myself) accurately, yet the results being post-rationalisations. From my own perspective, I relied on hunches, connections between facts "suddenly" becoming clear, a rather unstructured and unclear, sometimes "terrifying" process riddled with insecurities of "poking around in the fog" in order to understand what I was learning, make sense of it and proceed to a next step. The applied methodology

emerging quietly almost on its own in the background. Later, after completion, I would end presentations of my PhD Research with the statement: "Told as a story, my research appears pretty straightforward and top-down. In fact it was bottom-up and came together step-by-step over three years. The research process was a constant learning process." From that perspective, the written thesis did not describe in thorough, "honest" detail how new insights emerged, but made sense of it in post-rationalisation (Glanville 1999: 5). For example, even a meticulously kept journal would not reveal how exactly the grounded theory emerged in the analysis of interview data.

6. The following term, "systematic," is in my view the most problematic in the research of artists and designers. Systematic, meaning "acting according to a fixed plan or system, methodical." Following a fixed plan in practice-based design research contradicts, in my view, exactly the possibility of acting on new insights and diverging from a perhaps planned trajectory. It is this creative freedom that allows for new connections, experiences and discoveries. I believe it lies at the heart of research in the creative disciplines. Without it, we would be "drawing by numbers," while serendipitous and radically new discoveries would be less frequent. As a result, I think "systematic" may be relevant to the general overall structure or model[1] of the research process of PhD research but should be avoided in the active creative phases in which new ideas emerge and solutions are developed.

7. In this context, I ask where the "values" in scientific research might enter. In artistic research, they often are referred to, or better emerge, in a reflective chapter. Karl Popper asks why few scientists care to write about ethics and values:

> […] values emerge together with problems; that values could not exist without problems; and that neither values nor problems can be derived or otherwise obtained from facts, though they often pertain to facts or are connected with facts. (Popper 1976: 226)

8. I would say that it is here where a second-order cybernetics perspective might provide a valuable contribution to avoid the "view from nowhere" (Turnbull 2000: 221). When design research aims to answer a research question, a process that also involves looking at problems, then this should be linked to particular values held by the researcher, fundamentally informing the thinking and acting. However, these are rarely made explicit. If this happens this usually takes place in a reflective chapter towards the end of the written thesis.

9. I believe that Sweeting's emphasis on Karl Müller and Alexander Riegler's proposal (§§25f) linking second-order cybernetics and design creates a most promising direction for both disciplines. This might be especially so in view of current developments in design such as transition design (Irwin 2015), design for social change and user experience design. All three examples include theoretical models that inform acting, which may lead to designing intangibles, such as processes, involving (§25) complexity, adaptation and evolution, and (§26) self-reflexivity and the inclusion of observers (§30).

Open Peer Commentary: **Digital Design Research and Second-Order Cybernetics**

Mateus de Sousa van Stralen

1. Gordon Pask pointed out "it is easy to argue that cybernetics is relevant to architecture in the same way that it is relevant to a host of other professions; medicine, engineering or law" (Pask 1969: 494). Indeed, there are several publications about the application of cybernetics in design. In the target article, Ben Sweeting looks at this the other way around and proposes that design research can contribute to cybernetic thinking by suggesting that design research is not just a field influenced by cybernetics but is a form of second-order cybernetic practice. Sweeting relies on Glanville's work to underpin the strong relation of second-order cybernetics (SOC) to practice and design. Through his work, Glanville has shown that not only can cybernetics contribute to design, but that design can also inform cybernetics, understanding cybernetics and design not as separate entities but as a circular interwoven process of acting and reflecting, theory and practice. The discussion I put forward in the commentary is that Sweeting's arguments can be made even more explicit if we focus on a more specific form of design research that is based on digital processes – digital design – and look how it is practiced. The connections between digital design, design research and SOC can serve as bridge for a new generation of designers to access and incorporate radical constructivism in their reflections and actions.

2. In the last decade, there has been a growing interest in cybernetics amongst designers, especially young ones, driven by the increasing use of digital technologies in design. Computer programming and its promise of machine intelligence in the process of design,[1] manufacturing[2] or embedding it in the environment[3] are part of today's design practice. The development of the different digital processes and techniques was mainly motivated by transformations in praxis led by architects and designers trying to explore the potential of digital technologies in their work. As Neil Leach (2012) points out, much of the research in digital design was done outside the traditional academic environments. Designers had to develop their own software and building process to ensure the feasibility of their designs,[4] and many reached out to theories external to design to support their works.[5] But as Rivka Oxman (2006: 232) has noted, the impact of digital design on practices has resulted in a need for a revision of current design theories. Many research groups and designers have looked to cybernetics to create conceptual frameworks to guide research and development.

1. Among others, the following AI-based techniques are popular: neural networks, genetic algorithms, multi-agent systems, evolutionary architecture (Frazer 1995).
2. Topology optimization, digital fabrication, and self-assembling are examples of techniques in which computation is applied to the manufacturing process.
3. In interactive environments and relational architecture, computation is embedded in the environment to enable reactive, interactive and dialogical behavior. See, e.g., the works of Usman Haque, http://www.haque.co.uk, and Ruairi Glynn, http://www.ruairiglynn.co.uk.
4. See, e.g., the design companies Gehry & Partners and Zaha Hadid Architects.
5. The special issue of the London journal *AD* on "Folding in Architecture" (Lynn 1993) has several articles that exemplify how designers reached out to theories external to design to support their works.

3. Digital design research can be seen as a subcategory of design research, but given the impact of computation in designing and in production practices, it is evolving to become a unique field in design (Oxman 2006). In digital design, computation can be integrated in the total process of design, from the initial concept through to materialization, production and use. In this "digital continuum," as it is called by Branko Kolarevic (2003), design is directly connected to materialization, from the initial conceptual stages with rapid prototyping techniques, to the final object with digital fabrication processes and interactive systems. The connection between design and materialization, research and action indicates how the relations between design research and cybernetics can be even more evident in digital design. It is not a surprise that most examples of connections between cybernetics and design listed in §21 of the target article can be seen as examples of early digital design. Nicolas Negroponte's *Soft Architecture Machines* (Negroponte 1975) discusses computer-aided architecture related to machine intelligence in design. John Frazer's *An Evolutionary Architecture* (Frazer 1995) investigates form-generating processes by considering architecture as a form of artificial life. Glanville also had several articles related to digital design, such as "CAD Abusing Computing" (Glanville 1992) and "Variety in design" (Glanville 1994). Further evidence can be found by bringing the discussion of the concepts of self-reflexivity and the inclusion of the observer into the light of digital designing.

4. Sweeting discusses how self-reflexivity and the inclusion of the observer can be seen as important points of interconnection between design research, SOC and second-order science (SOS). Self-reflexivity is one of the central issues in digital design processes today. This becomes more evident in those practices where computation is inextricably part of the process, such as algorithmic and parametric design, in which the designer designs computational process to generate form. The design of the design process that generates form gives the idea that form is not "given," but "found." In the first case, data forces shape onto passive matter, and in the second case, matter and data interact and give shape. The idea of giving shape makes the connections between observer and process more explicit, as most designers are eager to claim their involvement in the process. That is why the inclusion of the observer does not seem to represent a problem in design. But in *form finding* this becomes more blurry, as questions can arise as to who is responsible for the design. This process, which is also called "emergence," leads to a false idea that computers themselves are generating autonomous objects. However, from an SOC perspective, the designer is also responsible for the final design because form is actually coded in the computer by the designer. The observer is included in a self-reflexive act of designing design.

5. Another point worth being discussed is the impact of the *digital continuum* in design. Digital fabrication enables designers to create short feedback cycles of designing, making and reflecting. In that context, practice-based research methods have become more widely used and accepted, as designers are now able to make high-end models and products in a fast and accessible manner through different iterative cycles. Either explicit or not, these feedback cycles can be seen as examples of cybernetic practice, which reinforces Sweeting's arguments.

6. In conclusion, Sweeting's target article positions design research as a contemporary variety of SOC and by doing so, establishes the connections between design and SOS, creating a circular relation where one can inform the other. The parallels between design research, SOC and SOS can be even more explicit in digital design. SOC and SOS can point towards the creation of an epistemological foundation to digital design, where self-reflexivity and the inclusion of the observer are central questions.

Open Peer Commentary: **Cybernetics Is the Answer, but What Was the Conversation About?**

Jose dos Santos Cabral Filho

1. Ben Sweeting's target article shows a genuine and welcome effort to amplify our understanding of the relationship between design and cybernetics. Sweeting explores in detail the intricacies of such a relationship, presenting a well-argued investigation into the possible links between the two fields of investigation and looking for a kind of mutualism, exploring the improvement that both parts can bring to one another. He does so by continuing Ranulph Glanville's lifelong enterprise of clarifying the intertwining of the two areas, an effort that was often made in unusual ways, escaping the conventional idea of applying cybernetics to design.

2. My collaboration in this open commentary is to suggest that if a radical consideration of the systemic nature of design were taken into account, the main arguments of the article could be constructed in an easier and simpler way. The question of simplicity here is less to attend the principle of Occam's razor and more to make the arguments even more compelling and, therefore, have a greater chance of extending their practical implications.

Design as an invitation to dialogue

3. There seems to be widespread consensus that to consider design under the principles of second-order cybernetics is mostly to acknowledge the inclusion of the observer and the conversation that originates from this acknowledgment. However, the bibliography on the subject shows that most researchers consider the inclusion of the observer to be restricted to the design process (the work of designers), and sometimes extended to the research into the design process (the work of researchers, academic or not). This applies to different researchers, such as Glanville and Donald Schön, and, in fact, it underlies the target article. Most of the time, the issue of the use of the designed object does not get much attention, as if the design role had ended with the creation of the object. Nevertheless, the consideration of the object and its use is not enough: if we understand the systemic nature of design in a radical way, we have to come to terms with the fact that the final product of the design chain is not an object but a system, in which the object is included. Acknowledging this will change, concomitantly, the design process and design process research.

4. Thus, if we want to push the idea of design research as a variety of SOC to its most interesting limits, we have to consider that what we design when designing is not merely an object but a larger system that includes the object (and in some cases, may even prescind from physical objects). That may seem common sense, obvious and self-evident, and, in fact, it is one way or another acknowledged by most designers, especially by architects (and it is surely indisputable amongst cyberneticians).

5. This matter was already highlighted by Gordon Pask in his seminal paper "The architectural relevance of cybernetics" (Pask 1969). It posits, amongst other things, that "architects are first and foremost system designers," a reasoning that can be easily extended to design in general. Pask's paper is certainly one of the most quoted papers in relation to cybernetics and architecture/design, especially due to its unexpected proposition that architecture is more relevant to cybernetics than the other way round.

6. However, Pask's assertion on the systemic nature of architecture is not fully taken into account by practitioners, if not even downplayed. Certainly, several researchers, particularly those with some sort of direct link to Pask himself, such as Hugh Duberly, Paul Pangaro, Usman Haque, and John and Julia Frazer, have all drawn attention to the groundbreaking aspect of Pask's contention, and have consistently tried to develop it further (§21). Dubberly & Pangaro (2015) even argue that this paper "anticipates Donald Schön's notion of design as conversation [...] and goes further than Rittel and others who described design as a cybernetic process" (Dubberly & Pangaro 2015: 10). For certain, Pask goes beyond Schön's notion of design as conversation, considering that Schön, even though he pushes the idea of design beyond mere problem solving, still regards the design process as somehow ending with the object. In this way, conversation, in Schön's view, ends up been a kind of soliloquy between the designer and his or her drawings, regardless of whether he or she is using drawings to articulate ideas and not just as a representation. However, it is undeniable that Schön's book *The Reflective Practitioner* (1991) turned out to be very influential and played a significant role in the general acceptance of design as conversational outside the circle of cybernetics.

7. Thus, on the one hand, we have a theoretical recognition of the importance of the systemic principle of design, and on the other hand, what we can term as a politically correct embracing of democratic intentions by designers. The problem is that despite this general and diffuse acceptance of a systemic approach, we are witnessing a continued and excessive focus on the design of non-systemic objects that is more and more tailored to meet the spectacularization of our lives and cities. In other words, we see not the use of a dialogical framework in the actual practice of design, but a dialogical discourse superficially applied to design. As a matter of fact, a dialogical discourse is a contradiction in itself, as discourse is opposed to dialogue, as the philosopher Vilém Flusser (2011: 83) reminds us.

8. The problem with a superficial adoption of design as conversation is that it can lead to sterile self-reflexive attempts such as Peter Eisenman's Cannereggio project, referred to in the target article (§28). On the one hand, it is for sure a meta-reflection on the design process and apparently it articulates an ingenious convergence of the three categories of design research – into, through and for (§§22, 30). On the other hand, its design scenario excludes so many layers of the concerns and stakeholders implied on that specific architectural design that it becomes a restricted conversation, a soliloquy so to speak, that ends up as a selfish and exhibitionist exercise, no matter how intellectually flamboyant and marketable it may be. In other words, it is not enough to be self-reflexive and simply engaged to explore the full potential of being an SOC observer. The question is not only about the engagement or detachment of the observer; it is not only about where we position ourselves as observers (§30) but also about how far we are willing to take the systemic approach, that is to say, it is about the extent and nature of the included observers invited to the dialogue.

9. Pask and Price, once more, have shown some possible paths to including the observer radically with their Fun Palace project – a collaboration with Joan Littlewood (§13). However, it is worth noting that the same contradiction regarding Pask's paper – praised but not fully taken into account – goes for the Fun Palace. It is widely reverenced in architectural magazines and at exhibitions but it seems to have had little practical impact on the production of contemporary architecture. The digital design trend of recent decades, for example, which is based on the design research of the 1970s, has promoted a change in practice from designing the object to designing the process of designing the object (designing design, form-finding, etc.). A radical move would change the focus on the object in

itself towards a systemic and relational scenario where the object exists in its full dialogical potential; that move, however, seems unattainable (or possibly, undesirable).

10. Even if we consider the development of so-called interactive architecture, the Fun Palace proposition is still far ahead of what we have achieved, in spite of the advances in digital technology at our disposal. It seems that, contradicting Price's famous dictum "technology is the answer, but what was the question?", technology is not the answer in our present situation. At least not technology outside an SOC framework. Perhaps we should bring Price's dictum up to date by saying: cybernetics is the answer, but what was the conversation about?

Conclusion

11. A significant advance in design towards a second-order level will come when designers embrace an all-encompassing systemic approach that will necessarily have the inclusion of the observer, at all possible levels, as its pivotal point. If the desire is to keep design and design research as a practical enquiry into openness, as Sweeting seems to aspire, designers must extend the conversational and recursive strategy used in the design process towards the creation of dialogical objects and the system in which they are inserted. To consider design within the complexities and seriousness of Pask's conversation theory would allow a radical rethinking of design in a way that it would necessarily become SOC in practice. Then, Glanville's assertion that "cybernetics is the theory of design and design is the action of cybernetics" (2007c: 1178; §22) would become unequivocal, and design, as well as design research, would be undoubtedly more similar to the tangible experimentation of first-order cybernetics, as the target article proposes.

Open Peer Commentary: **(Architectural) Design Research in the Age of Neuroscience: The Value of the Second-Order Cybernetic Practice Perspective**

Andrea Jelić

1. In the context of the increasing interest of neuroscience for architecture and, more broadly, evidence-based design, Ben Sweeting's target article offers a critical perspective for plausible positioning of design in such an interdisciplinary dialogue. Specifically, by understanding (architectural) design research as a contemporary variety of second-order cybernetics, an opportunity arises for tackling potentially crucial obstacles to future progress and the usefulness of neuroscientific investigations in an architectural context. Accordingly, the aim of this commentary is to highlight the value of the target article's view by examining its possible contribution to several crucial issues, including:

a. addressing concerns of prescriptive design solutions;
b. using the inherent second-order cybernetic structure of design research to question the roles of the architect-designer and the scientist in the context of experimental studies; and

c. indicating the need for and possibility of a new second-order science of interdisciplinary design research framed on the basis of cognitive science and phenomenology of architectural experience and design.

2. Before proceeding, it is important to contextualize the commentary's argument and motivations by sketching briefly the background and current efforts in the field dedicated to investigating the relationship between the mind, body, and built environment through a neuroscientific lens (for a comprehensive introduction, see Mallgrave 2011, 2013). On the one hand, a renewed interest in the experiential dimension of architecture and a turn toward human-centred design, and on the other, decades-long history of architectural psychology and environment-behaviour research have created conditions for a seamless opening of neuroscience-architecture dialogue. However, despite promising initial efforts, there is a lack of a systematic framework purposely aimed at defining and structuring the relationship between architectural design and scientific insights/evidence. It is in this light that the target article's cybernetic parallels between science and design are proposed as a direction for approaching this important issue.

3. Concretely, the continuity of ideas between cybernetics and design research as presented by the author (§22) establish potential interpretations for neuroscientific knowledge-architectural design connection at two levels:

- at the level of design research being exercised as a second-order cybernetic practice, and
- at the level of interdisciplinary design research as a second-order science.

4. Firstly, the essentially conversational and constructivist nature of the design process challenges directly any concern for developing evidence-based prescriptions for architectural solutions. In this sense, any (recurring) attempt to "scientise" design through neuroscientific methods and inputs – a genuine possibility in the age of neuroscience – can be countered effectively by bringing awareness of the cybernetic conditions governing design research into this interdisciplinary endeavour. Therefore, similarly to the capacity of the work of architecture only to trigger and not control the subject's experience (according to the enactive-embodied view, Jelić et al. 2016; see also Sweeting's hypothesis of architectural experience as facilitating second-order inquiry, Sweeting 2015a), neuroscientific inquiries into the experience of architecture primarily serve to shed light on design knowledge, to relate the intuitive decisions to spatial scenarios, and not to modify the design activity as such.

5. Secondly, in line with the theory of embodied cognition, architectural design is in itself an embodied process: it is hypothesized as being a neurological activity that always involves embodied metaphorical thinking and multi modal image-making (Arbib 2013; Mallgrave 2011). Indeed, reflecting phenomenologically upon one's own experience as a designer and based on (auto)biographical descriptions of the process by extraordinary practitioners (e.g., Zumthor 1999), it can be suggested that architects commonly have rather suggestive, lifelike, intensive (bodily) feelings when imagining the spaces they are designing, in resonance with imagined atmospheric qualities. Accordingly, a neuroscientific, or better yet, neurophenomenological investigation of the design process may bring forward the awareness about the bodily and emotional processes involved in (pre-)reflective experiences of "living" the designs. Hence, design research with reference to second-order cybernetics principles (§22) could help to distinguish the participant's di-

mension – how an exchange of different observational positions occurs (i.e., imagining experience from the position of the user and one's own as a designer), how such switching is incorporated into the conversation with the medium in which the designer works, and ultimately, in what manner such an observer's awareness could be introduced to teaching design and facilitating the learning process.

6. Following the target article's convincing argument for the necessary shift in understanding science as a design-like activity (§§13f), a concrete illustration can be offered in the context of the neuroscience-architecture inquiry. If design research is understood as a variety of second-order cybernetic practice, then this kind of interdisciplinary experimental work encounters a particular observer issue: who is a designer here – an architect or a scientist? Currently, the majority of neuroscientific investigations are one-sided, i.e., led and conceived primarily by cognitive scientists, with little or no support from the architectural side. In such a situation, designing experimental setups, involving the creation of architectural environments, is guided more by the requirements of scientific methods than by architectural purposes. Thus, the validity of resulting evidence can be questioned on the basis of its appropriateness and usability in design. For this reason, there is a need to strategize such an interdisciplinary endeavour by establishing a framework for the new second-order science (see, for instance, the proposal by Hugo Alrøe and Egon Noe 2014), which should include careful rethinking of the participants' roles in relation to their disciplinary perspectives, expertise, and corresponding impact on study outcomes. In other words, there is a need for self-reflexivity and differentiation according to the observer (architect or scientist), in the spirit of second-order cybernetics and science, as indicated in §15 and §30.

7. To illustrate further the parallels between design research and (second-order) cybernetics (§23), this last point considers the overlap between the notion of cognitive-science-*cum*-phenomenology providing a genuine second-order science (for a detailed account, see Vörös 2014) and the proposal by Sweeting (§26) of second-order science being a potential point of interchange between design and science. In the context of neuroscience-architecture dialogue, Sebastjan Vörös's argument can be transformed into a cognitive science-*cum*-phenomenology of architectural experience and design, where the latter refers to a longstanding tradition of architectural phenomenology and the above-mentioned phenomenological descriptions of architects' works and design thinking (classical examples including Holl, Pallasmaa & Pérez-Gómez 2006; Pallasmaa 2005). In parallel, current efforts to provide a systematic conceptual framework for the complex biocultural nature of architectural experience prevalently belong to the enactive-embodied understanding of cognition (see, for instance, Jelić 2015; Jelić et al. 2016; Rietveld 2016; Rietveld & Kiverstein 2014). Taken all together, a new second-order science of interdisciplinary design research can be conceived of as a conversational framework between the enactive-embodied approach and the phenomenology of architectural experience and design, which focuses on the interdisciplinary research itself. Thus, its aim is to identify and establish a plausible pathway of exchange between neuroscience and architecture – that is, to create a communicative space, a "*trans-domain*" where scientific and designerly research may converge (Jonas 2015a: 34). Accordingly, the architect's ways of knowing might be able organically to incorporate alternative approaches to life-world perspectives, in this case, one that is enactive-embodied and evidence-based, and thus strengthen in turn the architectural mode of structuring and representing the world.

8. Finally, the value of target article in the context of constructivist approaches more broadly, can be particularly emphasized in terms of its pertinence to addressing problems beyond cybernetics – more specifically, in the domain of enactivism as related to architecture – by indicating a way of structuring interdisciplinary research and thus tackling one of the key issues of design research in the age of neuroscience.

Author's Response: **Beyond Application**

Ben Sweeting

1. I wish to thank all commentators for their stimulating contributions. The first thing to note in response to these seven commentaries is the range of ground they cover, indicating the wide potential of the relation between cybernetics and design research to inform both fields. It is significant that many of the aspects raised by commenters are focused on core topics of cybernetic research: computing technology (**Mateus van Stralen**; **Christiane Herr**); cognition (**Andrea Jelić**); and, broadly, the relationship between research/theory and action/practice, which is a focus of **Herr** and **Michael Hohl**, and underlies the concerns of **Jose Cabral**, **Dai Griffiths** and **Tom Scholte**. As Karl Müller (2010) has noted, there is a need to focus on core topics in order to reinforce the coherence of radical constructivism (RC) and second-order cybernetics (SOC) as a research field. Müller's remarks could be taken as a call for a turn away from topics such as design that have been prominent in recent cybernetics. These commentaries, and the research to which they point, suggest that design may instead offer a focus in which a number of such core issues can be explored.

2. In this context, **Scholte**'s introduction to the work of Ann and Lawrence Halprin may be valuable even beyond the project of connecting cybernetics-inspired discussions in design and theatre studies (see also **Scholte**'s target article). Building connections such as this would seem to be a way to help broaden the relationship of cybernetics with both design and theatre beyond one of application, releasing their potential to explore central cybernetic concerns through practice (cf. Müller 2010: 36f).

3. Of the commentaries, those of **Griffiths** and **Cabral** put forward the most explicit questions, and I therefore concentrate on these below. In line with my approach in the target article, I have attempted to remain focused primarily on how issues raised in design can contribute to questions in cybernetics.

Ill-defined problems

4. **Griffiths** (§8) suggests that the account of design that I have given applies to a particular subset of design, whereas at least some other areas of design deal with well-defined problems. Some design tasks or components of design tasks are, indeed, characterised by more constrained problems than others. Yet even apparently clear and familiar design tasks regularly involve incomplete criteria or contestable premises, and a clearly-defined goal is no guarantee of a well-defined problem (cf. **Griffiths** §6). This is because design is always concerned with the new (target article §8), which is the case even when designers are not attempting to be especially innovative (that is, when we design a build-

ing, we are concerned with creating something new even when we stick to an established typology). This can be seen within the scope of the definition that **Griffiths** (§5) cites: the process of preparing a plan for constructing something is not solely a matter of setting out production information (the working drawings and specifications that will guide manufacture) but of devising what is proposed in these. This process involves forms of reflective, conversational activity whenever such a plan is considered in more than arbitrary terms (that is to say, when it is designed).

5. Take, for instance, some of the questions posed in the design of a new motorway (an example within the compass of engineering, and one to which Horst Rittel and Melvin Webber refer, Rittel & Webber 1973: 163). Different configurations of road junctions will be both better and worse according to different terms of reference. Even considering only the efficiency of traffic flow, there will be trade offs between congestion at different points in the road system. There are also many other relevant criteria, such as, for instance: safety, other road users, cost, construction sequencing, maintenance, noise pollution, air quality and impact on natural habitats. While these criteria are mostly easily recognisable, they are not all commensurable with each other, such that there is no one way to resolve definitively between them, nor is it possible to optimise against an overall goal without this being distorting. Further, the interactions between these different criteria and the limitations they set on each other in the specific situation that is at hand only become clear as particular solutions are developed, discussed and enacted. Taking a broader scope, one might also challenge the premises under which the project is advanced: having explored the likely consequences of the new motorway, we may take a different view on whether it is a worthwhile project and consider alternative options instead.

6. While such situations resist exhaustive analysis and conventional linear problem solving, designers deal with them as a matter of course and without regarding them as being problematic. In so doing, they develop and refine not just their design proposals but also the questions to which these proposals respond. Indeed, as Nigel Cross (2007a: 100) points out, designers treat even well-formed problems as if they are ill-defined, an approach that has the benefits of testing the assumptions that are given at the outset and searching for new opportunities.

7. **Griffiths** (§6) gives two counter examples – those of scientific and musical instruments – where questions are very tightly constrained. Indeed, these situations are so constrained that they might well not be considered as instances of design activity in that they respond to a plan rather than create one. The musical instrument example, which is perhaps better understood in terms of craft, is closely related to the existing tradition of musical performance in which each instrument must be usable. These constraints can, however, be understood as a result of a wider design process, one where the configuration of the musical instrument has co-evolved slowly over several generations together with the traditions of musical performance to which it is related (this is comparable in architecture to the development of a vernacular tradition). The development of scientific instruments can be thought of, similarly, as blurring with that of scientific experimentation itself, as is reflected in accounts of scientific practice (target article §10). What is learnt in experiments using the instruments generates new criteria for further experiments and so new or refined instruments. Thus we can think of this as one overall process, which we could characterise either in terms of science or design, encompassing scientific experimentation and the construction of the instruments that support this.

8. **Griffiths** (§8) asks the question of to what extent an SOC account of design can be convincing to those that do not share its epistemological position. I do not see this as a question of different design epistemologies but of different degrees of explicitness about the epistemology that is acted out in design, and different ways of making this explicit. What designers do in practice is not always what they describe themselves as doing, as discussed by **Herr** and **Hohl**. It is in retrospect that the paths taken seem clear and, as it is this clarity that is what designers need to communicate, the messy process by which this clarity is developed usually remains unremarked on. Making these sorts of processes explicit is a core concern of design research and something to which SOC can contribute. The purpose of this is not, as I see it, to reconfigure design practice in some specific way. Rather, articulating what would otherwise remain tacit helps maintain what is already special about design (including attitudes towards values, as raised by both **Herr** §2 and **Hohl** §§7f), something that can otherwise become lost.

9. This relation of SOC to design practice in terms of making the implicit explicit may, as **Griffiths** (§8) suggests, inform how SOC might be advanced more generally. Cybernetic processes are implicit in everyday life and, as with design, making these processes explicit reinforces what is special about them, which can otherwise become lost in the context of other concerns. Looked at in these terms, SOC's relation to practice is not limited to where its epistemological position is explicitly shared. It can enjoy a broad relation to practice in terms of implicitly cybernetic processes, while still contesting the ways in which particular practices are conventionally understood.

Material agency and viability

10. **Griffiths** points out tensions between RC and Andrew Pickering's (1995) account of material agency. As **Griffiths** (§11) notes, there is not necessarily a conflict here and it seems to me that such tensions can be defused, or at least sharpened to more precisely the points at issue.

11. This is supported by the case of design, which while constructivist in orientation is compatible with ideas of material agency, even if this was not emphasised in my account. This is both in terms of the media with which designers think and the technologies and industries with and in which they work:

- Media plays an active role in how designers work. It is important to how they deal with complexity (Gedenryd 1998), model the material and spatial (Sweeting 2011), and construct new possibilities (the process of sketching that Ranulph Glanville 2006a, 2007c emphasises is one that needs to be embodied in media of some kind). This includes the digital technologies discussed by **van Stralen**, as well as the more obvious materiality of the analogue. Accounts of the active role of instruments in science, such as that given by Pickering (1995), can be read as if referring to the design studio (target article §10).
- What is materially and technologically feasible is a crucial constraint on what designers propose. This is especially the case where designers try to use materials in forms to which they are particularly suited, as can be summarised by architect Louis Kahn's oft-quoted conversation with a brick – "You say to a brick, 'What do you want, brick?' And brick says to you, 'I like an arch.' And you say to brick, 'Look, I want one, too, but arches are expensive and I can use a concrete lintel.'

And then you say: 'What do you think of that, brick?' Brick says: 'I like an arch'."[1] As well as this material-focused approach, material agency can be seen in the way that technological changes have transformed the nature of material constraints (discussed by **van Stralen** §§2, 4), and it remains an important factor even where design approaches are focused elsewhere.

12. The principle move in RC is to change the orientation of epistemology from a concern with how we know (or do not know) about any real world beyond our experience, to a focus on this experience itself. This relocates epistemology to the realm of experience, in which (our experience of) the material is important to include (as is evident in design). While, therefore, RC can be contrasted with the material where this is meant in the sense of the real, there is no conflict between RC and our material experience. Indeed, the latter can be encompassed in the notion of viability, which is central to Ernst von Glasersfeld's account. RC is not a licence for unconstrained construction. Von Glasersfeld (1990) gives the example of not being able to walk through a desk, and thus being unable to maintain a viable idea of the world that would allow him to do this. This is an example of a material condition in which we experience epistemological, not just practical, resistance.

13. Von Glasersfeld sometimes referred to viability in terms of "fit." In RC, this is in the sense of "fitting with" or evolutionary fit, and so perhaps better phrased in terms of the elimination of the unfit. There is no sense of correspondence to the real and much room for contradictory explanations to be viable in our experience at different times. This is not to be confused with the athlete's notion of fit, of an idea becoming fitter and fitter in the sense of a closer match to the goal of the real. In this latter view, while it may still be acknowledged that we do not have access to the real, our experience is claimed to be a good guide to it in any case because of the constraints that are imposed on it, thus returning to a correspondence view of epistemology. The main point at issue here is, as I see it, not about material agency per se but whether this is understood in terms of the real or in the realm of experience, and about how this is then put to work epistemologically.

14. Similarly to what I have said above regarding the relation between SOC and design, I think that RC is agile enough to engage with the material and the performative across the "whole range of scientific and design activities" (**Griffiths** §11), while also contesting what is at stake epistemologically in these. Indeed, RC can help provide the honesty that Glanville (2014c) suggests will efface the differences between different research traditions (target article §14; and as expanded on by **Hohl**).

Designing systems

15. **Cabral**'s call for an increased focus on the systemic nature of objects is something that I support. The issue as I see it, and as **Cabral** (§3) points to, comes back to what, especially in architecture, is a surprising gulf between theories regarding how we understand, on the one hand, what is designed and, on the other, the process through which design occurs. Recent work has addressed this in part by seeing architecture in terms of its place within the building industry (Lloyd Thomas, Amhoff & Beech 2016). From the vantage point of SOC, there are further, more designerly opportunities for bridging between these areas. The work of **Jelić** is significant in this regard, establishing an account of architectural experience in commensurable terms to constructivist accounts of design

1. https://www.theguardian.com/artanddesign/2013/feb/26/louis-kahn-brick-whisperer-architect

practice. I have previously suggested there is potential in connecting conversational accounts of design with conversational accounts of architectural experience (Sweeting 2011), while in the context of the target article one can also understand particular examples such as the Fun Palace as being part of SOC enquiry not just resulting from it (**Cabral** §9; **Jelić** §4; Sweeting 2015a).

16. The building of such bridges does not, however, guarantee in what manner they will be crossed. In making the argument in the target article – that design is a form of SOC even where SOC is not explicitly referenced – it was important for me to refer to work in design beyond figures such as Cedric Price, Nicholas Negroponte and John Frazer, who were explicitly influenced by cybernetic ideas. My reference to Peter Eisenman is not therefore intended to validate his architecture but to point to the formal similarities between his work and second-order science (SOS) that are of interest whatever we think of his proposals. Indeed, the sort of critiques put forward by **Cabral** and others, such as that of Robin Evans (1985), may inform how SOS and SOC can be developed: as **Cabral** (§8) puts it, "it is not enough to be self-reflexive and simply engaged to explore the full potential of being an SOC observer." The question of how to design such systems is an open one, and a topic on which design research and cybernetics might collaborate.

Combined References

Alexander C. (1964) Notes on the synthesis of form. Harvard University Press, Cambridge MA.

Alexander C. (1984) The state of the art in design methods. In: Cross N. (ed.) Developments in design methodology. Wiley, Chichester: 309–316. Originally published in 1971.

Alrøe H. F. & Noe E. (2014) Second-order science of interdisciplinary research: A polyocular framework for wicked problems. Constructivist Foundations 10(1): 65–76. http://constructivist.info/10/1/065

Arbib M. A. (2013) (Why) should architects care about neuroscience? In: Tidwell P. (ed.) Architecture and neuroscience: A tapio wirkkala-rut bryk design reader. Tapio Wirkkala Rut Bryk Foundation, Espoo: 43–76.

Archer B. (1979) Design as a discipline. Design Studies 1(1): 17–20.

Ashby W. R. (1957) An introduction to cybernetics. Chapman & Hall, London.

Ashby W. R. (1991) General systems theory as a new discipline. In: Klir G. J. (ed.) Facets of systems science. Plenum Press, New York: 249–257. Originally published in 1958.

Baron P., Glanville R., Griffiths D. & Sweeting B. (eds.) (2015) Living in Cybernetics: Papers from the 50th Anniversary Conference of the American Society for Cybernetics. Special double issue of Kybernetes 44(8/9).

Boden M. (2006) Mind as machine, a history of cognitive science. Volume 1. Oxford University Press, Oxford.

Brand S., Bateson G. & Mead M. (1976) For God's sake, Margaret: Conversation with Gregory Bateson and Margaret Mead. CoEvolutionary Quarterly 10: 32–44. http://www.oikos.org/forgod.htm

Bremaud I. (2012) Acoustical properties of wood in string instruments soundboards and tuned idiophones: Biological and cultural diversity. Journal of the Acoustical Society of America, Acoustical Society of America 2012: 131(1): 807–818. https://hal.archives-ouvertes.fr/hal-00808347.

Broadbent G. & Ward A. (1969) Design methods in architecture. Lund Humphries, London.

Bédard J.-F. (ed.) (1994) Cities of artificial excavation: The work of Peter Eisenman, 1978–1988. Rizzoli International, New York.
Crease R. P. & Lutterbie J. (2010) Technique. In: Krastner D. & Saltz D. A. (eds.) Staging philosophy: Intersections of theater, performance, and philosophy. University of Michigan Press, Ann Arbor: 160–179.
Cross N. (1982) Designerly ways of knowing. Design Studies 3(4): 221–227.
Cross N. (2007a) Designerly ways of knowing. Birkhäuser, Basel.
Cross N. (2007b) From a design science to a design discipline: Understanding designerly ways of knowing and thinking. In: Michel R. (ed.) Design research now: Essays and selected projects. Birkhäuser, Basel.
Downton P. (2004) Studies in design research: Ten epistemological pavilions. RMIT University Press, Melbourne.
Dubberly H. & Pangaro P. (2007) Cybernetics and service-craft: Language for behavior-focused design. Kybernetes 36(9/10): 1301–1317.
Dubberly H. & Pangaro P. (2015) How cybernetics connects computing, counterculture, and design. In: Hippie modernism: The struggle for utopia. Walker Art Center, Minneapolis MN: 1–12. http://www.dubberly.com/articles/cybernetics-and-counterculture.html
Durrant A. C., Vines J., Wallace J. & Yee J. (2015) Developing a dialogical platform for disseminating research through design. Constructivist Foundations 11(1): 401–434. http://constructivist.info/11/1/401
Eno B. (2011) Composers as gardeners. Edge 11.10.11. https://www.edge.org/conversation/composers-as-gardeners.
Evans R. (1985) Not to be used for wrapping purposes: Peter Eisenman: Fin d'Ou T Hou S. AA Files 10: 68–78. http://www.jstor.org/stable/29543477
Feyerabend P. K. (1970) Against method. In: Radner M. & Winokur S. (eds.) Analyses of theories and methods of physics and psychology. Volume IV. University of Minnesota Press, Minneapolis: 17–130. Retrieved from http://www.mcps.umn.edu/philosophy/completeVol4.html
Feyerabend P. K. (1982) Science in a free society. Verso, London. Originally published in 1978.
Feyerabend P. K. (1993) Against method. Third edition. Verso, London. Originally published in 1975.
Fischer T. (2015) Wiener's prefiguring of a cybernetic design theory. IEEE Technology and Society Magazine 34(3): 52–59.
Fischer T. & Herr C. (2007) The designer as toolbreaker? Probing tool use in applied generative design. In: CAADRIA2004: Proceedings of the 12th international conference on computer aided architectural design research in Asia, Nanjing, China 19–21 April 2007: 367–375.
Fischer T. & Richards L. D. (2015) From goal-oriented to constraint-oriented design: The cybernetic intersection of design theory and systems theory. Leonardo Journal, in press. http://cepa.info/2299
Flusser V. (2011) Into the universe of technical images. University of Minnesota Press, Minneapolis.
Foerster H. von (1962) Perception of form in biological and man-made systems. In: Zagorski E. J. (ed.) Transactions of the I. D. E. A. Symposion. University of Illionois, Urbana IL: 10–37. http://cepa.info/1612
Foerster H. von (1992) Ethics and second-order cybernetics. Cybernetics & Human Knowing 1(1): 9–20. http://cepa.info/1742
Foerster H. von (ed.) (1995) Cybernetics of cybernetics: Or, the control of control and the communication of communication. Second edition. Future Systems, Minneapolis MN. Originally published in 1974.
Foerster H. von (2003a) Cybernetics of cybernetics. In: Understanding understanding: Essays on cybernetics and cognition. Springer, New York: 283–286. Originally published in: Krippendorff K. (ed.) (1979) Communication and control. Gordon and Breach, New York: 5–8. http://cepa.info/1707

Foerster H. von (2003b) Understanding understanding: Essays on cybernetics and cognition. Springer, New York.
Foerster H. von & Poerksen B. (2002) Understanding systems. Translated by K. Leube. Kluwer Academic, New York.
Frayling C. (1993) Research in art and design. Royal College of Art Research Papers 1(1): 1–5.
Frazer J. (1993) The architectural relevance of cybernetics. Systems Research 10(3): 43–48.
Frazer J. (1995) An evolutionary architecture. Architectural Association, London. http://www.aaschool.ac.uk/publications/ea/intro.html
Furtado Cardoso Lopes G. M. (2008) Cedric Price's generator and the Frazers' systems research. Technoetic Arts 6(1): 55–72.
Furtado Cardoso Lopes G. M. (2009) Gordon Pask: Exchanges between cybernetics and architecture and the envisioning of the IE. Kybernetes 38(7/8): 1317–1331.
Furtado Cardoso Lopes G. M. (2010) Pask's encounters: From a childhood curiosity to the envisioning of an evolving environment: Exchanges between cybernetics and architecture. Edition Echoraum, Vienna.
Gage S. (2006) The wonder of trivial machines. Systems Research and Behavioral Science 23(6): 771–778.
Gage S. (2007a) Constructing the user. Systems Research and Behavioral Science 24(3): 313–322.
Gage S. (2007b) How to design a black and white box. Kybernetes 36(9/10): 1329–1339.
Gedenryd H. (1998) How designers work: Making sense of authentic cognitive activities. Cognitive Studies 75. Lund University, Sweden. http://lup.lub.lu.se/record/18828/file/1484253.pdf
Glanville R. (1992) CAD Abusing Computing. In: Mortola E. et al. (eds.) CAAD instruction: The new teaching of an architect? 10th eCAADe Conference Proceedings, Barcelona: 213–224.
Glanville R. (1994) Variety in design. Systems Research 11(3): 95–103. http://cepa.info/2785
Glanville R. (1997) A ship without a rudder. In: Glanville R. & de Zeeuw G. (eds.) Problems of excavating cybernetics and systems. BKS+, Southsea. http://cepa.info/2846
Glanville R. (1999) Researching design and designing research. Design Issues 15(2): 80–91.
Glanville R. (2002) Second order cybernetics. In: Parra-Luna F. (ed.) Systems science and cybernetics. In: Encyclopaedia of life support systems (EOLSS). EoLSS, Oxford (Web publication). http://cepa.info/2708
Glanville R. (2004a) Appropriate theory. In: Durling D., de Bono A. & Redmond J. (eds.) Futureground: Proceedings of the Design Research Society international conference 2004. Monash University, Melbourne.
Glanville R. (2004b) Desirable ethics. Cybernetics & Human Knowing 11(2): 77–88.
Glanville R. (2004c) The purpose of second-order cybernetics. Kybernetes 33(9/10): 1379–1386. http://cepa.info/2294
Glanville R. (2006a) Construction and design. Constructivist Foundations 1(3): 103– 110. http://constructivist.info/1/3/103
Glanville R. (2006b) Design and mentation: Piaget's constant objects. The Radical Designist 0. Retrieved from http://www.iade.pt/designist/pdfs/000_05.pdf
Glanville R. (ed.) (2007a) Cybernetics and design. Special double issue of Kybernetes 36(9/10).
Glanville R. (2007b) Designing complexity. Performance Improvement Quarterly 20(2): 75–96. http://cepa.info/2694
Glanville R. (2007c) Try again. Fail again. Fail better: The cybernetics in design and the design in cybernetics. Kybernetes 36(9/10): 1173–1206. http://cepa.info/2464
Glanville R. (2009a) A (cybernetic) musing: Design and cybernetics. Cybernetics & Human Knowing 16(3–4): 175–186.
Glanville R. (2009b) A (cybernetic) musing: Certain propositions concerning prepositions. In: The black b∞x, volume III: 39 steps. Edition Echoraum, Vienna: 319–329. Originally published as: (2005) Cybernetics & Human Knowing 12(3): 87–95.

Glanville R. (2009c) A (cybernetic) musing: Design and cybernetics. In: The black b∞x. Volume III: 39 steps. Edition Echoraum, Vienna: 423–425. Originally published as: (2009) Cybernetics & Human Knowing 16(3–4): 175–186.
Glanville R. (2011a) A (cybernetic) musing: Wicked problems. Cybernetics & Human Knowing 19(1–2): 163–173.
Glanville R. (2011b) Introduction: A conference doing the cybernetics of cybernetics. Kybernetes 40(7/8): 952–963.
Glanville R. (ed.) (2012) Trojan horses: A rattle bag from the 'Cybernetics: Art, design, mathematic – A meta-disciplinary conversation' post-conference workshop. Edition echoraum, Vienna.
Glanville R. (2013) Radical constructivism = second order cybernetics. Cybernetics & Human Knowing 19(4): 27–42. http://cepa.info/2695
Glanville R. (2014a) Acting to understand and understanding to act. Kybernetes 43(9/10): 1293–1300.
Glanville R. (2014b) Freedom and the machine. Text of inaugural professorial lecture. University College London, 10 March 2010. In: Glanville R., The black b∞x. Volume II: Living in cybernetic circles. Edition echoraum, Vienna: 61–81.
Glanville R. (2014c) Why design research? In: Glanville R., The black b∞x. Volume II: Living in cybernetic circles. Edition Echoraum, Vienna: 111–120. Originally published in: Jacues R. & Powell J. (eds.) (1981) Design, science, method: Proceedings of the 1980 Design Research Society conference. Westbury House, Guildford: 86–94.
Glanville R. (2015) The sometimes uncomfortable marriages of design and research. In: Rogers P. A. & Yee J. (eds.) The Routledge companion to design research. Routledge, London: 9–22. http://cepa.info/2799
Glanville R., Griffiths D. & Baron P. (eds.) (2014) A circularity in learning. Special double issue of Kybernetes 43(9/10).
Glanville R. & Sweeting B. (eds.) (2011) Cybernetics: Art, design, mathematics – A meta-disciplinary conversation: Papers from the 2010 conference of the American Society for Cybernetics. Special double issue of Kybernetes 40(7/8).
Glasersfeld E. von (1990) An exposition of constructivism: Why some like it radical. In: Davis R. B., Maher C. A. & Noddings N. (eds.) Monographs of the Journal for Research in Mathematics Education #4. National Council of Teachers of Mathematics, Reston VA: 19–29. http://cepa.info/1415
Glasersfeld E. von (1995) Radical constructivism: A way of knowing and learning. Falmer Press, London.
Goodbun J. (2011) Gregory Bateson's ecological aesthetics: An addendum to urban political ecology. Field 4(1): 35–46. http://www.field-journal.org/uploads/file/2011 Volume 4/field-journal_Ecology.pdf
Gooding D. (1992) Putting agency back into experiment. In: Pickering A. (ed.) Science as practice and culture. University of Chicago Press, Chicago IL: 65–112.
Grand S. & Jonas W. (2012) Mapping design research. Birkhäuser, Basel.
Halprin L. (1969) The RSVP cycles: Creative processes in the human environment. George Brazillier, New York.
Herr C. M. (2011) Mutually arising abstract and actual. Kybernetes 40(7/8): 1030–1037.
Herr C. M. (2015a) Second order cellular automata to support designing. Kybernetes 44(8/9): 1251–1261.
Herr C. M. (2015b) The big picture: Connecting design, second order cybernetics and radical constructivism. Cybernetics & Human Knowing 22(2–3): 107–114. http://cepa.info/2468
Herr C. & Fischer T. (2004) Using hardware cellular automata to simulate use in adaptive architecture. In: CAADRIA2004: Proceedings of the 9th International Conference on Computer Aided Architectural Design Research in Asia, Seoul, Korea 28–30 April 2004: 815–828.

Herr C. & Fischer T. (2013) Systems for showing and repurposing: A second-order cybernetic reflection on some cellular automata projects. Journal of Mathematics and System Science 3(2013): 201–216. http://cepa.info/2323

Hohl M. & Sweeting B. (eds.) (2015) Composing conferences. Special issue of Constructivist Foundations 11(1). http://constructivist.info/11/1

Holl S., Pallasmaa J. & Pérez-Gómez A. (2006) Questions of perception: Phenomenology of architecture. William Stout, San Francisco CA.

Hollingsworth R. & Müller, K. H. (2008) Transforming socio-economics with a new epistemology. Socio-Economic Review 6(3): 395–426.

Irwin T. (2015) Transition design: A proposal for a new area of design practice, study and research. Design and Culture 7(2): 229–246.

Jelić A. (2015) Designing "pre-reflective" architecture: Implications of neurophenomenology for architectural design and thinking. Ambiances – International Journal of Sensory Environment, Architecture, and Urban Space: 11 September 2015. http://ambiances.revues.org/628

Jelić A., Tieri G., De Matteis F., Babiloni F. & Vecchiato G. (2016) The enactive approach to architectural experience: A neurophysiological perspective on embodiment, motivation, and affordances. Frontiers in Psychology 7. http://journal.frontiersin.org/article/10.3389/fpsyg.2016.00481

Jonas W. (2007a) Design research and its meaning to the methodological development of the discipline. In: Michel R. (ed.) Design research now: Essays and selected projects. Birkhäuser, Basel: 187–206.

Jonas W. (2007b) Research through DESIGN through research: A cybernetic model of designing design foundations. Kybernetes 36(9/10): 1362–1380.

Jonas W. (2012) Exploring the swampy ground. In: Grand S. & Jonas W. (eds.) Mapping design research. Birkhäuser, Basel: 11–41.

Jonas W. (2014) The strengths/limits of Systems Thinking denote the strengths/limits of Practice-Based Design Research. FORMakademisk 7(4): Article 1: 1–11.

Jonas W. (2015a) A cybernetic model of design research: Towards a trans-domain of knowing. In: Rogers P. A. & Yee J. (eds.) The Routledge companion to design research. Routledge, London: 23–37.

Jonas W. (2015b) Research through design is more than just a new form of disseminating design outcomes. Constructivist Foundations 11(1): 32–36. http://constructivist.info/11/1/032

Jones J. C. (1984) How my thoughts about design methods have changed during the years. In: Essays in design. Wiley, Chichester: 13–27. Originally published in 1974.

Jones P. (2014) Design research methods in systemic design. In: Sevaldson B. & Jones P. (eds.) Proceedings of the Third Symposium of Relating Systems Thinking to Design (RSD3). Oslo School of Architecture and Design, Oslo. http://systemic-design.net/rsd3-proceedings/

Knorr Cetina K. (1992) The couch, the cathedral, and the laboratory: On the relationship between experiment and laboratory in science. In: Pickering A. (ed.) Science as practice and culture. University of Chicago Press, Chicago IL: 113–138.

Kolarevic B. (2003) Architecture in the digital age: Design and manufacturing. Taylor & Francis, New York.

Krippendorff K. (2007) The cybernetics of design and the design of cybernetics. Kybernetes 36(9/10): 1381–1392. http://cepa.info/2463

Krueger T. (2007) Design and prosthetic perception. Kybernetes 36(9/10): 1393–1405.

Lautenschlaeger G. & Pratschke A. (2011) Don't give up! Media art as an endless conversational process. Kybernetes 40(7/8): 1090–1101.

Leach N. (2012) Parametrics explained. In: Leach N. & Yuan P. F. (ed.) Scripting the future. Tongij University Press, Shanghai. Reprinted as: Leach N. (2014) Parametrics explained. Next Generation Building 1(1): 8–15.

Littlewood J. (1964) A laboratory of fun. New Scientist 22(391): 432–433.

Lloyd Thomas K., Amhoff T. & Beech N. (eds.) (2016) Industries of architecture. Routledge, London.

Lobsinger M. L. (2000) Cybernetic theory and the architecture of performance: Cedric Price's Fun Palace. In: Goldhagen S. W. & Legault R. (eds.) Anxious modernisms: Experimentation in post-war architectural culture. MIT Press, Cambridge MA: 119–139.

Lynn G. (ed.) (1993) Folding in architecture. Architectural Design Profile 102: 8–15.

Mallgrave H. F. (2011) The architect's brain: Neuroscience, creativity, and architecture. John Wiley & Sons, Chichester.

Mallgrave H. F. (2013) Architecture and embodiment: The implications of the new sciences and humanities for design. Routledge, New York NY.

Mathews S. (2005) The Fun Palace: Cedric Price's experiment in architecture and technology. Technoetic Arts: A Journal of Speculative Research, 3(2): 73–91.

Mathews S. (2006) The Fun Palace as virtual architecture: Cedric Price and the practices of indeterminacy. Journal of Architectural Education 59(3): 39–48.

Mathews S. (2007) From agit-prop to free space: The architecture of Cedric Price. London: Black Dog.

Maturana H. R. (1990) Science and daily life: The ontology of scientific explanations. In: Krohn W., Kuppers G. & Nowotny H. (eds.) Selforganization: Portrait of a scientific revolution. Kluwer, Dordrecht: 12–35. http://cepa.info/607

Mead M. (1968) Cybernetics of cybernetics. In: Foerster H. von, White J. D., Peterson L. J. & Russell J. K. (eds.) Purposive systems. Spartan Books, New York: 1–11. http://cepa.info/2634

Medawar P. (1996) Is the scientific paper a fraud? In: The strange case of the spotted mice and other classic essays on science. Oxford University Press, Oxford: 33–39. Originally published in 1963.

Michel R. (2007) Design research now: Essays and selected projects. Birkhäuser, Basel.

Müller K. H. (2008) The new science of cybernetics: The evolution of living research designs. Volume I: Methodology. Edition Echoraum, Vienna.

Müller K. H. (2010) The radical constructivist movement and its network formations. Constructivist Foundations 6(1): 31–39. http://constructivist.info/6/1/031

Müller K. H. (2011) The new science of cybernetics: The evolution of living research designs. Volume II: Theory. Edition Echoraum, Vienna.

Müller K. H. (2014) Towards a general methodology for second-order science. Journal of Systemics, Cybernetics and Informatics 12(5): 33–42. http://cepa.info/2786

Müller A. & Müller K. H. (eds.) (2007) An unfinished revolution? Heinz von Foerster and the Biological Computer Laboratory (BCL): 1958–1976. Edition Echoraum, Vienna.

Müller K. H. & Müller A. (2011) Foreword: Re-discovering and re-inventing Heinz von Foerster. Cybernetics & Human Knowing 18(3–4): 5–16.

Müller K. H. & Riegler A. (2014a) A new course of action. Constructivist Foundations 10(1): 1–6. http://constructivist.info/10/1/001

Müller K. H. & Riegler A. (2014b) Second-order science: A vast and largely unexplored science frontier. Constructivist Foundations 10(1): 7–15. http://constructivist.info/10/1/007

Negroponte N. (1975) Soft architecture machines. MIT Press, Cambridge MA.

Nicolescu B. (2012) Transdisciplinarity: The hidden third, between the subject and the object. Human & Social Studies 1(1): 13–28.

Nowotny H., Scott P. & Gibbons M. (2006) Re-thinking science: Mode 2 in societal context. In: Carayannis E. G. & Campbell D. F. J. (eds.) Knowledge creation, diffusion, and use in innovation networks and knowledge clusters. A comparative systems approach across the United States, Europe and Asia. Praeger, Westport CT: 39–51.

Oxman R. (2006) Theory and design in the first digital age. International Journal of Design Studies 27: 229–265.

Pallasmaa J. (2005) The eyes of the skin: Architecture and the senses. John Wiley & Sons, Chichester.
Pask G. (1963) The conception of a shape and the evolution of a design. In: Jones J. C. & Thornley D. G. (eds.) Conference on Design Methods, September, 1962. Pergamon Press, Oxford: 153–167. Retrieved from http://pangaro.com/pask/pask conception of shape and evolution of design.pdf
Pask G. (1969) The architectural relevance of cybernetics. Architectural Design 39(9): 494–496. http://cepa.info/2696
Pask G. (1976) Conversation theory: Applications in education and epistemology. Elsevier, Amsterdam. Retrieved from http://pangaro.com/pask/ConversationTheory.zip
Pickering A. (ed.) (1992) Science as practice and culture. University of Chicago Press, Chicago IL.
Pickering A. (1993) The mangle of practice: Agency and emergence in the sociology of science. American Journal of Sociology 99(3): 559–589.
Pickering A. (1995) The mangle of practice: Time, agency, and science. University of Chicago Press, Chicago IL.
Pickering A. (2010) The cybernetic brain: Sketches of another future. University of Chicago Press, Chicago IL.
Popper K. R. (1976) Unended quest: An intellectual autobiography. Routledge, London.
Pratschke A. (2007) Architecture as a verb: Cybernetics and design processes for the social divide. Kybernetes 36(9/10): 1458–1470.
Price C. (2003) Re: CP. Edited by H. U. Obrist. Birkhäuser, Basel.
Ramsgard Thomsen M. (2007) Drawing a live section: Explorations into robotic membranes. Kybernetes 36(9/10): 1471–1485.
Rawes P. (2007) Second-order cybernetics, architectural drawing and monadic thinking. Kybernetes 36(9/10): 1486–1496.
Rawes P. (ed.) (2013) Relational architectural ecologies: Architecture, nature and subjectivity. Routledge, London.
Richards L. D. (2010) The anticommunication imperative. Cybernetics & Human Knowing 17(1–2): 11–24. http://cepa.info/925
Richards L. D. (2015) Designing academic conferences in the light of second-order cybernetics. Constructivist Foundations 11(1): 65–73. http://constructivist.info/11/1/065
Riegler A. & Müller K. H. (eds.) (2014) Second-order science. Special issue of Constructivist Foundations 10(1). http://constructivist.info/10/1
Rietveld E. (2016) Situating the embodied mind in a landscape of standing affordances for living without chairs: Materializing a philosophical worldview. Sports Medicine: 1–6.
Rietveld E. & Kiverstein J. (2014) A rich landscape of affordances. Ecological Psychology 26(4): 325–352.
Rittel H. (1972) On the planning crisis: Systems analysis of the "first and second generations." Bedriftskonomen 8: 390–396.
Rittel H. & Webber M. (1973) Dilemmas in a general theory of planning. Policy Sciences 4: 155–169.
Rittel H. & Webber M. (1984) Planning problems are wicked problems. In: Cross N. (ed.) Developments in design methodology. Wiley, Chichester: 135–144.
Rogers P. A. & Yee J. (eds.) (2015) The Routledge companion to design research. Routledge, London.
Schneidewind U. & Augenstein K. (2012) Analyzing a transition to a sustainability-oriented science system in Germany. Environmental Innovation and Societal Transitions 3: 16–28.
Schön D. A. (1988) Designing: Rules, types and worlds. Design Studies 9(3): 181–190.
Schön D. A. (1991) The reflective practitioner: How professionals think in action. Arena, Farnham. Originally published in 1983.
Scott B. (2003) "Heinz von Foerster – An appreciation" (Revisited). Cybernetics & Human Knowing 10(3–4): 137–149. http://cepa.info/2697
Scott B. (2004) Second-order cybernetics: An historical introduction. Kybernetes 33(9/10): 1365–1378.

Scott B. (2011) Explorations in second-order cybernetics: Reflections on cybernetics, psychology and education. Edition Echoraum, Vienna.
Shannon C. E. (1948) A mathematical theory of communication. The Bell System Technical Journal 27: 379–457. http://ieeexplore.ieee.org/stamp/stamp.jsp?tp=&arnumber=6773024
Simon H. A. (1996) The sciences of the artificial. Third edition. MIT Press, Cambridge MA. Originally published in 1969.
Snow C. P. (1961) The two cultures and the scientific revolution. Cambridge University Press, New York.
Spiller N. (ed.) (2002) Cyber_reader: Critical writings for the digital era. Phaidon Press, London.
Spiller N. (2006) Visionary architecture: Blueprints of the modern imagination. Thames and Hudson, London.
Sweeting B. (2011) Conversing with drawings and buildings: From abstract to actual in architecture. Kybernetes 40 (7/8): 1159–1165. http://cepa.info/1002
Sweeting B. (2014) Architecture and undecidability: Explorations in there being no right answer – Some intersections between epistemology, ethics and designing architecture, understood in terms of second-order cybernetics and radical constructivism. PhD Thesis, UCL, London. http://discovery.ucl.ac.uk/1443544
Sweeting B. (2015a) Architecture and second order science. In: Proceedings of the 59th Annual Meeting of the ISSS 1(1): 1–6. http://cepa.info/2843
Sweeting B. (2015b) Conversation, design and ethics: The cybernetics of Ranulph Glanville. Cybernetics & Human Knowing 22(2–3): 99–105. http://cepa.info/2845
Sweeting B. (2015c) Cybernetics of practice. Kybernetes 44(8/9): 1397–1405.
Sweeting B. & Hohl M. (2015) Exploring alternatives to the traditional conference format: Introduction to the special issue on composing conferences. Constructivist Foundations 11(1): 1–7. http://constructivist.info/11/1/001
Toffler A. (1970) Future shock: A study of mass bewilderment in the face of accelerating change. Bodley Head, London.
Trimingham R. (2008) The role of values in design decision-making. Design and Technology Education: An International Journal 13(2): 37–52.
Turnbull D. (2000) Masons, tricksters and cartographers: Comparative studies in the sociology of scientific and indigenous knowledge. Routledge, London.
Umpleby S. A. (2003) Heinz von Foerster and the Mansfield amendment. Cybernetics & Human Knowing 10(3–4): 187–190. http://cepa.info/1876
Umpleby S. A. (2014) Second-order science: Logic, strategies, methods. Constructivist Foundations 10(1): 16–23. http://constructivist.info/10/016
Umpleby S. A. & Dent E. (1999) The origins and purposes of several traditions in systems theory and cybernetics. Cybernetics and Systems 30(2): 79–103. http://cepa.info/2698
Upitis A. (2013) Alexander's choice: How architecture avoided computer aided design c. 1962. In: Dutta A. (ed.) A second modernism: MIT, architecture, and the "techno-social" moment. SA+P Press, Cambridge MA: 474–505.
van Ditmar D. F. & Glanville R. (eds.) (2013) Listening: Proceedings of ASC conference 2011. Special double issue of Cybernetics & Human Knowing 20(1–2).
Vörös S. (2014) The uroboros of consciousness: Between the naturalisation of phenomenology and the phenomenologisation of nature. Constructivist Foundations 10(1): 96–104. http://constructivist.info/10/1/096
Wiener N. (1950) The human use of human beings: Cybernetics and society. Eyre and Spottiswoode, London.
Zumthor P. (1999) Thinking architecture. Birkhäuser, Basel.

"Black Box" Theatre: Second-Order Cybernetics and Naturalism in Rehearsal and Performance

Tom Scholte

Introduction

1. Whether in print or in person, when highlighting the importance of James Clerk Maxwell's "black box" *gedanken-experiment* to his own work, Ranulph Glanville would often playfully gesture to the black box theatre (a simple, flexible, unadorned performance space with a flat floor and no proscenium arch) as decidedly not the kind of Black Box to which he was referring. It is with a similar mix of playfulness and earnestness (and as a small but heartfelt memorial to his enormous contributions) that I gesture back to the theatrical black box as precisely the place where that other kind of Black Box might be fruitfully investigated.

2. As a practitioner and teacher of acting and directing for the theatre, the theory explicated herein was inspired by the growth of my engagement with cybernetics to include not only my initial "first-order" concerns with the ways in which, through rigorously applied cybernetic heuristics, the Stanislavski system of acting can consistently generate "believable" performances (Scholte 2015), but also "second-order" questions regarding the mechanisms through which observers (audiences) assign this sense of "belief," as well as "meaning," to these performances. The result is the theoretic formulation below, which will be experimentally investigated and analysed over the next several years.

3. At all phases of the work as it has unfolded, the list of its second-order cybernetic implications has continued to grow and, along with it, my nascent belief that not only could second-order cybernetics bridge theoretical gaps between post-structuralists and cognitivists within theatre studies, but that naturalist theatre (along with related offshoots including Theatre of the Oppressed and psychodrama) could provide a hitherto untapped laboratory for the generation of quantitative and qualitative research pertaining to several dimensions of second-order cybernetics,[1] particularly cybersemiotics, which, as a result, might end up

1. One reviewer suggested that cybernetics, if it is indeed a science at all, is inductive rather than empirical and is, therefore, under no obligation to run laboratory experiments. This and other conceptions of the disciplinary status of the field, including Gordon Pask's eventual disavowal of the label "science" in favor of "applied epistemology" (Pask 1980b) and Lowell Christy's recent

better positioned to help dissolve onto-epistemological deadlocks between constructivists and realists of all stripes across the academy and beyond.[2] In my view, this represents a significant new approach to second-order research, indicating that, fifty years after its inception, the theoretical advances of second-order cybernetics are not only continuingly relevant, but, in vast areas of the humanities, still to be explored. My hope is that the theoretical considerations discussed here might inspire others to bring their expertise to this fledgling research program.

4. I will begin by outlining the current, hotly debated onto-epistemological deadlock in theatre studies that has been the backdrop against which my work has been carried out. I will then introduce the cybernetic fundamentals of the Stanislavski system of acting (the methodology most commonly employed in the rehearsal and performance of naturalistic plays). I will then further excavate the second-order cybernetic depths of naturalist theatre's methods of production and reception while proposing ways in which second-order cybernetics might be operationalized, through the application of elements of Gordon Pask's conversation theory (CT), as a cybersemiotic intervention in the aforementioned dispute within theatre studies as well as other similar outstanding questions across the humanities and social sciences. Conversely, it is my hope that the employment of the naturalist theatre as a "research site" might help the field of second-order cybernetics further refine its theoretical foundations lending credence to Ross Ashby's assertion that "the discovery that two fields are related leads to each branch helping in the development of the other" (Ashby 1956: 4).

The impasse in theatre studies

5. A "Cognitive Turn" has swept through theatre studies, along with many other disciplines within the humanities, during the opening decades of the 21st century. The previous era of high postmodernism had been a dangerous time for any scholar willing to make a statement even remotely resembling some kind of universal truth-claim, and this had been the case particularly for university-based teachers of the Stanislavski system of acting, grounded, as it is, in 19th century notions of the "self" and the "truth of the human spirit" (Stanislavski 2008: 19).

6. In terms of theatrical practice as a whole, the genre of realism, the performance mode most often conflated with the theoretical commitments of naturalism, and where Stanislavski's methods have traditionally been most at home, had been hit particularly hard. Considered "cranky, weary, old-fashioned, prescriptive, middle-class, and formulaic," realism is described by Amy Holzapfel as "one of the most berated dramatic forms within contemporary theory," and as commonly considered "a conservative force that reproduces and reinforces dominant cultural relations" (Holzapfel 2014: 5).

comment to me that the field's difficulties establishing an institutional home for itself are due, in part, to the fact that it is actually an "anti-discipline," certainly deserve consideration.

2. There is, of course, nothing like a total embrace of cybersemiotics amongst cyberneticians, but its insistence on the role of observer-dependent onto-epistemology in the interpretation and description of both symbolic and non-symbolic phenomena makes it sufficiently compatible with other strands of second-order research to be of wide experimental usefulness across the field.

7. Regarding the Stanislavski system in particular, an early paper by Rhonda Blair vividly diagnoses its American descendent, "the Method," as a similar casualty of the critical theoretic onslaught; pilloried for its reification of "a non-existent "self" and accused of "colluding in mechanisms of representation that serve the ends of decadent late capitalism," its "ahistorical, noncritical, sentimentalized or sensationalized view of experience" is said to participate in a dubious "humanist project of universalizing experience" that is "inherently patriarchal and misogynist" (Blair 2000: 202).

8. But, as the new millennium dawned, a cluster of theatre scholar/practitioners, frustrated by the fact that "few scholars in [the] field have asked the advocates of Bordieu, Lacan, and Derrida [...] to justify the epistemology of their theories on scientific grounds" (McConachie 2008: 9), grabbed hold of cognitive science as a means of "defying the sceptical paralysis that characterizes much of postmodern thought" (Rokotnitz 2011: 5) and rebuilding our ability to "trust performance." Undergirding this view is the belief that "[e] mbodied modes of reception and perception are those that do not require logical analysis for their verification; their presence and effects are made manifest in the body" and that this bodily apprehension and confirmation provides us with "truths of which we may be sufficiently satisfied in order to trust with confidence" (ibid: 2).

9. Regarding Stanislavski's position in this reappraisal, the early days of the movement saw Blair offer the thoroughly isomorphic relationship between the Method's terminology and that of the computational theory of mind as a potential arrow in the quiver of the theatrical cognitivists.

> Behavior that is adaptive is designed to accomplish or prevent something or to achieve a goal, which is *action* in the Method's sense. 'Opportunities' and 'constraints' defined by 'other people's behavior' are analogous with [the Method's] 'objectives,' 'obstacles' and 'given circumstances.' From this perspective, key parts of the Method are fundamentally Darwinian and biological, not necessarily Romantic, modernist, humanist, psychological, Western, and spiritual. If this theory is correct, it means that the appeal of the method is due not necessarily (and only) to the hegemony of various kinds of realism and habits of representation and mimesis but to the fact that this is how we function organically. From this perspective, self and behavior are grounded in our being as physical organisms while also being affected by culture and conscious choice. (Blair 2000: 206)

10. The obvious short step from the computational theory of mind to the goal-directed and feedback-controlled conception of cybernetics, and its place within a unified theory of human behavior (see Grinker 1958), could, potentially, position the field to play its own role in the cognitive turn and its insistence that...

> while doubt is necessary for rigorous analysis, and essential for the advancement of intelligent inquiry, art and drama can also teach us of other forms of knowing that are no less valid, rigorous, or true. (Rokotnitz 2011: 12)

11. But, while the notion that art and drama may provide access to "other [...] no less valid [...] forms of knowing" is appealing, the notion that such forms of knowing might be "true" speaks of a desire for the kind of ontological certainty that is the hallmark, and blind-spot, of first-order science (and, for that matter, first-order cybernetics). And it would, in my view, be most unfortunate if this blind-spot were to prove the very undoing of the cognitive turn, as the movement offers an opportunity to confront certain persistent phenomena that have not been satisfactorily explained away by a strictly social constructivist semiotic approach.

12. Many theatre practitioners continue to work with notions of "truth" in performance, along with a sustained belief that representational drama in the realist/naturalist tradition can contain genuine epistemic goods. Arguably, a large percentage of mainstream theatre audiences would agree, at least tacitly, with this notion. Countless acting students are routinely stunned by the sudden emergence of "truthful" behaviour under imaginary circumstances when their colleagues first begin to effectively execute the cybernetic principles of the Stanislavski system. Peer reflections inevitably include such comments as: "It was great to hear you use your real voice instead of your acting voice," "I've never seen your movements and your gestures seem so natural," "It was just like watching you being yourself in that situation except the words weren't the words you would use." But, of course, these accounts have long been inadmissible as "evidence" in the postmodern court of opinion since words such as "real," "natural," and "self" have been effectively stricken from the record. They are, of course, similarly suspect in second-order cybernetic discourse. So what are we to make of the persistent phenomena attested to daily in classrooms and professional rehearsal halls throughout the Anglo-American and European world for more than a century? What can we say about this difference between acting that is commonly called "good" and that which is called "bad" and that seems to have something to do with a kind of relationship to the behaviour we recognize from our everyday observations and that is fairly dependably produced by the application of cybernetic principles to the world of "make-believe"?

13. Semiotically-minded theatre scholars such as Ric Knowles have already pushed back against the "proselytising […] excesses of [the cognitive turn's] still early days" and declared the phenomenological "lack of mediation" it vouches for to be "a theoretical ideal rather than a practical possibility" (Knowles 2015: 83–87). Interestingly, Knowles's defence of semiotics' alleged "unscientific errors" also gestures towards its integration with "biological structuring" in the work of Jakob von Uexküll, a scientist whose biosemiotics have also played such a pivotal role in Søren Brier's cybersemiotic formulation. This seems to set the stage nicely for a second-order intervention that might provide a productive "middle way" between these two contesting camps in theatre studies.

Stanislavski's cybernetic system

14. Scholte (2015) examines, in some detail, the essentially cybernetic vision at the core of the Stanislavski system of acting, first highlighted by Robert Cohen, who begins his explanation with the most often cited example of cybernetics in action: the thermostat on a regulated furnace that is, of course, "calibrated to go on when feedback reveals the room temperature is too low, and to go off when feedback reveals the opposite" (Cohen 1978: 33).

15. Cohen then points out that, in order to provoke her own organic responses on a moment-to-moment basis in performance, the actor must remain responsive to the feedback emanating from her fellow actors that will let her know whether she, like the regulated furnace, is getting closer to, or further from, the "objective" that she, in orthodox Stanislavskian practice, has chosen for her character to pursue throughout the scene (i.e., soliciting a declaration of love).

> This feedback includes not only conscious dialogue and conscious non-verbal signals like winks and shrugs but also the unconscious and autonomic responses of our physiology; behaviours of which we (as humans) are skilled observers and interpreters. They are all part of the feedback of information which makes us adjust our behaviour toward the satisfaction of our intentions – and sometimes to adjust our intentions as well. (1978: 36)[3]

16. Guided by this feedback, characters create what engineers would describe as "closed loop" control systems as they mutually seek to influence each other's behaviour. The mental discipline to remain genuinely engaged in such loops within imaginary circumstances lies in direct correlation to the seeming "naturalness" of performance that is the hallmark of great Stanislavskian, or Method, acting.

Active Analysis: An ordinary language description

17. Nowhere did Stanislavski harness the power of his cybernetic approach to acting more productively than in the rehearsal methods known as his "later legacies" including Active Analysis (AA). Bella Merlin outlines the process of AA as follows:

> First of all, the actors read the scene. Second, they assessed the facts of the scene… What is the event? What are the inciting objectives and counter-objectives? […] The third stage consisted of the actors improvising the scene using their own words, incorporating any of the facts that they could remember. […] Following the improvisation, the actors reread the scene and compared it with what they had just experienced. They noted which facts were retained and which were forgotten, and whether the inciting incident took place. Rehearsing […] consisted of repeating this four-stage process. […] With each new improvisation, the actors strove to add more details of events, language, and images… The fifth and final stage involved memorizing the scene… In fact, if the improvisational work had been successful, they found that the scene had virtually 'learned itself.' (Merlin 2003: 34f)

18. This initial process is meant to last for the first two thirds of the total allotted rehearsal time. In the final third, the director engages in the more traditional work of formalizing ("blocking") the final staging of the show; setting the actors' physical movements, tempos, and rhythms so that "the words of the dialogue, contained in the text, are scattered through and inscribed in the time and place of the stage" in such a way as to "make the deep meaning of the dramatic text tangibly evident" (Pavis 1998: 364). The more or less conventional gestural, proxemic, illocutionary, and temporal semiotic codes she employs to do so comprise what has become known as the "performance text" as distinct from the "dramatic text" comprised of the author's written word (Elam 1988: 3). In shaping this text, the director initiates an additional control loop that remains open when she is satisfied with what is unfolding on stage and that she closes periodically in order to modify the performances to her satisfaction.

19. According to Merlin, the fact that the actors participating in AA "[were] starting from *themselves*" allowed them to "kick-start the creative process into action" in a manner that pays off in "profound effects both in rehearsal and in performance" as "the sense of improvisation is carried all the way from first preview to last night" requiring "no 'creative'

---

3. Of course, unlike the furnace, as an organism that learns and also exercises "theory of mind," the feedback to which the actor/character responds will be continuously contextualized by the moment-to-moment predictions she is making regarding the other characters' behavior.

force' or impossible demands" (Merlin 2003: 35). After observing Stanislavski's Moscow Art Theatre in rehearsal, Norris Houghton corroborates this claim indicating that the powerful connection between actor and character forged in early rehearsals remains unbroken by the director's later interventions and that "once it is achieved, the director can in a fairly short time add shape to the performance and not disturb the actor in so doing" (Houghton 1962: 80f).

Beginnings

20. In May of 2015 I undertook an experiment, in collaboration with Alan Kingstone of the UBC Department of Psychology, to explore the potential impacts of the AA rehearsal method (seldom, if ever, employed in professional North American theatre practice for reasons discussed in Scholte 2010) upon audiences. Two actors rehearsed and performed David French's naturalist drama *Salt-Water Moon* under my direction, employing the AA rehearsal process. The play portrays an encounter on a single evening in August of 1926 in which a young man returns to his hometown in rural Newfoundland to try and win back the girl he left behind when he suddenly fled to Toronto without a word a year previous. Following the completion of the first two thirds of the process, audiences were invited to view four performances of the entire piece in which the actors adhered to the author's written text while the performance text was still allowed to unfold autopoietically each night based solely on the actors' emergent and self-organizing cybernetic response. The final third of rehearsal, in which I, as director, fixed an allopoietic performance text, was then completed followed by a further four performances for invited audiences. At all eight performances, data on the audience's qualitative appraisal of the "truthfulness" and "believability" of the actor's performances,[4] as well as overall satisfaction with the performance, was gathered through the use of Likert scale questionnaires. Additionally, data regarding audience interpretation of the piece as a whole was gathered using a method Kingstone had previously developed and employed in a study of film audience responses (see Coleman et al. 2013).

21. Participants are asked to press a button on a mobile clicker when the action is found to be "meaningful." Observers press and hold down the key and let go when the moment ends. Any segment of action during which multiple participants held down their buttons simultaneously is deemed a moment of "convergence." We hypothesized that audiences would indicate a greater sense of "believability," with fewer moments and lower levels of "convergence" around "meaning," in the autopoietic performances. The difference in degree of "audience satisfaction" with these two models remained an open question at the very heart of this experiment, the answer to which would lead to further hypothesizing as to the expectations and desires of theatrical audiences.

22. Detailed analysis of the copious data generated by this experiment is on-going and will, ultimately, be presented in a separate article co-authored with my collaborator, Dr Kingstone. But it was during the rehearsal of the production itself that the second-order cybernetic underpinnings of the AA process came into sharper focus for me, suggesting a

4. While the dialogue in certain plays may be more "heightened," or stylized, than everyday speech (even within the realm of naturalism), the indicators of "believability" found in vocal tone, body language, and dynamics remain constant. We are particularly impressed when a gifted actor achieves such an effect in, for instance, a performance of Shakespeare.

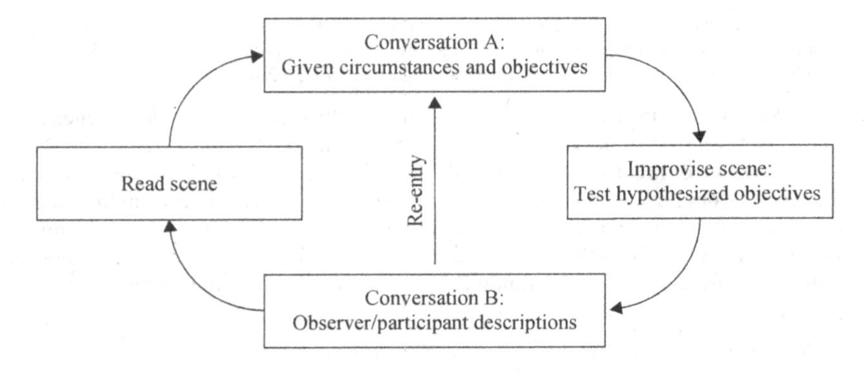

Figure 1. Active Analysis: A conversational theoretic diagram.

potential new direction for this type of inquiry that could, through the additional deployment of elements of CT, shift future experiments onto a firmly cybersemiotic footing. The explication of this notion, which will make up the remainder of this article, must begin by re-examining the process of AA itself through a distinctly second-order cybernetic lens.

Active Analysis: A conversation theoretic description

23. Figure 1 illustrates the circular nature of the first two thirds of the AA process and indicates the recursive re-entry of observations as it cycles through its successive iterations. We will examine each component of this process in some detail, highlighting the second-order cybernetic considerations implicit in their procedures and suggesting ways in which they may be made more explicit, and analytically useful, through the application of CT.

24. When a production team of actors, designers, and director begins its work, its members must confront a formal object that is both a result and embodiment of second-order observation: the playwright and his/her text. Before examining the subsequent operations taking place within the rehearsal process, it is necessary to spend some time framing the naturalistic play text itself in second-order cybernetic terms.

The naturalistic play as a Black Box

25. In Glanville's view, Maxwell's Black Box *Gedankenexperiment* provided "one of the greatest inventions of human thinking, because it allows us to develop understandings of the universe without having to claim we know what it's made up of or that we have access to actual mechanisms, emphasizing that our descriptions are descriptions, not to be confused with actuality." He defines this "fantas[tical]" (i.e., wholly imaginary) conceptual tool as follows.

> In its simplest form, the Black Box is taken to be a virtual input/output machine. A signal is input, the output observed, and then another signal (usually the most recent output) is input, producing another output – and so on. What is observed is the behaviour of this in-

> put/output machine. The collection of observations is itself observed, so we may construct a pattern. When we have constructed such a pattern, we are inclined to say that we know what the Black Box does, and that it is now 'white'. (Glanville 2012: 421)

26. Swept up by the explanatory promise of late 19th century Darwinian science (particularly as developed in the writings of Hippolyte Taine (Pickering & Thompson 2013: 15)), the program of literary naturalism (still, arguably, the most prominent dramatic genre of the present day) sought to bring a rigorous analysis of heredity and environment to bear upon works of the imagination in the hopes that they might play a role in diagnosing psychological and social pathologies. The movement's most prominent early proponent, novelist, essayist and playwright, Emile Zola, expressed the movement's aims in the following terms:

> [T]o possess a knowledge of the mechanism of the phenomena inherent in man, to show the machinery of his intellectual and sensory manifestations, under the influences of heredity and environment, such as physiology shall give them to us, and then finally to exhibit man living in social conditions produced by himself, which he modifies daily, and in the heart of which he himself experiences a continual transformation. (Zola 1893: 649)

27. Of particular note here, from a cybernetic perspective, is the invocation of the term "mechanism" to denote some set of functions within man responsible for generating the phenomena that are "give[n]" to the observer through behavioral effects in the "physiology" of the individual(s) under observation, as well as the acknowledgement of a circular rather than linear causal relationship between the modifications of both man and his "social conditions." The latter insight indicates a nascent systems perspective inherited directly by Zola from his primary scientific role model, Claude Bernard. Pre-echoes of what might be characterized as a proto-systems theoretic viewpoint can be found throughout Zola's naturalist manifesto and even foreshadows notions of both biological and social autopoiesis (Zola 1893: 650).

28. As an example of the "experimental novelist" at work, Zola points to his countryman Honoré de Balzac's portrait of the character Baron Hulot in his 1846 work, *Cousin Bette.*

> The novelist starts out in search of a truth.[...] The general fact observed by Balzac is the ravages that the amorous temperament of a man makes in his home, in his family, and in society. As soon as he has chosen his subject he starts from known facts, then he makes his experiment, and exposes Hulot to a series of trials, placing him amid certain surroundings in order to exhibit how the complicated machinery of his passions works. (Zola 1893: 647)

29. Zola's employment of the term "experiment" to describe the operations performed by the author is obviously dubious. There is clearly no actual individual who is being placed within "certain surroundings" in order that his subsequent behaviours may be observed. What the author is providing, rather, is a hypothesised pattern of behaviour that an individual possessed of certain proposed psychological mechanisms would, in the author's view, be likely to exhibit. Zola, essentially, acknowledges as much.

> The idea of experiment carries with it the idea of modification. We start, indeed, from the true facts, which are our indestructible basis; but to show the mechanism of these facts it is necessary for us to produce and direct the phenomena; this is our share of invention, here is the genius in the book. (Zola 1893: 647)

30. Despite the overblown sense of ontological certainty explicit in Zola's language, the idea that the "indestructible basis" of this imaginative elaboration are the "true facts" previously "observe[d]" by the author in the course of his societal interactions and that, furthermore, the elaboration itself is intended to "show the mechanism of these facts" renders the naturalist novel or play (as well as each of the characters with which it is populated) a type of Black Box – again, as defined by Glanville – invoked by the author, who, having observed an "unclear [...] mess" that, nonetheless, appears to be "some action which [one] might be able to call behaviour," and applied as an "ordering concept" allowing "the unclear chaotic mess" to be "(re)constructed as behaviours associated with an input-output machine, where the machine is the Black Box (the home of the mechanism" (Glanville 2009a: 153).

31. The Black Box that is the naturalistic play is now confronted by the production team in a further process of second-order analysis implicit in Conversation A.

Conversation A

Play analysis as cybernetic explanation

32. While the traditional Western theatre artist would not be likely to conceive of or articulate her work in these terms, I suggest that it is this very process of Black Box "whitening" on both the level of individual characters and the social system that they participate in constituting that motivates the entire enterprise of naturalistic theatre vis-a-vis the "deeper meaning" that the director is charged with "inscribing in the time and space of the performance."

33. The dominant Western model of play production that evolved (along with the emergence of the position of director itself) in the second half of the 19th century, and is still taught in many Master of Fine Arts Directing programs, including the one in which I teach, begins with an intensive "pre-production" period of textual analysis. Confronted with the behavioural descriptions that constitute the playwright's text, the director embarks on what is clearly a process of "cybernetic explanation" as articulated by Gregory Bateson, in that it is always "negative" and considers

> what alternative possibilities could conceivably have occurred and then ask[s] why many of the alternatives were not followed, so that the particular event was one of those few which could, in fact, occur. (Bateson 1987: 405)

34. Various textbooks on the subject of play direction have laid out overlapping categorical templates for organizing the information contained in the play and that Stanislavski grouped under the heading "given circumstances." The composite picture that emerges from this analysis can also be classified under another term. Returning to Bateson, "the course of events is said to be subject to *restraints*, and it is assumed that, apart from such restraints, the pathways of change would be governed only by equality of probability" (Bateson 1987: 405f). Actors and directors are similarly charged with the task of close textual analysis, leaving no potential restraint unaccounted. It is these restraints, including the internal mechanisms of the characters, that will govern the process described by David Ball, in which the beginning of a play presents a portrait of stasis (Claudius on the throne

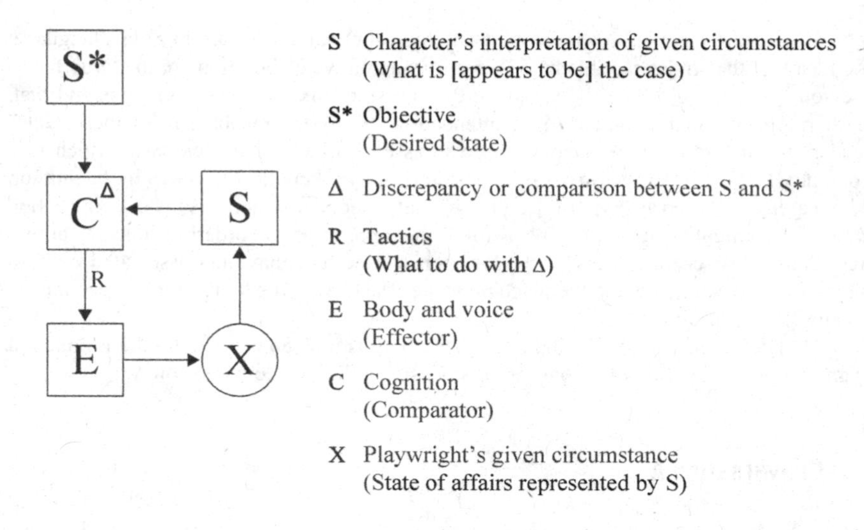

Figure 2. Character as basic feedback control system.

and Hamlet sulking silently), the stasis is disrupted by a moment of intrusion (the ghost of Hamlet's father commands his son to avenge his murder) and, by the play's end, a new stasis is established (Fortinbras on the throne and the stage littered with the corpses of Hamlet, Laertes, and Gertrude as well as the previously slain Polonius, Ophelia, Rosenkrantz and Guildenstern) (Ball 1983: 19–24). The example of Hamlet, while by no means naturalistic in the strict sense, does illustrate well a play's action as the self-reorganization of an autopoietic social system (a la Luhmann) following a substantial perturbation. The amount of ink spilled speculating upon the particular ways in which the internal mechanisms of that play's central Black Box (i.e., character) constrain the paths of that reorganization outstrip the amount spilled in the name of cybernetics by considerable orders of magnitude; and in the realm of theatre production, the structure of such mechanisms is the essential purview of the actor.

The naturalist actor as radical constructivist

35. Figure 2 adapts Jim Rebitzer's diagram of a basic feedback control system to illustrate the cybernetic conception of the actor/character in performance (Rebitzer 1995: 47; the categories from Rebitzer's original legend are included in parentheses). The existence of playwright's given circumstances and character's interpretation of given circumstances as separate categories points to the assumption of a distinct, subjective onto-epistemology for each character; the construction of which it is the actor's task to trace carefully.

36. Of the many templates on offer to assist in this task, I regularly recommend to my acting students one developed by Anita Jesse as particularly thorough. It guides the actor to answer, from the character's perspective, questions about their particular social background, relationships, attitudes, desires, fears, etc. based upon textual evidence. The

Place	The environment the characters are in	structural coupling
Character Biographies	The events in the past that shape the characters	ontogeny
Events	The changes that affect the behaviour of the characters	perturbations
Intentions	The pictures of the future that drive the present action of the characters	S*
Relationships	The thoughts about others that calibrate the behaviour of the characters	S

Table 1. Katie Mitchell's given circumstances in second-order cybernetic perspective.

final, and most encompassing, category of given circumstances in Jesse's template is titled: "What is my point of view, or how does the world work?" It is characterized as follows:

> The *point of view* is the character's belief system, the frame of reference in which the character operates. Each of the convictions that make up the belief system should be anchored in specific action. For every element of a person's belief system there is an event that either inspired the belief or at least corroborates the point of view. People don't consciously choose to *invent* convictions; they interpret their experiences as proof that 'this is how the world works.' (Jesse 1996: 136. Italics in the original)

37. In Piagetian/Glasersfeldian constructivist terms, this "point of view" contains, or is the summation of, the entire repertoire of conceptual and behavioural schemes that have proven viable over the course of the character's imagined ontogeny and have subsequently been internalized according to the innately conservative phylogenetic tendency, articulated by Maturana, of repeating what has worked in the past as the surest means of maintaining equilibration and, ultimately, the continued realization of autopoiesis (Glasersfeld 1980: 77f). In fact, the very climax of a good percentage of naturalist plays is the moment when the central character either succeeds or fails at making a necessary accommodation of the sort von Glasersfeld defines[5] (Ibsen 2015). It is, therefore, incumbent upon the actor to scour the playwright's description for reports of those formative ontogenetic episodes in the character's "back-story" (again, understood as having taken place in the hypothesised "possible world' of the play) as well as the assimilatory and accommodating schemes employed and on display in the present action. Jesse lays strong emphasis upon the need for the actor to "internalize" rather than "show" this conceptual repertoire; to get "behind the eyeballs of the character" as Cohen puts it, so that it will guide spontaneous behaviour through the recursive iterations of improvisation leading to performance.

Director and actors in "strict conversation"

38. The director will, of course, also have completed a similar process of investigation of the given circumstances. Table 1 adapts the template of renowned British director Katie Mitchell (2009:10), incorporating second-order cybernetic terminology drawing from

5. E.g., Hedda Gabler's fatal inability to conceive of Eilert Lovborg as anything other than a kind of Dionysian übermensch "with vine leaves in his hair" and to accommodate successfully his sordid end: accidentally shot through the bowels in a tawdry whorehouse.

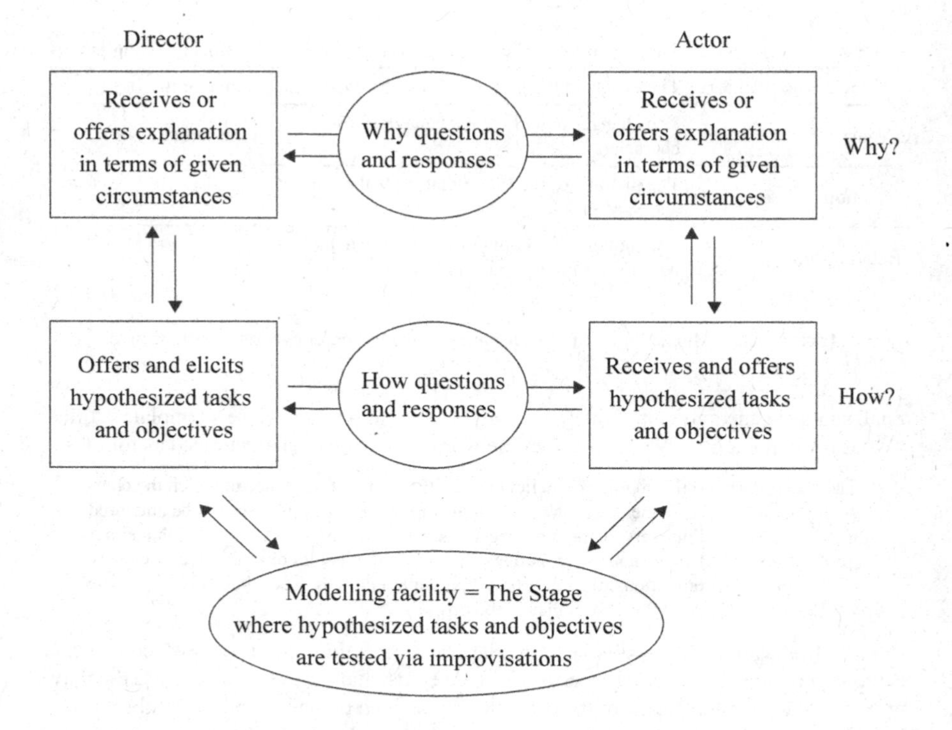

Figure 3. Skeleton of conversation A (after Scott 2011).

Maturana and Varela and referring back to the previous figure of the character as feedback control system.

39. Conversation A, in which the director and actors share the fruits of their analyses, refine and expand their findings, and, ultimately, formulate scene objectives to test through improvisation, qualifies as what Bernard Scott would call a "strict conversation." Figure 3 adapts Scott's "skeleton of a conversation" from its initial description of a teacher/student interaction through the lens of CT (Scott 2011: 310) to the particularities of the director/actor relationship. In the context of Pask's CASTE educational application, the modelling facility is the space (either literal or virtual) in which projects testing the student's level of mastery of the topic are completed. For our purposes, the modelling facility is the stage, as this is where the collaborative formulation of given circumstances and objectives will be tested for their efficacy in actor improvisation, observations of which will re-enter subsequent iterations of Conversation A. But before making that move, I propose our first application of CT in order to leave, in the words of its creator, a "residue" of the conversation that can be employed for analytical purposes after the fact.

40. As a depiction of an uninterrupted encounter between two characters, *Salt-Water Moon* provides an ideal play upon which to experiment with the graphical expression of conversation as "concept sharing" envisioned through CT. The analytical process executed

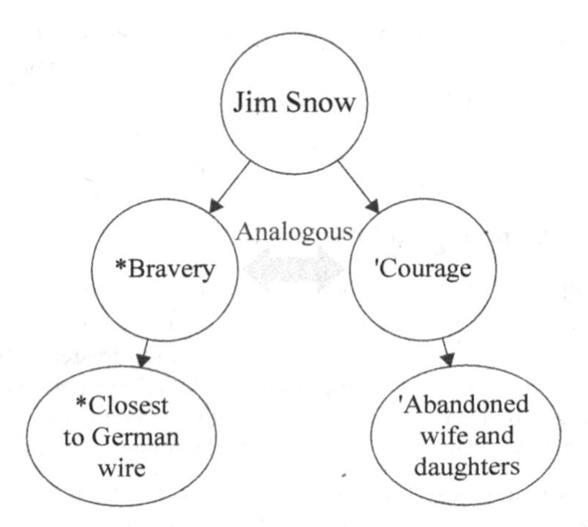

Figure 4. Entailment structure showing Jacob and Mary's operationalized concepts regarding the topic, Jim Snow. * = Jacob's concepts, ' = Mary's concepts.

by actors and director described above deals quite specifically with the identification of stable concepts within the participants (characters). Figure 4 adapts some of Pask's techniques for the construction of entailment structures in order to diagram a conversation between two participants, the result of which, if agreement is reached, some, or all, of the conceptual procedures belonging to each may now be shared by both (Pask 1980a). This adaptation reflects the concept-sharing executed in the following passage of text from the play, regarding the character Mary's late father who was killed at the battle of Beaumont-Hamel in the First World War.

> Jacob: Go on with you. Jim Snow was a brave man. The one the stretcher bearers found closest to the German wire that night when they went out to collect the dead.
>
> Mary: Yes, and a lot of good it done, his courage. He left behind two daughters and a wife who can't look after us. (French 1988: 40)

41. The entailment structure in Figure 4 establishes a schematic overview of the contesting ontologies at play in this interaction in which the concepts "brave" and "courage" are assumed to be analogous (this could, of course, be contested) while the differing entailed descriptions produce an ambiguous and jarring counterpoint. Interestingly, neither character contests the other's description allowing them fully to resonate in tension within the conceptual repertoire of the audience who, like the characters themselves, are free to take on or reject parts or all of both descriptions or combine them in novel ways. This basic entailment structure can, of course, be elaborated to express the conceptual relations between all of the topics of conversation between the two characters throughout the course of the play. That one entailment represents the conceptual operations of a hypothesized independent young man in the summer of 1926 and the other a hypothesized dependent young female housekeeper of the same time period seems ripe for the kind of social constructivist analysis intended to trouble and problematize the naturalization of such concepts

and the unremarked erasure of the contingency of their production. This is what makes the composition of such entailment meshes during Conversation A a critical component of the cybersemiotic investigation I am proposing to take place after the passions of performance have cooled.

The stage as modeling facility

42. Robert Cohen's notion that "humans are skilled observers and interpreters" of "autonomic responses" of the "physiology" of other humans commits theatre practitioners working in the Stanislavskian tradition to a biosemiotic epistemology characterized in Brier's cybersemiotic framework as "the genuine semiotic level belonging to all living systems." The "paralinguistic [...] sign games" of this level function below that of "social language systems" and serve as the scaffolding upon which they are built (Brier 2008: 390).

> It is obvious that what we call language games arise in social contexts in which we use our minds to coordinate our wilful actions and urges with fellow members of our society. Some of these language games concern our conceptions of nature as filtered through our common culture and language. But underneath that, we also have emotional and instinctual *psychological sign games.* For humans these function as unconscious paralinguistic signs, such as facial expressions, hand gestures, and body positions that originate in the evolution of species-specific signification processes in living systems. (Brier 2008: 395)

43. I posit that it is the spontaneous, non-premeditated emergence of these sign games amongst actors engaged with each other cybernetically, and their subsequent "emotional and instinctual" recognition by observers, that constitute the very phenomena categorized by acting students (and, arguably, the majority of mainstream theatre audiences) as "truthful" or "natural" and gestured to by the theatrical cognitivists as phenomenologically unmediated embodied forms of knowing.

44. In his critique of the Stanislavski system of acting, postmodern performance theorist Phillip Auslander rejects the notion that "the actor's self precedes and grounds her performance and that it is the presence of this self in performance that provides the audience with access to human truths." He asserts, instead, that "[t]he act of signification produces its own significance" and that there is "no presence behind the sign lending it authority" (Auslander 1998: 30). From the cybersemiotic perspective, however, the "presence behind the sign" is the actor's very embodiment itself. Its fluctuating autonomic and unconscious behaviors, while unconcerned with consciously signifying anything, engaged cybernetically within imaginary circumstances, enter the realm of semiosis only when indicated as distinctions by conspecific observers sharing a mutual structural coupling. Nowhere in this process, for either actor or audience, do we find the "reification" of a "non-existent autobiographical self" as alleged by Auslander and other theatre scholars of a similarly postmodern bent. It is also noticeably absent from any of the instructions Stanislavski bequeathed to his followers.[6]

---

6. Stanislavski's error, according to Auslander, is that he "treats the subconscious as what Derrida shows it is not; a repository of retrievable data, as in his famous metaphor of the house through which the actor searches for the tiny bead of an emotion memory." Citing Derrida's reading of Freud, Auslander reminds us that "the making conscious of unconscious materials is a process of creation, not retrieval" and that "[t]he unconscious is not a source of originary truth – like lan-

45. Appealing to interdisciplinary arguments "for the importance of embodiment in semiosis," Brier goes on to invoke a second-order cybernetic concept that will play a central role in the theory of theatrical "meaning making" presented below.

> Ethology and embodied metaphor theory have both discovered that the conception of a sign as standing for something for somebody in a particular way is controlled by releasing mechanisms that connect motivation, perception, and behaviour/action in one systemic process, as von Uexküll described in his *Funktionskreis* and which Heinz von Foerster refers to as perceptual *Eigenvalues*. (Brier 2008: 393, italics in the original)

46. Hypothesized given circumstances, objectives, tasks, and obstacles from Conversation A "pass the test" in the modelling facility if their operationalization within the improvisation generates behavioural eigenvalues across the group of observers present (including the actors second-order observations of their experience after the fact) that are isomorphic with eigenvalues observed in off-stage daily living, seem appropriate to the behavioural descriptions in the text, and in which the director feels reasonably confident that, in subsequent performative iterations, they will generate similar eigenvalues across new groups of observers; namely audiences. At that stage, it is probably most appropriate to describe them, after von Foerster's article introducing the concept, as eigenbehaviours (Foerster 2003a). These emerge, in the words of Bruce Clarke, when "a multiplicity of mutually reinforcing observers maintains relationships and states of stable cross-systemic resonance [...] at both the biological and social level" (Clarke 2009: 46). While the director's (and, subsequently, audience's) observations, interpretations, and assessments of the plausibility of the character's behaviour will be influenced by the "theory of mind" that they have developed for each of the individuals under observation, such behaviour could also be illustrated on a purely symbolic semiotic level by highly calculated performances of obvious "theatricality" bearing little or none of the "spark of genuine life" sought by Stanislavski and his artistic heirs. The "emotional and instinctual psychological sign games" enumerated by Brier, plus the haptic effects of organic, spontaneous vocal tonality and dynamics borne of felt, rather than feigned, emotion, all held to be pre-linguistic, pre-conceptual, and pre-symbolic by the proponents of the cognitive turn, are the exclusive domain of the biological eigenbehaviour.

47. At this stage in the rehearsal process, it is essential that the director's relationship to the emergent behaviour remain an "open" rather than "closed" loop in that she does not seek to intervene and correct the behaviour if it is discordant with either her theory of mind

---

guage, it is subject to the vagaries of mediation" (Auslander 1998: 26). While Stanislavski does, indeed, invoke the metaphor quoted above, Auslander completely distorts the spirit in which it is offered. Stanislavski fully acknowledges that a memory is not an exact replica of the event to which it refers and tells us that there is an unconscious and automatic selectivity involved over time lending "poetry to memory" and rendering it "clearer, deeper, denser, richer in content and sharper than reality itself" (ibid: 206). More importantly, he actively *discourages* his students from relying upon the deliberate, repeated reconstruction of autobiographical memories as the foundation for "truthful" performance (ibid: 207). He warns them not to "imagine for a moment" that they "can retrieve a feeling that has gone forever" and admonishes them to "give up the idea of hunting old beads – they are beyond recall" (ibid: 216). Instead, Stanislavski exhorts his charges to allow ever new and evolving emotional overlaps with their characters to arise, as much as possible, in spontaneous response to the actual sensory "stimuli" available to them in the moment of performance itself.

regarding the characters or on the level of eigenbehaviours isomorphic with observed daily life. For Stanislavski, it is essential to remain a "director of the root" rather than a "director of the result" (Cole & Chinoy 1976: 109). In other words, she must not interrupt the unfolding improvisation nor even, following the performance, ask the actors for specific behavioural alternatives in the next iteration ("faster/slower, louder/softer, more angry, more seductive, with a pause here, etc."). She must not even compare the actors' emergent behaviour with some desired ur-performance she has imagined but rather must obey the imperative of director, Ann Bogart, to watch "without desire" (Bogart & Landau 2004: 31) and allow only the "plausibility" and "believability" of the scene, regardless of its actual dynamics, to guide her response.7 Even then, she must not express those responses in terms of "right" or wrong" but seek only to address those moments and passages that seem unsatisfactory by reverse-engineering alternative behaviour through further mining the onto-epistemic mechanisms of the characters and the constraints manifest in their relationships and environment to which they must constantly adjust. This will be the business of Conversation B.

Conversation B and the return to the text

48. The only important point that needs to be raised here regarding the move through Conversation B and the beginning of the next recursion with the return to the text is the importance of maintaining the strict separation between the two steps. Conversation B is entirely concerned with distinctions and descriptions of what took place during the previous improvisation. The subsequent return to the text provides the opportunity to refine or indicate further distinctions within the playwright's descriptions of the interaction system with which the improvisation is intended to be, in ever increasing degrees through the process of recursion, isomorphic. These distinctions will now meet the re-entry of the distinctions drawn in Conversation B in the subsequent iteration of Conversation A. It is here that non-isomorphic irregularities will be identified and adjustments in the onto-epistemologies at work through the characters in the action of the play will be proposed as a means of generating new eigenbehaviours. Beyond the first iteration of Conversation A, the onto-epistemologies in question may not only be those of the characters but of the actors as well.

Conversation A: Subsequent iterations with re-entry

49. Confronted with those moments when "the given material fails to stimulate you sufficiently and you must search for something which will trigger an emotional experi-

7. I learned this lesson the hard way a number of years ago directing a company of actors of wildly varying levels of experience. When I finally got so frustrated trying to get them to do what I wanted in a repeatable fashion that I completely gave up, I was suddenly free to see what they were *actually* doing, free from an ongoing comparison with how I thought their rhythms and dynamics should sound and where I believed all the pauses and transitions should be, which, to my delighted, and humbling surprise, more than adequately satisfied the two essential criteria specified above. Critics particularly praised the production for its acting and I have never been the same director since.

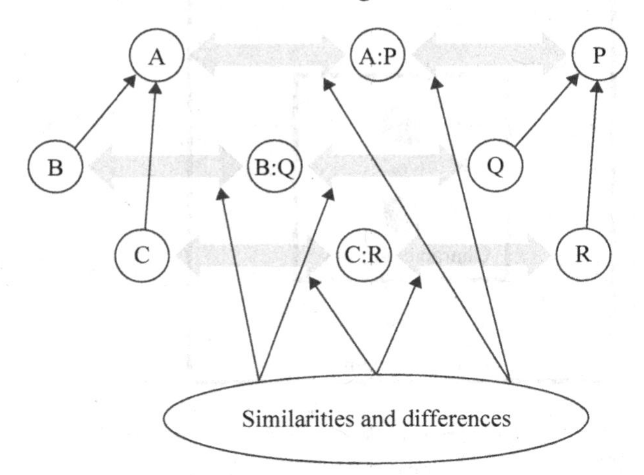

Universe I: Manuela's behaviour	Universe II: Actress's behaviour
A Manuela's shame	P Actress' shame
B Fraulein von Bernburg	Q Lynne Fontanne
C Torn chemise	R Soiled panties

Figure 5. Substitution as analogical entailment (after Scott 2011).

ence… and send you into the immediate action of the play," Stanislavskian actor/director/teacher Uta Hagen extends the range of inputs available to the actor through her development of the substitution technique.

> A young actress working on the part of Manuela in *Children in Uniform* was having difficulty with the moment when Fraulein von Bernburg, the teacher she loves and admires confronts her with her torn chemise and says, 'This will never do!' Manuela must react with deep shame and humiliation. The actress could not make this moment meaningful. Neither the garment nor the actress playing the teacher seemed to matter enough to her. Accidentally, I supplied her with a stimulating substitution for both teacher and chemise. I said, 'What if Lynne Fontanne had a pair of your soiled panties in her hand and showed them to you?' The actress turned beet red, snatched the chemise from her Fraulein von Bernburg and hid it frantically behind her back. (Hagen & Frankell 1973: 35)

50. Substitution has become a widely known and practiced technique over the last half century plus and the exchange described above is exactly the sort of thing that might take place when the results of an improvisation are being compared with the playwright's descriptions in an iteration of Conversation A with re-entry. It is also a technique to which CT can be applied through the expression of an entailed analogy as presented in Figure 5, again adapted from Scott (2011: 319). The composition of such entailment structures at this point in the AA process would provide us with yet another object of second-order cybernetic study with particularly rich possibilities for analysis.

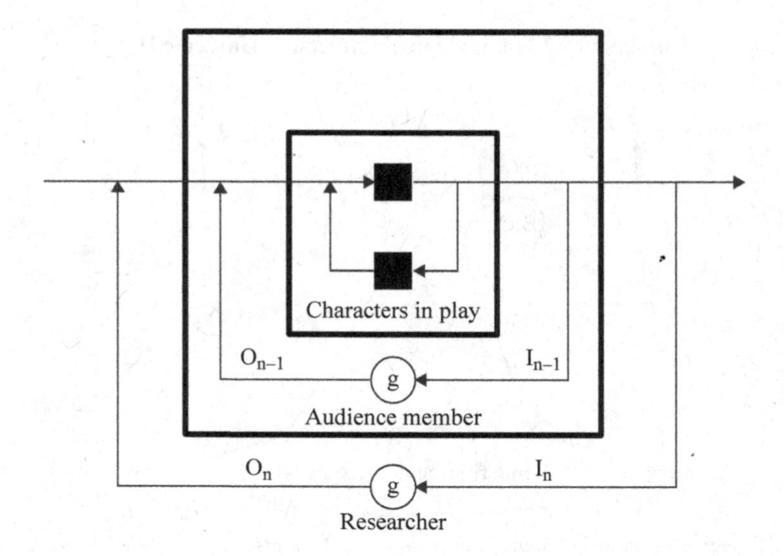

Figure 6. Cybersemiotic theatrical laboratory as nested black boxes.
O: output, I: Input, g: stable observations across recursions (eigenbehaviours).

51. At work in the example above is a complex constellation of social signification through the dual lenses of both the character's and actor's conceptual operations regarding such topics as hierarchical social status and the female body and its traces (to name only two of a great possible number), which, when executed simultaneously, give rise to the biological eigenbehaviours of blushing, snatching, and hiding likely to be "read" cybersemiotically by audiences as signs of shame; biological eigenbehaviours triggered by elements of the character/actor's "semiosphere" that are quite likely to have been deeply influenced by the impingement of societal concepts around normativity, deviance, reward, and sanction in their cognitive formation. Much detailed analysis of the complex interplay between biologically and socially constructed semiosis employing the full arsenal of critical theory could potentially be facilitated through a combination of audience questionnaire data, audience clicker data, and the conversational "residue" of entailment structures illustrating the conceptual operations at work across the range of nested Black Boxes that are the observer/participants in the production and reception of these eigenbehaviours in performance.

Performance

Inside every theatre there are *n* Black Boxes trying to get out

52. Figure 6 applies Glanville's diagram of recursive Black Boxes to the naturalistic theatrical cybersemiotic laboratory I am proposing. To the audience, the interaction system depicted in performance is a Black Box that, as Glanville indicates, is made white within as

the Black Boxes nested inside it (the characters) make descriptions of each other (Glanville 1982). These descriptions guide the emergent dynamics of cybernetic interaction between them. Continuing to follow Glanville's recursive logic, audience members plus performance now constitute new Black Boxes that are whitened within as individual members draw their own distinctions regarding the behaviour under observation and describe its mechanisms to themselves through the act of semantic and semiotic interpretation. These acts of distinction and interpretation can be considered second-order operations performed upon emergent eigenbehaviours (both biologically and socially/symbolically constructed) and carry the requisite awareness of contingency (to greater and lesser degrees across the various audience members) to qualify as observations of a genuinely second-order (Luhmann 2000: 54–101). The final Black Box (and subsequent whitening) we encounter in this discussion is that of the researcher (myself) seeking to describe the mechanisms of interpretation at work amongst the audience as individuals, in total, or on average by performing second-order operations of distinction and interpretation on the data collected through the questionnaires and clickers. If, over its recursive iterations, the nested systems described above stabilize on particular descriptions that hold across their boundaries, they assume the status of eigenbehaviours for which it is possible, as Glanville says,

> to have an apparently fixed, shared, social value, or to be what are thought of as 'facts' The black box model does not, thus, preclude e.g., science. (Glanville 1982: 6)

It is critical, however, that we do not fall into an error similar to that committed by the hard core of the cognitive turn and mistake the word "science" for "objective ontological truth" rather than continuing to view it as a powerful and useful description that continues to work only until it does not and is always circumscribed by the closed operations and bounded rationality of the observer/participants.

Performance texts: autopoietic vs. allopoietic

53. While celebrating the epistemological latitude provided by the Black Box, Glanville reminds us how essential it is that we never relinquish our constructivist perspective on its use.

> It is important to note the essential contribution of the observer in all this: (s)he brings in the Black Box, operates it, observes, and creates the pattern that is taken to be symptomatic of the mechanism within. (Glanville 2012: 421)

54. Once a play enters the production process, the most empowered observer is, traditionally, the director. It is easy to see how, without keeping Glanville's admonition alive, she can be easily seduced by the belief in one's own powers of "objective" analysis so prevalent in first-order science (and first-order cybernetics for that matter). This is not surprising given that virtually every canonical text employed in the teaching of play directing (explicitly naturalistic or otherwise) reinforces the notion that the director's primary function is to use all of the theatrical resources at her disposal to make tangibly visible her interpretation of the play's "deeper meaning" (see e.g., Dean & Carra 1989; Hodge 1994; Sievers, Stiver & Kahan 1974) or, in our terms, her own particular whitened version of the Black Box. This positions the director as a kind of "privileged observer" who will, as Niklas Luhmann describes "observe his emerging work in anticipation of its observation by others" and, despite the fact that

> there is no way of knowing how others [...] will receive the work through their consciousness [...] he will incorporate into the work ways of directing the expectation of others. (Luhmann 2000: 40)

At work here, perhaps tacitly, is precisely the kind of "ontological thinking" described by Krzystof Matuszek:

> Ontological thinking presupposes the existence of a privileged observer endowed with authority, who describes and explains reality in the only correct and binding (for everyone) way [...] The privileged observer instructs those who stray, corrects their mistakes and guarantees the final convergence of all observations. (Matuszek 2015: 204)

55. While the likelihood of completely achieving such a feat might be slim, this description vividly captures an ideal of the theatrical director that many postmodernists within the field of theatre studies suspect, with good reason, to be the, more or less, unconscious *modus operandi* of traditional theatre practitioners and, particularly in the case of realism/naturalism, view with the deep suspicion illustrated in the opening section of this article.[8] On the second-order cybernetic front, such a procedure can also been seen to "violate" both Heinz von Foerster's ethical imperative to "always act so as to increase the number of choices" as well as Larry Richards anticommunication imperative stating that, if we "desire the new" we should "compose asynchronicity" (Richards 2010: 13f). A desire to confront these issues and assumptions played a crucial motivating factor in the experiment described above and inspired me to specifically investigate the impacts of both autopoietic and allopoietic performance texts in the first place; an investigation that I intend to continue with the even sharper second-order cybernetic focus provided by the application of CT.

56. When watching an autopoietic performance the audience observes an interaction system for which, as Richards puts it, the initial description manifest in the formal language of the script (dialogue and stage directions) provides a "skeleton of constraints to guide action" that can then "*trigger* a dynamics of interaction that can lead to new distinctions" (Richards 2010: 15). The potential development of a cybernetically grounded naturalistic theatre less hermeneutically coercive than is traditionally the case is one of the research strands to be pursued as this work continues.

Future directions

57. In reference to an fMRI study performed on improvising jazz and rap musicians at Johns Hopkins, Clayton Drinko comments that "[t]he outward focus that improvisation demands allows the intuitive and creative brain centers to flourish, while drastically inhibiting self-censoring regions." While it may be difficult to interpret such scans with explanatory exactitude, at the very least "it appears that the brain is operating quite differently while improvising than while performing memorized scores" (Drinko 2013: 96). And, while the promise of the "mirror neuron" theory (an eagerly embraced and essential cornerstone of the cognitive turn) has increasingly come under fire (see, e.g., Hickok 2014), it may still indicate some degree of "contagion" of mental states capable of creating

8. Moreover, the more-or-less continuous "closed loop" control exerted by the director working in this manner quickly begins to sap the genuine responsiveness and creativity of the actors; not to mention their very enthusiasm for the project itself.

a palpable sense of difference in an audience if, as Merlin promises "the sense of improvisation is carried all the way from first preview to last night" (see §17) of a play rehearsed through the cybernetic practice of AA.

58. Hans-Ulrich Gumbrecht suggests that we challenge the primacy of the hermeneutic imperative at the heart of "Western" culture and, instead, "conceive of aesthetic experience as an oscillation (and sometimes as an interference) between 'presence effects' and 'meaning effects" (Gumbrecht 2004: 2). In an essay taking up Gumbrecht as well as George Spencer-Brown and Niklas Luhmann, Edgar Landgraf outlines a psychic mechanism within the observer of an improvised performance that might engender a sense of "identification" powerful enough to account for such interference.

> Identification here would mean that temporarily one would no longer observe the action from a distance – that is, draw one's own distinctions to observe the event – but instead would 'embody' the process. That is, to experience a performance as such, one has to experience the logic of the distinctions drawn and the operations performed as one's own. The point is that such an identification with the performance is not symbolically mediated, as Adorno would have it, but rather results from a (nevertheless cognitive) identification with the self-programming of the form-creating process. (Landgraf 2009: 194)

59. Gumbrecht and Landgraf's theories identify one thread of the complex interplay between biologically and socially structured semiosis that I intend to follow as I continue this work. At the same time, I intend to investigate Bertolt Brecht's claims that the kind of identification described above, and in which Stanislavski's theatre most deliberately traffics, militates directly against the awakening of social consciousness and subsequent activism and calls for a vehemently anti-naturalistic theatre capable of encouraging a genuine second-order awareness of the contingency of dominant societal structures.

Conclusion

60. In this article I have posited that an analysis of naturalistic theatre's processes of meaning-making filtered through the constructivist ontological agnosticism of second-order cybernetics offers a productive middle way forward for those on both sides of the social constructivist/embodied cognitive realist divide, within and beyond theatre studies. Once employed, we need not be paralyzed by Knowles admonition that "we risk policing the 'appropriate,' 'normal,' or valued characteristics of elements of the world and of humanity each time we say of a representation, 'yes, I recognize that'" (Knowles 2014: 4) or that giving some credence to our embodied responses to naturalist performance dooms us to being "what the French Marxist philosopher Louis Althusser calls 'hailed,' or 'interpellated' into an ideological system" (ibid: 33) that we are not still free to deconstruct employing Marxist, feminist, queer, or critical race theory to name only a few of the discourses which would remain fully at our disposal.

61. If we can re-conceptualize naturalist theatre (and perhaps all representational art) as dealing in the generation of eigenbehaviors rather than laying claim to objective portraits of "reality" and examine those eigenbehaviors through a cybersemiotic lens employing both quantitative data around meaning and believability and conversation theory reflecting the operationalized conceptual schemas of characters, actors, and, ultimately, audiences (through either pre- or post-show group facilitation or software assisted compi-

lation), and that, with regard to their emergence, denies neither the influence of culturally constructed language games nor bioconstructive couplings between embodied cognitive agents and certain invariant (but only indirectly accessible) aspects of their environment, we can commence to tease out the complex interplay between "biological and social level" eigenbehaviors in a manner free of the fundamentalist excesses of both totalizing cognitive-scientific objectivism and paralyzing postmodern scepticism providing great potential benefit to scholars and practitioners on both sides of such theoretical divides. The theoretical formulation and experimental procedure described in this article were undertaken as a hopeful first step in this direction.

Open Peer Commentary: **Audience and Autopoiesis**

Bruce Clarke & Dorothy Chansky

1. Tom Scholte introduces his directorial experiment as part of a research agenda concerned with "the ways in which [...]the Stanislavski system of acting can consistently generate 'believable' performances [...] but also [...] questions regarding the mechanisms through which observers (audiences) assign this sense of 'belief' as well as 'meaning' to these performances" (§2). To open, or "whiten," the Black Box of such theatrical phenomena, Scholte adapts Ranulph Glanville's elaboration of Norbert Wiener's evocation of James Clerk Maxwell's invention of the idea of the black box to denote any system for which we can make external observations of its behavior – for instance, that inputs of one sort produce outputs of another sort – while lacking knowledge regarding the internal processes by which these behaviors are produced. In Glanville's summary, Maxwell posited the notion of the black box…

> in order that he could justify the building of functioning descriptions (i.e., in his case, equations) that accounted for the observed behavior of some phenomenon when the workings of that phenomenon were not clearly visible. (Glanville 1982: 1)

Glanville noted that Maxwell was previously the author of a fictive "demon," and we would add that along with the invention of the Demon as a kind of "supernatural" observer, Maxwell's thought experiment also placed his Demon within a box-like apparatus, a sort of proto-black box for the Demon to inhabit and within which to work its white scientific magic.

2. Maxwell conceived his famous Demon, you will recall, in order to "pick a hole" in the second law of thermodynamics. The drift of closed physical systems toward increasing entropy due to the leveling of energy differentials could be counteracted, Maxwell surmised, if only one could insert into that system an agency capable of discriminating high- from low-energy particles and sorting them into separate containers. Thus the Demon's box was partitioned into two chambers. The Demon would do its sorting among randomly hot and cool molecules using an aperture in the partition to gather hot molecules into one chamber and cool molecules into the other. The Demon would thus defy the second law by lowering the entropy of the system, restoring a heat differential from which work could be extracted, not by the application of additional energy, but simply by the ordering effect of

its ability to operate upon an internal observation of the system contained in the box as a whole (Clarke 2001: 103–110).

3. Maxwell's Demon comes down to the history of cybernetics not only as the genius presiding over the transformation of the quantity of entropy in Boltzmann's statistical mechanics into the quantity of information in Claude Shannon's information theory, but also as a precursor to the second-order cybernetic prescription to position the observer *within* the system to be observed. One may pick up this thread in Heinz von Foerster's 1960 essay "On Self-Organizing Systems and Their Environments," where von Foerster puts not one but two Maxwell's Demons to work observing the ordering and disordering processes constituted by adding environmental considerations to the observation of self-organizing systems (Clarke 2009). In this regard, Scholte describes a directorial experiment in which...

> audiences were invited to view four performances of the entire piece in which the actors adhered to the author's written text while the performance text was still allowed to unfold autopoietically each night based solely on the actors' emergent and self-organizing cybernetic response. (§20)

4. Scholte's second-order cybernetic analysis of theatrical matters through the lenses of Gordon Pask's conversation theory and Niklas Luhmann's autopoietic theory draws upon a conceptual account well-funded by its inheritance of this line of systems discourse. With Scholte's guidance, one also recovers the theatrical and performative elements latent in systems theory's long history of constructing models in order to explore systems too complex to seize in their totality, but experimentally reducible to key parameters whose dynamics can be fruitfully observed. Every theatrical representation "constructs *a* reality" which is then rendered to reconstruction by multiple observers. Scholte's essay is illuminating, particularly in walking its reader through the recursive observational dynamics constituted by both the construction of the play through the rehearsal process that transforms a script into a theatrical work and the performance of that work ultimately presented to the gaze and response of an audience. With regard to the constructive work of the rehearsal, Scholte borrows Pask's concept of the "modeling facility," which for his purposes "is the stage," the theatrical Black Box,

> as this is where the collaborative formulation of given circumstances and objectives will be tested for their efficacy in actor improvisation, observations of which will re-enter subsequent iterations. (§39)

5. It is also the case that we hit some speed bumps along the way. Scholte's account of how the director got out of the way of the players for a while as they – well – *played* with and in the text gave way to an opportunity for experimenting with audience participation/feedback. Scholte's first audience(s?) saw four performances that arguably were still rehearsals, in that the director withheld formal interaction with the actors' autopoiesis. Later audiences were invited to four performances of the director's "fixed [...] allopoietic performance text" (§20). If we understand correctly, four of the performances featured actors treating the full text via what felt right and responsible to them, while four other performances featured the same text then sculpted by the director's blocking (creation of set movements, timing, and stage pictures). We would like to know more about how these audiences were constituted. For the autopoietic performances, "audiences were invited"; for the allopoietic ones, there were "invited audiences" (§20). It's unclear

to us whether "invite" is being used the same way in these two instances. Were random people welcome to show up, or was the presence of particular people solicited?

6. George Pierce Baker's early experiments with audiences as respondents for the work of his fledgling playwrights at Harvard in the 1910s had membership requirements and stipulated that people could be dropped from the roster if they missed too many performances (Chansky 2004: 97–106). Did the two sets of four audiences comprise the same people? The statement "Audiences were invited to view four performances" at first glance said to us that they had to commit to four viewings – something that could reveal much about the differing choices actors might make while still in that allopoietic phase. For instance, "by the third time around, I wasn't clicking as long or as hard," or, arguably more interestingly, "my heart raced when she turned upstage on that line – *so* unexpected after what she'd done the first three times." Minus such a control mechanism, are we to assume that this experiment takes all audiences to be interchangeable? We are mindful here, too, of Joseph Henrich, Steven J. Heine and Ara Norenzayan (2010) calling most research studies that use North American undergraduates as respondents flawed if they are meant to talk about humans, either as cognitive or as sociosemiotic beings. These authors used the acronym WEIRD as shorthand for the usual Western, Educated, Industrialized, Rich, and Democratic suspects.

7. Who then, exactly, was in Scholte's audience? Does the experiment assume that audiences are self-selecting members of the imagined community that more or less "gets" realistic theatre? If they are not – if they come from other cultures, or know little about theatre, or do not like theatre, will this experimental protocol fall apart? Does the requirement for them to be "conspecific observers sharing a mutual structural coupling" (§44) load the deck to excess? One hoary platitude has it that ninety percent of effective directing is casting. Should the director also get to cast the audience? Or, from another perspective, should a director be required to take seriously a response – via clicker, questionnaire, or what used to be called voting with one's feet – from a cohort ill-equipped to grasp his or her allopoiesis? (This may be irrelevant in the case of realism, but it is a definite theatrical impasse in work in other aesthetic modes.)

8. Be that as it may, for the performance studies scholar also at home in the praxis that Scholte investigates, the mainstream, text-driven theatre practice built around mimetically constructed characters, he does nail the description of a conundrum, even as his answer to the problem may suggest little to directors and actors unschooled in the language of second-order cybernetics. To wit, he pinpoints an impasse within his field between the abstract linguistic-cultural idealism of the post-structuralists and the newer neo-Darwinist adaptationism of the "cognitivists." In rough terms, actors and directors who would embrace Scholte's suggestions might readily understand his "goal-directed and feedback-controlled" AA rehearsals (§10) as a generative period based on the crude idea that "acting is reacting" bundled with responsible textual analysis, a generous dose of improvisation, and the director getting out of the way of the players for a while as they – well – *play* with and in the text. Further, we cannot imagine a practitioner who would not welcome the idea of "applied epistemology" (Footnote 1) – the idea that we know via embodied doing – nor any who would refuse the idea of gently and generously taking care not to mistake "the word 'science' for 'objective ontological truth' rather than continuing to view it as a powerful and useful description that continues to work only until it does not" (§52). The concept of something "working" (unless it does not) is so common a shorthand in theatre that it might deserve a separate analysis of its own.

9. By drawing out von Foerster's concept of eigenvalues (a topic explored at some length in Glanville 1982) as expounded in Søren Brier's discourse of cyber semiotics and, above all, in numerous variations by Louis Kauffman (1987, 2005, 2015), however, Scholte does demonstrate the potential of second-order cybernetics to deconstruct cognitivistic truth-claims regarding "phenomenologically unmediated embodied forms of knowing" (§43). In short, whereas the postmodern critique on the one hand lacks an account of systemic operationality and the cognitivist critique on the other hand lacks an account of recursive sociality, a robustly eclectic deployment of second-order cybernetics loops individual bodyminds and their semiotic mediations into recursive social circuits in a manner that can account for their creative success if not predict the spontaneous content of their operations. Thus we find Scholte's approach to be an effective reminder of the breadth of second-order cybernetic conceptuality as it has developed since the 1970s and an apt demonstration that this vocabulary provides an analytical repertoire adequate in this instance to the complexity and manifold stages of the theatrical phenomena being described, from the construction in rehearsal to the delivery in performance of the play.

Open Peer Commentary: "Truthful" Acting Emerges Through Forward Model Development

Bernd Porr

1. One of the aims of Tom Scholte's target article is to re-introduce the stigmatised word "truth" back into the discourse of theatrical practise and also constructivism. This has been (from my point of view) successfully achieved by using the rehearsal process as devised by Constantin Stanislavsky as a constructive example. Central to this approach is improvisation, where the actors base their actions on the internal goals of their characters and start to interact. If successful, the director, the audience and, thus, observers, as a result report the behaviour of the actors to be "truthful." The article bravely goes beyond the postmodernist notion that "truth" needs to be avoided at all costs and successfully removes its stigma.

2. While the article succeeds in making "truth" credible again, it could have also benefited from being ambitious on the front of open vs. closed loop. This could have easily been taken into account as well because from the cybernetic point of view, the target article is not just about closed loops but also about open loops. However, as with the word "truth," "open loop" is also often frowned upon in constructivism, which traditionally demands that descriptions are based on closed loops, recursions and the observation of loops by other loops. In this commentary I remind the audience of the concept of the *forward model*, which is a well-established construct in second-order cybernetics and control theory (Palm 2000). This concept is implicitly woven into the main text and my aim is to make it explicit in this commentary.

3. First of all, we need to define "forward model." An ideal forward model is an open loop controller that no longer needs feedback to arrive at a desired outcome (Palm 2000). If we refer back to the well-trodden territory of the thermostat, then a thermostat action using a forward model will not require its feedback path because it knows the exact temperature

change in advance when switching on/off the heating. It would notice a change from the desired state and then would switch on/off the heating without making any comparison of the achieved result with the desired result. Another example is a chef who knows exactly how much salt needs to be added to a soup without tasting it afterwards. The chef is able to achieve the perfect taste because he/she has operated in closed loop mode many times before but no longer needs to do it because he/she has a forward model.

4. One might argue that we will not need forward models. It is of course possible to live without developing any forward models in our lives but this is a risky strategy. A purely reactive feedback system is always at the mercy of the environment, hoping that its requisite variety will always be sufficient when reacting against disturbances. The rabbit hopes to be fast enough all its life to escape all attackers. However, animals – and in particular humans – develop a multitude of forward models to pre-empt what is going to happen. This can only be achieved through learning, which step-by-step develops forward models through experience on top of feedback loops (Porr & Wörgötter 2002, 2005). Even if these forward models fail from time to time and the feedback loops need to kick in overall, the agent has developed models of its environment. This does not mean that the agent knows everything about its environment, but it has understood its own closed loops. With that knowledge, the agent knows how to avoid unexpected surprises. In the worst case, these might kill the agent. However, they could be just a situation where the agent enters a cocktail party with a room full of strangers. This leads us to the special case of human–human interaction, where two or more people try to develop forward models of each other.

5. What happens if agents develop forward models of each other by interacting with each other? This is what Niklas Luhmann calls "double contingency" (Luhmann 1984). It is mastered by creating mutual forward models to achieve a high degree of certainty. For example, bakers often talk about recipes or theatre practitioners about the rehearsal process and not baking recipes. Here, learning develops forward models of the other person because the other person (alter) disturbs the closed loop processes of the first person (ego) and vice versa. It is important that both persons start off from their personal closed loops and that if they do not learn they just see each other as mutual disturbances (think again of the cocktail party with a room full of strangers). Only because they develop forward models do they actually create a closed loop system that spans through both of them (Porr & Di Prodi 2014). This is an important step and is often overseen because the people themselves become open loops (!) because they no longer need their own personal feedback loops. For example, when talking about the weather, the response of the other person is highly predictable. The person has developed a forward model of the other person in terms of the topic of weather. This can be termed as a theory of the other person's mind.

6. Now we can go one step further and observe a conversation of two people, for example in a pub. It is important that the observer has developed her own forward models of conversations in the past as described above. The observer can observe and perhaps join into the conversation because of her forward models. This will work more or less seamlessly, depending on the topic and shared experiences, but it will be just part of the everyday operations in our environment.

7. Now, observing acting is a special case in contrast to observing people interacting in everyday situations. The main text is spot on that the actors and the director need to find out what goals (or in control theory, desired states) the different characters want to achieve and that then, through the technique of improvisation, this will be tried, tested and evolved. Again, this can be understood in terms of forward models: at the start of the improvisation,

the actors have a very limited or perhaps no forward model of the other actors' goals or closed loop behaviour. However, the two actors then learn to predict what the other actor is going to achieve so that their mutual uncertainty is reduced, in the sense of Luhmann's reduction of double contingency. If the improvisation has been successful (very similar to the everyday conversations), the actors will mainly act in open loop using their forward models by knowing what the other actor is trying to achieve. This is in stark contrast to reading out lines, which require very little predictive power and, thus, no forward model. An observer who watches the improvisation (or the director) should then be able to compare their forward models to that of the two or more actors on stage. If there is a reasonable match, then this is perceived as being "truthful" in the sense that there are similarities of forward models developed by both the actors and the audience.

8. Scholte's article also has wider implications because improvisation imitates everyday double contingency reduction and acts as a convincing demonstrator/simulator of how everyday communication emerges. The actors face a similar challenge to somebody entering the aforementioned cocktail party with a room full of strangers. Again, here, forward models need to be developed to engage in meaningful conversations.

9. As a final remark, I would like to draw attention to film, where certain directors use improvisation not just to shape the acting but as a tool for developing the story as such (as done by Mike Leigh for example). Another example is the recent film "Victoria," which indeed feels very "real." This has been achieved by just prescribing inner goals for the protagonists in the form of a treatment that they then use to improvise the action. Even in more traditional environments, film is usually developed as a two-stage process where first, a treatment is written, which often describes the characters' goals, and then a script based on the treatment is evolved.

10. Be it film or theatre, improvisation should be at the heart not only of the rehearsal process but ideally also of the story development itself.

Open Peer Commentary:
Naturalism in Improvisation and Embodiment

Edgar Landgraf

The introduction of the chorus was the decisive step with which war was declared openly and honourably against any naturalism in art. (Nietzsche 2008: 28)

1. The "essentially cybernetic vision" (§14) Tom Scholte locates at the core of the Stanislavski system of acting (or Method acting) is most apparent in Constantin Stanislavski's use of improvisation. Improvisation demands a particular mindset, the attentiveness to one's surroundings and the willingness to stay "engaged in [feedback] loops within imaginary circumstances" (§16). Improvisation serves as a tool to evoke spontaneity and immediacy in acting, qualities that promote the semblance of naturalness and authenticity. These qualities are retained even after a scene has been memorized and has undergone "formalization" by the director.

2. Stanislavski's use of improvisation resonates with one of the central theses of my book *Improvisation as Art: Conceptual Challenges, Historical Perspectives,* namely that "improvisation is best understood as a particular mode of *staging* art that shares properties common to various individual arts and fulfills many of the expectations we have for the arts in general" (Landgraf 2011: 11). My argument is based on the recognition that it is impossible to decouple improvisation fully from structure and repetition. Yet, despite following a deconstructive logic, I wanted to be able to account for the effects that are at the base of the continued association of improvisation with notions of spontaneity, immediacy, and inventiveness, effects recognized by practitioners as much as by their audience. Like Scholte, I drew on second-order cybernetics to circumvent the impasse between postmodern skepticism (expressed, for example, by Jacques Derrida's proclamation that improvisation is impossible) and the essentializing tendencies adopted by many practitioners of improvisation (e.g., Derek Bailey), who find the practice of improvisation in its "immediacy" utterly incompatible with theoretical reflection.

3. Stanislavski's association of improvisation with naturalism most closely parallels the many comparisons of improvisation with conversation. In these comparisons, conversation is idealized as a form of free and spontaneous social interaction where the participants are thought to be able to act naturally and authentically. As in improvisation, this freedom is the result of long practice, of experience and situational familiarity, and is something that not everyone will learn to master equally well. A comparative analysis can further underline the social constructedness of "naturalism," even for conversation. Merely consider, for example, how aristocratic etiquette in pre-modern times encouraged quite different conversational practices than the bourgeois emphasis on sensibility that became dominant in the course of the eighteenth century and put strong emphasis on non-verbal signs (blushing, feinting, gestures, etc.) in and outside the theatre. As I explored more extensively in my book, the expectation that interlocutors of a conversation engage in an open exchange of ideas ("stay engaged in feedback loops") is itself a modern development. Again, in feudal societies, by comparison, conversation was highly regulated and improvisation, too, did not aim for naturalism or inventiveness. The latter is apparent in the use of masks in the *Commedia dell'arte*, its repeatable plot structures, or the rhetorical corsets that defined and made predictable the practice of improvisation in older theatre and poetry traditions (see Angela Esterhammer 2008 and Beatrix Müller-Kampel 2003 for examples).

4. This brief historical comparison should serve as a reminder of the constructedness of our contemporary views on the naturalness of conversation, improvisation, and acting. More specifically, I want to argue that the comparison between conversation and improvisation suggests that today the appearance of "naturalness" must itself be linked to the recognition of open and dynamic processes of social interaction. In cybernetic terms, nowadays actors and audiences alike have learned to read the emergent qualities of autopoietic performances as markers of authenticity and naturalism. This is why Scholte's turn to Gordon Pask's conversation theory for his quantitative analysis promises to be highly productive. Scholte might also want to include Keith Sawyer's work. Sawyer draws on Erving Goffman in his description of processes of "collaborative emergence" that he observes in improv theatre dialogues. The performances Sawyer analyses demonstrate how from contingent beginnings, an unforeseen story can emerge within seconds, a story with a defined plot, novel characters, specific times and places, and so on (Sawyer 2003: 41–43). In this regard, improv theatre can serve as a model for how to *stage* the absence of planning to create effects of spontaneity, immediacy, freedom, and authenticity.

5. To look at conversations and dialogues as a model for autopoietic and self-steering processes is also to acknowledge that such processes develop not only around linguistic utterances, but consciously and subconsciously involve many other elements, such as gestures, facial expressions, perception of place and time, timing, the surrounding atmosphere, etc. These are all factors that help determine what will be perceived, but also what an actor will experience and "embody" as natural and authentic behavior. Paskian cybernetics allows us to understand such experiences as mediated through our body's sensory apparatus. This means admitting pre-conscious and pre-symbolic modes of cognition and communication, while nevertheless maintaining that such experiences are acquired, are the product of cultural training (socialization and education), and reflect culturally determined expectations and interpretive patterns.

6. The cultural dimension is often shortchanged by the humanities' recent turn to the cognitive sciences. In my essay "Form aand Event," I take to task the "naturalism" of theories of embodiment that employ the term "embodiment" as a foundational concept, as evidence for an evolutionary-physiological determination of our being-in-the-world. If we look at current debates surrounding the schools of thought associated with posthumanism, we notice on a much broader scale the renewed desire for ontological or physiological foundations, for a return to a realism that is not plagued by the specters of postmodernism. Cybernetics is not immune to this quest, even a cybersemiotically versed thinker such as Søren Brier adopts embodiment as a central concept from where to develop a notion of reality that is thought to exist – and exist with definable qualities – independent of its (semiotic re)construction by an observer.

7. Brier argues that "although we have rightly abandoned the notion of 'objective reality' in second-order cybernetics, we should not give up the notion of a partly independent 'outside reality'" (Brier 2008: 92). From the perspective of Niklas Luhmann's operational constructivism – which is more consequential, and asks us not to lose sight of the social involvement (communication) in our construction of knowledge – this "outside," too, has to be understood as the "inside" product of its observation, whereby the primary observer for Luhmann is not a subject, mind, or body, but the system of communication. The argument is simple. Anything we (individual minds) observe and any meaning we might attach to any observer, be it the mind, the body, a subject, or communication, will have to draw on the operations of the system of communication to do so. This applies also to our observations of reality. Accordingly, when Luhmann defines reality, for example, as resistance to arbitrariness, he still insists that reality remains the construct of system operations:

> When reality should still be understood as resistance against arbitrary thematization – and what other concept of reality would we still have – then we must be dealing with resistance of signs against signs, language against language, communication against communication. That is: we are dealing with recursively formed complexity. In its continuing operations, the system tests, so to speak, against self-produced uncertainty and self-produced resistance what, from moment to moment, it can treat as eigenvalue. (Luhmann 1997: 1126f, my translation).

8. I will not be able to do justice to Brier's concept of cybersemiotics here, nevertheless, I want to suggest that Brier's "realism" fails to give due credit to the system of communication's role in defining the mind's ability to observe and attribute meaning to what it observes, including what it observes as physiological capabilities and needs that developed as the result of evolutionary processes. As Scholte notes, the point is not to ignore the reali-

ties science offers in its first-order observations, but to remind oneself that they, too, are observer-dependent and ultimately have to draw on and operate within the system of communication. Only by retaining an eye on both, observer and observed, can we overcome what Scholte describes as the "onto-epistemological deadlocks between constructivists and realists" (Scholte §3) and avoid the "fundamentalist excesses of both totalizing cognitive-scientific objectivism and paralyzing postmodern skepticism" (§61).

9. More specifically, then, rather than naturalizing embodiment, drawing on Luhmann's systems theory, we can observe how the body is accessible by our nervous system (the sensory apparatus) that interacts (is structurally coupled) with the mind (consciousness understood as an emergent phenomenon) that in turn is conscious due to its structural coupling with the system of communication, its ability to use language for the reproduction of its elements (perceptions, thoughts, feelings). Structural coupling is responsible for a high degree of stability in the interaction between these systems – what Scholte examines as eigenbehaviors – a stability that on the level of interaction between nervous system and psychic system is the result of long-term, evolutionary processes. We can think of consciousness or cognition or feelings as embodied in this sense and do not have to question the value of research that investigates the biological, physiological, chemical, semiotic, and other parameters that enable such a system to function and maintain relations to other systems and its environment. Such research will no doubt strengthen our understanding of the biological foundations of our mind, and of our sense of being-in-the-world. But one should not neglect what nothing demonstrates more clearly than acting itself: that the signification processes that are associated with embodied modes of cognition and experience, as well as their meaning, exist separate from these experiences, and can always be faked, simulated, feinted, and manipulated.

10. How can we conceive the role of theatre and, more broadly, of art in the context of these observational practices? I quoted Nietzsche above, who sees naturalism as a principle enemy of the theatre. Nietzsche cites theatre's most famous character, Hamlet, in support of his thesis that art responds to the fundamental need for illusion (the need to "turn those thoughts of disgust at the horror or absurdity of existence into imaginary constructs which permit living to continue" – Nietzsche 2008: 29). Scholte reads the current return to naturalism as a reaction to a crisis of meaning, as representing a backlash against "paralyzing postmodern skepticism" (§61). It is certainly not surprising that a time that sees itself confronted with overwhelming natural and technological challenges and finds itself disillusioned about its political maneuverability, might put an increased stake into science and feel the need to emphasize the reality of the threat rather than concern itself with language games, rhetoric, performativity, or ideological contradictions (which appeared more meaningful in the aftermath of WWII).

11. It is legitimate to ask what the role of art and theatre is under these circumstances. Does art want to insist on expressing the reality of the looming social and environmental crises and their effects on contemporary society? Or does art remain committed to purely aesthetic criteria and thus aim to preserve its autonomy, or at least a certain distance from the immediate concerns of its time? From a systems theoretical point of view, there might be a middle ground. If we adopt the theory of modern society's functional differentiation, we will have to acknowledge that a social subsystem such as art is part of, but also limited in how it might affect society at large. Which raises the question of the specificity of art's mode of communication vis-à-vis that of other social subsystems. If we take "illusion" to be central to art and theatre, we can distinguish art in terms of it not being bound by what

other social systems construct as their external reality. This freedom allows art to cite and recontextualize other societal discourses and thus make apparent the constructedness of the reality to which they appear to respond. Put differently, art in modern society "naturally" invites second-order observation on society's observations, instilling a sense of contingency and freedom, what Scholte, in a Brechtian spirit, calls a "genuine second-order awareness of the contingency of dominant societal structures" (§59).

12. Whether naturalism in rehearsal and performance contributes or undermines this particular function of art, however, will again depend on the observer, on the audience, and on its ability to recognize naturalism in art as staged. Scholte's research project holds the promise to further our sensibilities in this regard, by examining more closely *how* naturalistic effects are produced, what signs, behaviors, gestures, and so on accompany communicational exchanges and create the impression of naturalism in today's culture – and thus invite second-order observation also of naturalism itself.

Open Peer Commentary: **Opening the Black Box of Minds: Theatre as a Laboratory of System Unknowns**

Lowell F. Christy Jr.

1. Heinz von Foerster loved magic, with its sleight of hand and how one's perceptual field is directed. Magic was of interest but his passion was how the sleight of mind in perception undermines the realist's claims to independent reality and objective tenets of science. In 1971 when I first met him in Cuernavaca, Mexico, where he was collaborating with Gordon Pask, Humberto Maturana and Douglas Engelbart over the future of the technologies arising from cybernetics, he was fascinated by the Eastern spiritual traditions of the koan. The koan is a question/riddle that cannot be solved by remaining in the same paradigm that framed the question. For von Foerster, second-order cybernetics was a koan requiring making the mind larger than the problem at hand and the facts given. Tom Scholte's answer to the koan of second-order cybernetics requires going to another level by moving upstream from the paradigm of science into the pre-conceptual, pre-perceptual world of art and theatre.

2. Scholte's target article "Black Box Theatre" offers a refreshing alternative. With the example of theatre as an investigative tool in second-order cybernetics delving into the epistemological and ontological levels of systems, Scholte offers a research approach and agenda leading out of constructivism's "infinite linear circularity"[1] Instead of waging

1. "Infinite linear circularity" arose out of a discussion in July 1971 between Engelbart and Pask in Cuernavaca, Mexico, which I attended. Pask was lecturing on conversation theory and doing one of his famous doodles of a double mirror picture of the human head, within a human head within a human head. Engelbart said that Pask's conversation theory reminded him of a fatal software programming mistake of infinite do-loops that went nowhere. "Infinite circularity" was a 1960s substitute term for continuous do-loops creating programming mistakes where there is no resolution except continuous processing. Engelbart added the term "linear" to infinite circularity because he wanted to emphasize the difference between vicious cycles, where learning occurred only within the bounded rationality of Pask's "teach-back," versus virtuous interactions, which

an assault on the paradigm of privileged observer knowledge and strong epistemological reductionism of science, Scholte opens the possibility of a research methodology based on the formation, conformation and deformation of the social life of information.

3. What is radical in its very hypothesis is that beyond the level of command, control and objectivity, new research methods can test regulatory principles in the forming of information and the bounded contexts of communication. Meaning, identity, relationship, ecological insights could emerge in a carefully designed "system of systems" research effort. Von Foerster's central theme "How we know what we know?" (Segal 2001: 1) requires moving from worrying about truth to intelligence, from cognition to perception and from shattering the fantasy of an objective reality to deciphering the logic of interacting systems. The task is not to know what is *in* the black box but the structure of the box and its linkages, feedback loops and transformative processes of signs and symbols. Only by revealing systems of feedback loops, their consequences over time and the structural patterning of vicious versus virtuous cycles could von Foerster's deep concern for reigning in the "monsters of reason" (ibid: 2) set constructivism on a new course.

4. Scholte boldly proposes that naturalist theatre "could provide a hitherto untapped laboratory for generation of quantitative and qualitative research pertaining to several dimensions of second-order cybernetics" (§3). His "hope is that the theoretical considerations discussed here might inspire others to bring their expertise to this fledgling research program" (ibid). Scholte points to a different way forward for epistemological constructivists by investigation of the perceptual interactions leading to the formation of the image or narrative through which the world emerges. Using the clash of meaning in the interactive theatre of observers and the observed, the regulatory functions within and between systems could be revealed and studied. Instead of focussing on constructivism itself, this direction of research is asking about the meta-rules and regulators of systems and their consequence in individuals, groups, cultures and beyond.

5. What is the unique configuration and constitution of these perceptual lens and feedback loops? Organized systems of interactions of observers and observed are present in theatre. The arts operate at the pre-linguistic, pre-conceptual and pre-symbolic levels, where science appears excluded by its very act of operating on formed information and its reliance on the analytic droppings of data. Scholte makes the case that regulatory principles forming our fields of intelligence, perception could be probed. Probing the circuitry of connectedness and dependence within and between bounded systems offers the opportunity to move beyond the strait jackets of knowledge/no knowledge into learning to learn.

---

address complexity by creating evolutionary intelligence. This was an epistemological clash between Engelbart, who championed the co-evolution of human mind and machine that seeks collective intelligence processes versus Pask's conversation theory, which emphasized learning to make knowledge explicit. Scholte relies on Pask's conversation insights but must emerge from Pask's inward spiral feeding upon existing "knowns" if a results-oriented research agenda is to be achieved. Engelbart's and, more recently, Francis Heylighen's work on bootstrapping processes of emergence of ever more complex adaptive systems (Heylighen 2013) is important in constructivist research. The productive link between Pask, Engelbart and Heylighen's work is their common pursuit of higher levels of relationships or meta-structures (scaffolding), which a reflexive theatre of theatre could investigate. Gregory Bateson's emphasis on co-evolutionary structures of minds would assist researching higher levels of learning of systems of systems of second-order cybernetics.

6. In his target article, Scholte draws from von Forester's eigenbehaviors, Pasks' conversation theory and Bateson's cybernetic epistemology to investigate the interrelations and ecology wherein the world and the observer co-create each other. Scholte understands that "[t]he logic of the world is the logic of the description of the world"; (Segal 2001: 1) in doing so, he illuminated a path that bypasses the trap of accounting for the black box of cognition and the totality of our mental faculties. Calling for research not on cognition but on the field of intelligence emergent between the black boxes of perceptions and cognition, Scholte states that theatre provides an ideal place for understanding the structuring of our dreams of reality in an environment that starts with the "willing suspension of disbelief" (Coleridge 1817: Ch. XIV).

7. Scholte relies on Søren Brier's bio-cybersemiotic framework. Brier (2008) proposes a semiotic level that belongs to all living systems, and that serves as the scaffolding for "social language systems" (§42). Emergence of images and linking back to perception requires weaving of information based on differences making a difference to the whole system. Identification of the meta-patterns and how they constrain lower level functions would be a research objective of perceptual weaving and its holistic economy of relations. How information becomes entangled (develops) within the whole scaffolding as well as the potential for disentangling or "reverse engineering" complexity is a distinct possibility.

8. For a theatre of theatre, Scholte ends his target article with this thought about the nature of the organization (read regulator) of information, communication that co-creates itself. Teasing out the complex interplay between eigenbehaviors requires the re-conceptualization of...

> naturalist theatre (and perhaps all representational art) as dealing in the generation of eigenbehaviors rather than laying claim to objective portraits of 'reality' and examine those eigenbehaviors through a cybersemiotic lens employing both quantitative data around meaning and believability and conservation theory reflecting the operationalized conceptual schema of characters, actors and ultimately audiences. (§61)

9. The question remains of how to deal with black boxes of the unknown and whether there is a researchable pathway to super-sensible knowing and learning? Scholte's insight is that by researching processes (eigenbehaviors) through cybersemiotic and conversation theory in the terrain that is pre-scientific, we can complete the revolution of second-order cybernetics. The task of second-order cybernetics was to turn epistemology on its head. The question now is building on the constructivist revolution. What is tantalizing is that he points not to the secrets (the things) inside the black box but to relationship structures, understood both quantitatively and qualitatively, of the feedback loops of perception, deception and conception in the interactions between actors, audience and script.

10. Scholte's method simply states there is a field of inquiry that can not only investigate itself but can learn about patterns of organization and relational symmetry in the very perceptual and conceptual processes that create stability of objects. Theatre of theatre opens the possibility of investigating eigenbehaviors as tokens of processes. These processes are the relationships, feedback loops and interactions between the black boxes of minds. This research would answer questions not just about what is present between the black box of minds but importantly the development potential for change and transformation in the circuitry of information and communication through which we organize the world. A theatre of theatre research agenda would document the processes in the development of structures of increasing complexity and improbability.

11. The black box of unknown inner principles of change and stasis in living systems operates as a token of processes giving rise to fully formed objects. When we consider a pattern of patterns or cybernetics of cybernetics or the proposed theatre of theatre that fold back on themselves, we find there is a foothold for analysis. Louis Kaufman in his paper on von Foerster's eigenbehavior writes,

> Such concepts appear to close around upon themselves, and at the same time they lead outward. They suggest the possibility of transcending the boundaries of a system from a locus that might have been within the system until the circular concept is called into being, and then the boundaries have turned inside out. (Kauffman 2003: 73)

12. Ranulph Granville makes a similar point in his 1979 paper delivered at the London Cybernetics Society, "Inside Every White Box are Two Black Boxes Trying to Get Out," where he concludes:

> There comes, then, a point at which formal (artificial) systems, as we understand them, are limited by the distinction between level and meta-level. In terms of our (level distinctive) logics – themselves artificial systems – this distinction is sacrosanct. [...] Thus, the black box model [...] requires not only this change but also as one means for the establishment of eigen-behavior and hence, objects, the observer's ability to 'step outside' or transcend levels. (Glanville 1982: 9)

13. The movement away from objects that are socially constructed and a world where everything is relative does not mean the world of mind simply floats on the whim of the dominant class or prevailing paradigm. Like the relative in a family, where faces bear a striking wholeness of relationship (like Wittgenstein's family resemblances), what constitute the structures are it's underlying differences. Identification of those patterns that connect and disconnect would be the heart of the research effort.

14. Intelligence, perception, morphogenesis and in-formation are all eigenbehavior delineating processes that move upstream from science's world view to where objects are not objects of study but indications of processes – "the concatenation of operations upon themselves" (Kaufman 2003: 73) The unit of analysis is the two-way interactive co-creation of "organism plus environment" forming a structural coupling (Bateson 1999). The theatre of theatre would investigate structural coupling where the phase (transformative) space between meta-system scaffolding and operations, between organism and environment can not only be identified and studied but is regulatory functioning viewed.

15. Scholte's value-added proposition in creating a theatre of theatre second-order laboratory points towards a new, non-trivial paradigm. This paradigm requires heavy theoretical lifting and the design of new experimental questions and tools. Whereas the scientific method is analytic probing of the discrete *variables*, second-order methods seek the dual processes of the *whole* that

- *sustain* and provide stability over time, and
- have the capacity to *change*.

Change has many dimensions, including structural (morpho-genesis), change in ideas, habits, responses in behavior or in symbolic representations (ideo-genesis), or change niche/ environment (eco-genesis). All three areas could constitute second-order research. The change from seeking variables in the slice and dice analytic of science and the unification of holistic systems design due to changes in its circuitry is a significant shift in the way we think.

16. A voice from theatre is a strange place to listen for how patterns of organization and relational symmetry emerge, are sustained and evolve or dissolve. Black box theatre points in an important direction. Second-order rigor probes the circuitry of the entangled relationship of observers and the observer's co-creation. A method and tool of investigation emerges out of a hierarchy of levels of analysis incorporating the system of observer and observed. Establishing a theatre of theatre laboratory is fully consistent with the reflexive cybernetics of cybernetics, linguistics of linguistics, logic of logic, learning of learning or meaning of meaning.

17. Scholte's article stands on its own as a strong argument that black box theatre research offers a new way of thinking about thinking with a potential to change how we approach knowledge, intelligence and evolutionary design of the world. In seeking constructivist foundations, Scholte's idea is that aesthetics and art may provide deeper insights into the nature of epistemological circuits, their closure as a system and their consequences. Improving the way we think requires entering into the pre-perceptual and pre-conceptual domain of the arts. When asked in Mexico in 1971 about the future of cybernetics, von Foerster gave me a book, *The Dream that Was No More a Dream: Search for Aesthetic Reality in Germany, 1890–1945* (Kinser & Kleinman 1969). Witnessing Nazi Germany's Berlin during WWII, von Foerster's pioneering work in constructivism and cybernetics was driven by the epistemological bankruptcy of reason, rationality and its handmaiden, science. Countering the tyranny of certitude required tilting at science, truth and reality, but for von Foerster, Bateson and Kenneth Boulding (cf. his 1956, *The Image*), the future required probes into art countering construction of images, symbols and language that produced the "monsters of reason" (Segal 2001: 2).

18. Theatre is uniquely positioned to provide methods and tools to understand consequences of differing configurations forming perception and conception. The koans embedded in what von Foerster accomplished in raising the specter of second-order cybernetics requires new experimental design and the heavy lifting of theory. Repeating what the pioneers of cybernetics and constructionism said means we roll their koans up a hill only to have them crash down. Beyond the mountains of truth/no truth and relativity lies a field of images, narratives and relationships with consequences. I will meet you there.

Open Peer Commentary: **Does Second-Order Cybernetics Provide a Framework for Theatre Studies?**

Albert Müller

1. I understand the intentions of the author, Tom Scholte, as an attempt to create a systematic relationship between Stanislavski's naturalistic theatre, his own theatre work and cybernetics, especially second-order cybernetics, with a focus on a well-known key concept of cybernetics, the black box.

2. It is always welcome to see transdisciplinary cybernetics used either as a theoretical framework for the solution of a set of special scientific problems or entangled with other scientific disciplines. In fact, this applies not only to the sciences but also to various applied fields, including systemic therapy or consultancy schools, and, as it the case here, to vari-

ous fields of the arts. Among many others who supported such an agenda were Heinz von Foerster, Gordon Pask, Ernst von Glasersfeld and Ranulph Glanville.

3. Already in an earlier publication, Scholte (2015) tried to establish this relationship between cybernetics and the undoubtedly ingenious theoretician and practitioner of theatre, Stanislavski – who, besides Bertold Brecht, must be considered the most influential revolutionary in 20th century theatre. For an historian this opens up the problem of anachronism or the problem of a possible anachronistic fallacy (Fischer 1970) because Konstantin Stanislavski, who passed away in 1938, could definitely not have been aware of the development of cybernetics. This took place in the 1940s and the 1950s, mainly carried out by Norbert Wiener, Warren S. McCulloch and the Macy Group, and the British cyberneticians, including Ross Ashby. Since Stanislavski's major works were published in the 1930s, we need to wonder what kind of relationship Scholte tries to establish, for it cannot be a relationship of influence or mutual influence. Even though I am aware of the fact that Heinz von Foerster, Ernst von Glasersfeld and Gordon Pask as well (and probably others) read some of Stanislavski's books, I cannot assume that he had any influence on the development of early cybernetics for, as far as I know, there is no literature in early cybernetics making any reference to Stanislavski.

4. Besides this (for a historian) obvious criticism, there are other issues one needs to look into. One of them is the question of intellectual economy (or, if you will, Occam's razor *sensu* Hahn 1980): Of course, there are obvious parallels between Stanislavski and cybernetics – and Scholte tells us very interesting details about that – but is it actually necessary to adopt (second-order) cybernetics in order to understand, to explain, let alone to develop his conception of theatre, i.e., Stanislavski's system?

5. Another example in Scholte's article, depicted in Figure 6, is the reformulation of a quite conventional theatre situation (with "characters in play," "audience member" and "researcher") as nested black boxes in the sense of Glanville (2012: 447) but with an additional time variable. There is no doubt that the concept of the black box is a fundamental theoretical instrument in the history of cybernetics. While it would be possible to demonstrate that there were predecessors, it is clear that the first full description and discussion of this concept goes back to chapter 6 of Ashby's *An Introduction to Cybernetics* (Ashby 1956). In Ashby's handwritten *Journal*, http://www.rossashby.info, we find a first entry concerning this concept in the year 1951. Glanville (2012: 42) suggested it was possible to trace the general idea of the black box back to James Clerk Maxwell. A similar suggestion was made by Heinz von Foerster when he used Maxwell's demon in his thought experiments related to his work on "self-organizing systems and their environments" (Foerster 2003b). In any case, the black box has been one of the traditional concepts of (first-order) cybernetics that has often been used innovatively in new contexts, Scholte's article being one of them.

6. Glanville (2009c, 2012), in some ways, broke with the traditions of cyberneticians' black box thinking and went considerably beyond it. One of his central innovative ideas was to ascribe to the black box the quality of being "whitened." By being "whitened," the black box becomes a white box. This clearly transcended Ashby's conception of a black box, which would always remain a black box, never to be opened and only to be hypothetically ascribed a specific function by an observer (or the experimenter coupling himself to the box, in Ashby's 1956: 87 terminology). In Glanville's terms, "whitening" the black box refers to the building of a circular system as a new whole that includes the black box and the observer, who provides a functional description of the black box. Glanville's approach

takes into account that different observers may come up with different functional descriptions. The whitening of the black box also whitens the observer but the circular system they are forming appears again as a black box for a second observer. With this reformulation of Ashby's "Problem of the Black Box" (Ashby 1956: 86), Glanville turned the originally first-order cybernetics concept of the black box into a second-order cybernetics concept and made it a universal *epistemological* tool. But was it meant to be applied beyond epistemological questions, questions of what we can or cannot know, questions that were also formulated in von Glasersfeld's radical constructivism (Glasersfeld 2007)? I do not think so. Was it meant to be used for applied research, including social and psychological, and for problems emerging and being studied in theatre studies of the type Tom Scholte is doing? I have my doubts. The view that a concept, a theory is beautiful does not necessarily mean that it matches certain problems better than other or older concepts and theories. However, this does not mean that concepts and theories from second-order cybernetics cannot be successfully used – it must be carefully decided in which context they can be applied.

7. With his target article, Scholte announces a research program accompanying his theatre work that could last for years. In particular, this ongoing work could influence both theatre and research and might very well lead to lasting changes in concepts and theories as well. We shall remain curious.

Open Peer Commentary: **A Theatre for Exploring the Cybernetic**

Ben Sweeting

1. While Tom Scholte has concentrated on ways in which cybernetics can inform theatre, the connections that he has developed between the two fields are significant for being not ones of application but, rather, overlap, where cybernetic processes are seen to be being enacted within an already established set of practices. Scholte's bridge building is, therefore, suggestive of further possibilities, opening up a new avenue for exploring how cybernetics may be understood in terms of action rather than theory, and so as an active research tradition rather than one form of worldview amongst others. This is highly relevant to the context of this volume and previous concerns in *Constructivist Foundations* with second-order science (Riegler & Müller 2014).

2. One point of comparison for Scholte's target article is with the development of similar connections between cybernetics and design, such as in the work of Ranulph Glanville (e.g., 2007). This commentary is not the place to work through the various connections that can be made between design and theatre via cybernetics (a study that would be in the spirit of cybernetics' original trans-disciplinary agenda). However, reflecting on the parallels between Scholte's account and the invocation of design in cybernetic literature suggests ways in which the connections that Scholte has explored may be further developed.

3. Scholte concentrates on the underpinning that cybernetics can offer to processes in theatre, for instance in moving beyond the theoretical impasse that he describes (§§5ff), and ways in which the use of ideas from cybernetics such as entailment meshes may en-

rich those processes (§§40f). There are several areas of design where, similarly, cybernetics can provide theoretical support, particularly as regards interactive technology (e.g., Spiller 2002) or the relation between design and research (e.g., Glanville 2015; Jonas 2007, 2015). Glanville's analogy between cybernetics and design is, however, notably two-way: "cybernetics is the theory of design and design is the action of cybernetics" (Glanville 2007: 1178). That is, design contributes back to cybernetics, such as where second-order cybernetics is understood in terms of the cybernetic practice of cybernetic ideas (Sweeting 2015), and where the overlaps between cybernetics and core aspects of design practice have allowed designers to contribute to cybernetics through their tacit understanding of such processes, rather than via theory (on this see also my contribution elsewhere in this volume; Sweeting 2016).

4. Similarly, given the parallels that Scholte has suggested, and his quotation from Ashby (§4), we might expect ideas from theatre to inform or challenge ideas in cybernetics as much as vice versa – to provide a theatre, as it were, in which to explore the cybernetic. If the relations between cybernetics and theatre have not yet been explored in as much depth as those between cybernetics and design, there are, as with design, a number of clear parallels in existing work that can be drawn on. These include Heinz von Foerster's (2003c: 325ff) concerns with magic; the performance events that have long been part of the conferences of the American Society for Cybernetics (Richards 2015); and Andrew Pickering's (2010) interpretation of British cybernetics as what he refers to as "ontological theatre," where ideas are explored through their staging in experimental devices or other forms of practice. Central in Pickering's account is the work of Gordon Pask, who is also a key reference for Scholte. Scholte's concern with Pask stays close to the formal aspects of conversation theory, which he uses to make connections with the Stanislavski method (§§23ff). This is similar to the way that Glanville draws on Pask in building bridges between cybernetics and design (Glanville 2007, 2009b). Pask's oeuvre, however, suggests further possibilities for building the relationship between cybernetics and theatre. Pickering (2010) emphasizes the performative qualities of Pask's devices, through which he embodied his ideas in order to explore them in a way not unlike Scholte's (§§42ff) account of the stage as a modeling facility. Most explicitly, Pask was directly engaged in the theatrical, most notably with the development of the Musicolour device with Robin McKinnon-Wood (Pask 1971) and his substantial collaboration with avant-garde theatre director Joan Littlewood and architect Cedric Price on the Fun Palace project during the 1960s (Mathews 2007). By building on these connections, together with the analogies that Scholte has developed, theatre and cybernetics can offer each other mutual support in much the same way as cybernetics and design.

5. Theatre provides a rich territory in which to explore epistemological and cybernetic ideas, and the laboratory that Scholte (§52) proposes is one such exploration. The varied ways of configuring the relationship between performers and those they perform to, and the possibility of interactive or self-reflexive arrangements, also offer a number of other possibilities. Even in conventional formats, theatre is a significantly interactive medium, compared to, say, film, because of the way that actors respond to the way that the audience responds to them (this is Pask's starting point in his collaboration with Littlewood[1]). Theatre therefore offers the potential for staging different epistemological relations that

1. See Pask's unpublished report "Proposals for a Cybernetic Theatre" produced on behalf of Littlewood's Theatre Workshop & Pask's own System Research as part of the Fun Palace project. A

can be explored by participating in them from different observer positions: for instance, whereas Figure 6 shows a straightforward hierarchy, the audience or researchers may also find themselves within a play being observed by the characters, and so on.

6. In this light, it is interesting that it is not clear where second-order cybernetics, with its concern with observer inclusion, would sit vis-à-vis the debate between naturalistic and anti-naturalistic approaches to the theatre that Scholte briefly mentions (§59). Both approaches are concerned with observer inclusion: on the one hand, an anti-naturalistic approach explicitly articulates our presence as observers and agents in the social setting of the theatre; on the other, it is in the naturalistic approach where we are caught up within the flow of the constructed world of the performance, identifying with characters and their situations. Whereas second-order cybernetics is often presented in simple opposition to first-order cybernetics, theatre's modeling of observer relations offers possibilities for exploring nuances of how our presence in our observing is configured.

Open Peer Commentary: **The Many Varieties of Experimentation in Second-Order Cybernetics: Art, Science, Craft**

Laurence D. Richards

1. Tom Scholte expresses concern that second-order cybernetics (SOC) is being marginalized within mainstream academia. The implication seems to be that if SOC was recognized as a legitimate and mainstream approach to science and design, it could contribute significantly to many types of human endeavor. Scholte proposes that the theatre could provide a laboratory for experimenting with ideas in SOC as a way to add some legitimacy and demonstrate value. I find this to be a novel and intriguing proposal and encourage its further development. Scholte has the unusual combination of expertise in theatre studies, directing and cybernetics necessary to pull it off. I am unsure how many people with these abilities and interests there might be; perhaps Scholte's work will stimulate more interest. In this commentary, I wish to question the prospects for, and even the desirability of, pushing SOC into "mainstream" academia.

2. SOC is distinguished by the new questions it asks, not by the answers it might supply to current questions. Its legitimacy lies in the logic(s) embedded in these questions and the desirability of the consequences of exploring the questions further. Its method is deductive. Looking for empirical support for cybernetics concepts in current systems is not of value in responding to questions about systems that do not yet exist, but that might be desirable if they did exist. The activity of designing and exploring new systems invokes the realm of the un-decidable question – questions only we can decide, questions of desirability. Artistic performance, as in the theatre, provides a vehicle for creating new systems and then experimenting with them. The form of experimentation, however, may not be in the tradition of the scientific experiment, where empirical results are used to support or oppose

copy of the document is archived in the Cedric Price Archive at the Canadian Centre for Architecture, Montreal, reference: DR1995:0188:525:001:009.

pre-formulated hypotheses and theories; on the contrary, the more appropriate experimentation might be in the form of "playing" with the dynamics of interactions and relations. Opportunities for playing with dynamics reside in the composition of the script/score; in the interactions among the actors, between actors and director, between performance and audience and among audience members and others; and in the scheduling of a performance as an event among other events.

3. Experimental composition and experimental theatre are common subjects in university programs in the fine and performing arts. In fact, the movement arts, and all the arts, are experimental. Ideas are tried, consequences explored and new ideas generated. The controlled experiments of science, on the other hand, use the word "experiment" in a different way: scientific experiments are intended to prove or disprove an explanation of a current phenomenon. The controlled aspect of these experiments requires the specification of a current system in which the explanation will be tested. The idea of a craft merges art and science and adds action. For the craftsperson, the repetition of an activity – that is, learning through doing – could be regarded as a continuous process of experimentation: through practice, the craftsperson develops her craft.

4. There is, of course, an art, science and craft involved in all these activities, but it is the focus on experimentation in the sciences that increasingly gives legitimacy to an academic field of inquiry. Cybernetics has always been trans-disciplinary, even anti-disciplinary, in approach, treating all systems (existing or imagined) as potential subject matter. The approach of SOC involves art, science and craft together, simultaneously and without bias. Mainstream science is disciplinary, empirical and oriented toward questions that can be decided through observation and controlled experiment. SOC does not belong in the mainstream and is not likely to carve out a place for itself there; rather, it provides an epistemology for an entirely different system of inquiry, one that focuses on the observers/listeners themselves, their ways of thinking, their desires and their interactions with each other. The hope of SOC is in the prospect of a new way of thinking and talking about our world, our society and ourselves.

5. Why focus on experimenting with the "dynamics" of human interactions and relations when speaking of SOC? Three features unique to cybernetics come to mind.

Dynamics and relations

6. Cybernetics attempts to address two domains of inquiry simultaneously: the domain of dynamics (experience) and the domain of relations (explanation) (Richards 2010, 2013). Bridging the domains requires the observer/listener to select a clock by which observations are to be made and explanations formulated. I use the word "clock" to speak of a way of sampling a dynamics. In science, the commonly accepted tradition is to select an external standard clock (years, weeks, days, hours, minutes, seconds, milliseconds and so on – equal increments based on revolutions of the earth around the sun, rotations of the earth on its axis or other fixed reference points, which self-referentially already require the selection of a clock in order to call them "fixed"), with the only choice being how often to record observations of states or structures in the phenomena being observed. For example, observing the growth of a culture in a petri dish requires a decision about whether to record observations every day, every hour, every ten minutes or some other interval (sampling rate), so that causal relations among the variables under consideration can be inferred from the changes in state or structure observed. Different sampling rates produce different rela-

tions, with different consequences for science, for the scientist and for the world that accepts the results. In the arts, many possible clocks (or conceptions of time) are employed or invoked, often without explanation or justification by the artist. For example, spray painting carries a different conception of time than splattering with a brush. Music plays with conceptions of time in the performer and the audience. Artists accept neither the regularity of time nor the desirability of the standard clock, preferring to challenge accepted notions of time. Invoking multiple conceptions of time can create an out-of-synch-ness among composer, performer and audience – a situation of conflict to be resolved and an opportunity for new ideas to emerge. This is the role of the arts in society (Richards 2010). SOC deals explicitly with the choice of clocks, placing responsibility for the consequences on the observer/listener.

Recursion

7. Cybernetics deals with recursive processes and closure in the dynamics of operations of systems, rather than with the whole systems and open systems approaches more common in the sciences and humanities. SOC suggests that focusing on the dynamics of operations of systems – patterns of changes – can throw light on the human predicament in ways no current science does, but that it does so by including the observer/listener, and their selection of (a) clock(s), in the system of interest.

Conversation

8. Cybernetics is enacted in conversation: a particular dynamics of interaction in a language such that the dynamics moves from an asynchronicity (a friction, conflict, contradiction, disagreement, being on a different plane, being out of synch) towards synchronicity (including agreement or agreement to disagree). This dynamics is the realm of SOC.

9. I have often talked of the cybernetician as a craftsperson in and with time (Richards 2016). All artists manipulate time; the cybernetician does so thoughtfully and deliberately. For scientists to deviate from the standard clock in their research would be to insert, deliberately, the observer and the observer's desires into the system being observed. This would be a new science, one where the theatre and other arts could become a playground for research. At present, this conception of science is so far removed from what is accepted that it makes little sense to push the SOC agenda onto it. SOC will become appreciated by the desirability of the consequences realized when people employ this way of thinking – namely, a reduction or elimination of violence. I use the word "violence" to speak of any action that reduces the participation of some by eliminating their choices and alternatives. I regard the reduction or elimination of violence as a consequence most of humanity (even if not all) could agree on as desirable and therefore what experiments with SOC need to demonstrate. The theatre is a place to practice and hone this craft.

10. I would also like to make the case that SOC implies an approach to experimentation with language that is different from traditional approaches. Specifically, treating signs and symbols as fundamental units of analysis in the study of language, as in semiotic research, does not recognize them as objects generated by the very language being studied. SOC recognizes languaging as a process (the coordination of the coordination of action), with the language produced then serving as a medium through which the dynamics of a conversation can happen. If there is to be a unit of analysis in SOC experimentation, it

should be the entire conversation. In theatre, the conversations could be those that actors or directors have with themselves – namely, those that generate thinking; those that occur on stage as modulated by a script or score; those between actors and between the actors and director in preparation of a performance; and those between actors/directors and audience, or between audience members, or between audience members and others not in attendance. In all cases, the opportunity is to play with the dynamics of interactions and relations. Characteristics to observe include amplitude, speed, frequency, rhythm, emphasis, pivots, events and, of course, synchronicities and asynchronicities, among others – anything that would distinguish a pattern of dynamics.

11. In conclusion, Scholte lays out for us a challenge: let us advance SOC by doing it. Our ability to generate significance through scientific experiments on minds, societies and the world in general is limited by current conventions and resources and complicated by constant change in conditions, factors and desires. The theatre (and all the arts) offers the opportunity to create micro-worlds where these complications can be accounted for and experimented with, without the same constraints of convention and resources that limit the traditional sciences. Current best available knowledge can be applied and the artist's skills brought to bear on the creation of a performance, while also applying the craft of the cybernetician. The results will speak for themselves. I look forward to hearing about the experiments, if not participating in them.

Author's Response: "Playing With Dynamics": Procedures and Possibilities for a Theatre of Cybernetics

Tom Scholte

1. I must begin this response with an expression of deep and sincere gratitude to the eight authors of the open peer commentaries and to the editors of this journal for facilitating their rich contributions to this project. The questions and comments they have provided have not only helped clarify my reflections on the work as it has proceeded thus far but have also played a pivotal role in inspiring and orienting its direction moving forward. My response will be organized along these two thematic lines under the categories "procedures" and "possibilities." In the context of this response, the term "procedures" is meant to denote both the particular research protocols queried by **Bruce Clarke** & **Dorothy Chansky** and the embodied cognitive operations of actors and audiences interrogated by **Edgar Landgraf** and **Bernd Porr.** Possibilities will reflect upon **Larry Richards**'s and **Ben Sweeting**'s suggestions for future directions of the overall project as well as **Lowell Christy**'s endorsement of, and **Albert Müller**'s objections to, its conceptual foundations.

Procedures

2. **Clarke & Chansky**'s request for clarification regarding the research protocols employed in the experiment described in §20 of the target article is well-grounded given the significant potential ramifications of the details in question. To clarify, the description of

the two sets of performances (autopoietic and allopoietic) in §5 of their commentary is, indeed, accurate. The term "invite" in both instances refers to an invitation to "random people to show up" (ibid.) to a single performance put out through the social media outlets of the Theatre and Film and Psychology Departments at UBC as well as personal invites from the investigators (Alan Kingstone and myself) and actors. Suspecting that the social media invitations were most likely to solicit the participation of university students, the personal invitations (targeted toward non-theatre specialists) were intended to introduce some diversity into our subject pool. Admittedly, the majority who attended would still fall within the categories of the WEIRD acronym (§6). They are also assumed to be relatively interchangeable in their competency as spectators of naturalistic/realistic theatre given its position of dominance within the field and the ubiquity of its modes of actor performance across the adjacent fields of film and television. As the authors indicate, more experimental aesthetic modes would likely introduce new potentially confounding influences (§7) but this was not a concern in this instance.

3. As a description of the cognitive operations underpinning the kind of everyday social behavior that naturalistic theatre seeks to simulate, **Porr**'s insistence on the appropriateness of forward modeling and open control loops provides a welcome conceptual refinement that is entirely compatible with the Stanislavskian notions outlined in the target article. Indeed, many moments of heightened drama occur precisely when the forward model that a character is depending on is perturbed and "these forward models fail [...] and the feedback loops need to kick in" (§4). Furthermore, as referenced previously in Scholte (2015), Stanislavskian teacher/practitioner Uta Hagen tells us that, in life, while "[w]e never know what the next moment will be [...] we always have expectations about it." "Utter spontaneity" onstage becomes possible when the actor can learn to "suspend knowledge of what [is] to come by unearthing the character's expectations" and let those expectations collide with what actually takes place "in the moment" (Hagen & Frankel 1991: 128). The forward model/open loop conception of the character's operations also seems compatible with an earlier cybernetic conception that proved useful in rehearsals for our experimental production of *Salt-Water Moon*: Warren McCulloch's postulated "redundancy of potential command" summarized by Gordon Pask as describing a set of "goal-directed subsystems" that "compete for dominance" with the command position "shift[ing] from time to time in a way that favors the subsystem currently in possession of the most relevant information [...] from the environment (or from the aggregate of subsystems, or both)" (Pask 2011: 528).

4. Extending **Porr**'s cocktail party scenario, an individual in attendance in order to seek a potential mate can fairly effortlessly engage in small talk relying on open loop forward models while the "search for mate" subsystem is highly attentive to feedback as he or she surreptitiously scans the room for potential candidates. In this sense, the subsystem currently in command may be engaged in a closed loop while the other subsystems necessary to continue to function successfully in the environment can "get by" on open loop forward models. That is, until the individual distractedly commits a conversational faux-pas and their partner responds in a perturbing fashion. At that moment, the "social damage control" subsystem will seize control, temporarily overriding the forward model of the conversational partner and closing the loop until the situation is satisfactorily stabilized and "search for mate" can resume the command position. Again, such moments are often the stuff of drama; or perhaps, more often, comedy.

5. Regarding the film-based improvisational processes referred to by **Porr** (§9), both varieties have played a major role in my own artistic practice as I have been an actor/co-

creator on three feature films developed in the mode of Mike Leigh (*Dirty*, 1998; *Last Wedding*, 2001; *Crime*, 2007) and a trilogy of films that, like *Victoria,* were entirely improvised based on objectives and given circumstances in a collaboratively developed treatment (*Mothers&Daughters*, 2008; *Fathers&Sons*, 2010; *Sisters&Brothers*, 2011). However, with a focus on circular causal interactions and increased observer agency for the audience, the level of directorial/editorial control in terms of shot selection, pacing, rhythm, size, etc. and the linearity of traditional editing strategies, I have not found the medium of film to be as fruitful an arena for the type of investigation I am carrying out as live theatre. Of course, it is possible to push against predominant, mainstream cinematic techniques and provide more room for the elements in which I am interested within the filmic medium, and it may yet find its way into this program of research. Certainly, improvisationally-generated films manifest performance modalities that are markedly different from traditional author-centred works and, like **Porr**, I would attribute this difference to the wider margins within which cybernetic self-organization can take place facilitated by these processes.

6. **Landgraf**'s incisive problematization of our notion of naturalness resonates productively with **Porr**'s observation that "improvisation imitates everyday double contingency reduction and acts as a convincing demonstrator/simulator of how everyday communication emerges" (§8). In this vein, **Landgraf**'s previous deployment of the manner in which Judith Butler "draws on the concept of improvisation to describe a person's practice of enacting his or her identity in terms of a complex interaction between individual doing and societal constraint" (Butler 2004: 16) in the monograph to which he refers (§2) provides a useful lens through which to view the "simulation" of the "everyday" that a second-order cybernetic conception of the naturalistic theatre might illuminate. In interrogating cultural norms around gender, Butler describes the manner in which women are expected to make manifest behaviorally a socially constructed model of sanctioned femininity as an "improvised performance in a scene of constraint" (Butler 2004: 1). While I do not wish to diminish, in any way, the depth, power, and specificity of the manner in which such mechanisms impact and oppress women in particular, I agree with what I take to be **Landgraf**'s suggestion that the notion of "improvised performance in a scene of constraint" can be usefully extended to describe the inherent performativity of all manner of social identity lying, mostly unconsciously, at the center of our daily living. **Landgraf** tells us that Butler's vision of the highly circumscribed agency emerging when "individual 'doing' intersects with the particular social constraints it engages":

> [R]everses the causal link between doer and deed, giving primacy to the 'doing' rather than the doer. She asks us to conceive of agency (or, more precisely, of the mere appearance of agency) as the *result*, not the source, of a continued, improvised practice. In this regard, improvisation's object of invention – the 'thing' created – is the improviser herself; for Butler, improvisation marks the simultaneous composition and performance of the 'doer'. (Landgraf 2011: 18)

7. The creation of "character" by an actor in the Stanislavskian mode is no different. This is why actors training in this tradition are constantly urged to let go of any notion of "character" as a consistent entity outside themselves that they must understand, describe, and then seek to recreate, and rather, simply to be "themselves" inside the given circumstances (i.e., constraints) of the text on both the level of the societal milieu represented by the fiction and the stylistic tendencies of the particular genre or author that shape the text itself. There is no "character," only a pattern of actions described by an observer as a

"character." (Hence my invocation of the "black box" a la Ranulph Glanville.) What has made the Stanislavski system of acting the default "best practice" in the generation of nuanced and "life-like" performance is that it is these very mechanisms that lie at its heart. That the "life" these performances are "like" reflects and conforms to social constraints that are different, but just as constructed, as earlier historical periods is a point well taken. This overlap between the actor's work and the mechanics of "everyday double contingency reduction" identified by **Porr** is precisely what makes the naturalistic theatre such a rich field for second-order observation.

8. **Landgraf** critiques Søren Brier's commitment to softening the constructivist stance somewhat by embracing the "notion of reality that is thought to exist – and exist with definable qualities – independent of its (semiotic re)construction by an observer" (§6). Nevertheless, I remain sympathetic to Brier's position and persuaded by his argument that, without some degree of external invariance, the generation of an *eigen-object* would not be possible (Brier 2008: 105). It is enough to acknowledge that our access to this external environment is indirect and our descriptions of it the inferential constructions of closed systems. Even Ernst von Glasersfeld himself comes close to admitting as much when he speaks of the "obstacles" and "constraints" within our environment with which we "clash" even if this does not tell us "what the obstacles *are* or how a reality consisting of them might be structured" (Glasersfeld 1995: 73, italics in original). My embrace of Brier's softer position does not represent a capitulation to the tendency in the contemporary humanities, rightly identified by **Landgraf**, to embrace embodiment as a "foundational" position from which to fend off constructivist skepticism of various stripes. At the very heart of my proposal is the intention to interrogate the manner in which the *eigenbehaviors* **Landgraf** eloquently describes as "stability that on the level of interaction between nervous system and psychic system is the result of long term, evolutionary processes" (§8) are hijacked and repurposed to serve hegemonic power structures that rely, largely, on the symbolic arsenal within the system of communication. Even though, as **Landgraf** points out by way of Niklas Luhmann, "[a]nything we observe [...] will have to draw on the operations of the system of communication to do so" (§7) Luhmann is also open to the idea that systems theoretic functional analysis

> can clarify 'latent' structures and functions – that is, it can deal with relations that are not visible to the object system and perhaps cannot be made visible because the latency itself has a function. (Luhmann 1995: 56)

Luhmann goes on to point out that "[t]he more starkly a system is hierarchized, the more clearly do forms whose latent function is to protect hierarchy's need for latency stand out" and that

> [c]onsciousness can undermine social latencies when it forces communication, and communication can sabotage psychic latencies, especially in the form of communication of a person who is defined as seeking to protect and conceal personal latencies. (ibid: 336f)

(Perhaps this is the manner in which systems boundaries are "turned inside out" in the passage from Louis Kauffman cited by **Christy** in §11). I count the undermining of such latencies among the cybersemiotic theatre's possibilities and certainly among the desires I have for its use.

Possibilities

9. While I have great respect for **Müller** as an esteemed historian and theorist of cybernetics, I fear that, in this instance, he has completely misconstrued my main arguments. At no point in the target article or in the previous one that he mentions (Scholte 2015) do I make the claim that Stanislavski could have been influenced by cybernetics. When one looks at the dates of the two bodies of theory in question, as **Müller** has done, it is obviously a question of basic mathematics. Neither do I suggest that the early cyberneticians were influenced by Stanislavski. As **Sweeting** cogently observes,

> the connections that [I have] developed between the two fields are significant for being not ones of application but, rather, overlap, where cybernetic processes are seen to be being enacted within an already established set of practices. (§1)

That Stanislavski's independently and previously developed theory of human behavior, completely absent the influence of what would become cybernetics, could so closely mirror the latter is precisely the point. As to **Müller**'s question regarding the appropriateness of my redeployment of Glanville's black box as "epistemological tool" (§6), I am not sure what kind of issue observer-dependent distinctions, beginning with the naturalist author, through the naturalist actor and director and, finally those of "believability" and "meaningfulness" indicated by audiences of naturalistic theatre, would be if not epistemological; an issue with which the "other and older concepts" **Müller** suggests might "match better" have been inadequate in coping as the target article seeks to detail. It is also not my intention to "adopt (second-order) cybernetics in order to understand, to explain, let alone to develop [my] conception of theatre, i.e., Stanislavski's system" (§4) or to "replace" Stanislavski's original formulation with cybernetics. Rather, I am suggesting that, if the implicit cybernetics of the naturalistic theatre (including Stanislavski's approach to its performance) is made explicit, it can facilitate *new* uses of the naturalistic theatre as an instrument of cybernetic inquiry. But perhaps I have unwittingly created some of this confusion regarding my intentions myself by mingling second-order cybernetics with empirical experimental psychology as it has been traditionally practiced. This is an error that might be well avoided in light of the persuasive corrective offered by **Richards** and augmented by further analysis from **Sweeting.**

10. **Sweeting**'s citation of "Pickering's (2010) interpretation of British cybernetics as what he refers to as 'ontological theatre,' where ideas are explored through their staging in experimental devices or other forms of practice" (§4) is particularly apt vis-a-vis this research project. The autopoietic generation of a performance score by actor/characters engaging in what Stanislavski describes as goal-seeking behavior can be read as an analogue of Ross Ashby's multi-homeostat setup, which "stages for us a vision of the world in which fluid and dynamic entities evolve together in a decentered fashion exploring each other's properties in a performative back and forth dance of agency" (Pickering 2010: 106). The director of such a performance finds himself in a position similar to Ashby, whose

> atomic knowledge of the individual components of his machines and their interconnections [...] failed to translate into an ability to predict how aggregated assemblages of them would perform. Ashby just had to put the units together and see what they did. (ibid: 108)

(Remembering, of course, the circumscribed nature of "agency" thematized by Butler, mirrored by the fact that, while unpredictable, the behaviour of the multi-homeostat setup is, at bottom, deterministic on a mechanical level.) **Sweeting**'s assertion that a cybernetic investigation of theatre might work together with related trends in cybernetic design research in "opening up a new avenue for exploring how cybernetics may be understood in terms of action rather than theory, and so as an active research tradition rather than one form of worldview amongst others" (§1) takes on an increased poignancy when contextualized by the experimental vision sketched by **Richards** and in conversation with **Müller**'s theoretical reservations.

11. **Richards** captures the spirit of my enterprise thus far as a proposal that "the theatre could provide a laboratory to experiment with ideas in SOC (second-order cybernetics) as a way to add some legitimacy and demonstrate value" but questions "the prospects for, and even the desirability of, pushing SOC into "mainstream" academia." He suggests instead that "SOC is distinguished by the new questions it asks, not by the answers it might supply to current questions" and that a more "appropriate form of experimentation" for the theatrical laboratory that I am proposing "may not be in the tradition of the scientific experiment, where empirical results are used to support or oppose pre-formulated hypotheses and theories;" but rather "in the form of 'playing' with the dynamics of interactions and relations" (§2). This is an appealing and accurate description of the kind of activity I most definitely have in mind; however, my current interest is less "in responding to questions about systems that do not yet exist, but which might be desirable if they did exist" (ibid) than it is in shedding light on pathologies in existing systems through cybernetic analysis of the processes of their representative embodiment and observation in naturalist drama; as well as potential pathologies in the processes themselves. (Of course, over the arc of an extended research project, the latter may well end up facilitating the former.) This is why I have confined my focus to more traditional theatrical forms and, thus far, not engaged with the experimental theatre lineage of Pask referred to by **Sweeting** (§4), as rich and valuable as it certainly is. Similarly, it has led me to side-step the excellent work on purely improvised performance by Keith Sawyer alluded to by **Landgraf** although the themes emergent across the commentaries have led me to reconsider him as an important potential source of insight into **Landgraf** and **Porr**'s overlapping conceptions of improvised identity and "everyday double contingency reduction" discussed above. The same could be said of the "frame analysis" work of Erving Goffman and its foundational role in the rich literature of symbolic interactionism. **Richards's** admonishes that "SOC will become appreciated by the desirability of the consequences realized when people employ this way of thinking" (§9) and to just get on with "advance[ing] SOC by doing it" in "micro-worlds [...] without the same constraints of convention and resources that limit the traditional sciences" (§11). This is a powerful and persuasive message that I am considering deeply as I ponder the next phase of this work. Such a micro-world may indeed be a more effective venue in which to "demonstrate the value of cybernetic thinking" while, perhaps, also avoiding some of the philosophical features of the current project that **Müller** finds disconcerting. But what might be sacrificed through disengaging with mainstream science in areas such as psychology, with its extensive literature on areas of cognition that have long been a part of cybernetic inquiry? Is the emergence of "a new science, one where the theatre and other arts could become a playground for research" such a long shot (**Richards** §9) that we should embrace and entrench our position on the margins with deliberate purpose? Perhaps. But another strand of cybernetics-inspired theory, touched upon by **Christy**, offers a language through which a long-ago called-for but as yet undeveloped science might be explored: Kenneth Boulding's eiconics or science of the image.

12. Around the middle of the last century, two books emerged from Stanford's Center for Advanced Study in the Behavioral Sciences; the second, George Miller, Eugene Galanter and Karl Pribram's *Plans and the Structure of Behavior* (1960), offered as a direct and complementary response to the first, Boulding's *The Image* (1956). Boulding's book laid out his conception of the Image as "what [one] believe[s] to be true; [one's] subjective knowledge. It is this Image that largely governs [one's] behavior" (Boulding 1956: 7f). The Image is subject to the impact of "messages" that "consist of *information* in the sense that they are structured experiences. *The meaning of a message is the change which it produces in the image*" (ibid: 7, italics in original). Each individuals image is composed along the following dimensions:

- Spatial (location in space)
- Temporal (the stream of time and his place in it), relational (universe as a system of regularities)
- Personal (individual in the midst of persons, roles, and organizations)
- Value (scales of better or worse)
- Affectional/emotional (various items imbued with feeling or affect)
- Division of conscious/unconscious/subconscious, certainty or uncertainty/clarity or vagueness, reality or unreality (degree of correspondence between image and "outside")
- Public/private scale (degree shared by others) (ibid: 47).

The role of "value" as well as "fact" in this subjective knowledge structure is critical.:

> At the gate of the image stands the value system demanding payment. This is as true of sensory messages as it is of symbolic messages. We now know that what used to be regarded as primary sense data are in fact highly learned interpretations. We see the world the way we see it because it pays us and has paid us to see it that way. (ibid: 50)

13. Miller, Galanter, and Pribram pick up this conceptual thread and offer their notion of a "plan" as "any hierarchical process in the organism that can control the order in which a sequence of operations is to be performed" and link them to Boulding's Image in the following reciprocal manner.

> Changes in the Images can be effected only by executing Plans for gathering, storing, or transforming information. Changes in the Plans can be effected only by information drawn from the Images. (Miller, Galanter & Pribram 1960: 18)

14. Before concluding his book, Boulding calls "half tongue in cheek" for the founding of a new science he dubs "eiconics," to study "the formation of images, the impact of messages, and the consequences of images for behaviour" (Boulding 1956: 172). Such a science would be intrinsically designed to operate on a second-order level of analysis as "[w]e can examine consistency, coherence, survival value, stability, and organizing power in the image, because the image can investigate the image" (ibid: 174). If "a new science" in which "the theatre and other arts could become a playground for research" is to coalesce, the framework of eiconics developed over these two books might be a place from which to begin. Why Boulding chose a stance of only half-seriousness to float this conception can, of course, only be a matter of speculation. My guess is that it served as a pre-emptive defensive maneuver by a man who knew only too well how both radical and unformed his thinking was in this particular instance and who feared the derisive scorn of the mainstream

academics with which he was more regularly in contact. Moving towards this endeavor from a position of artistic practice rather than economics, I consider myself fortunately less vulnerable to the kind of professional price Boulding might have been asked to pay and am unashamed in suggesting that we take up his proposal in earnest. (In fact, the authors of the second volume acknowledged that, traditionally, artists were the ones best equipped to carry out such a project (Miller Galanter & Pribram 1960: 214). The expansive vision in which this proposal is enveloped is articulately captured in **Christy**'s commentary and I can add little to it other than my humility in the face of its vast aspirations. Whether this new field would retain the identifier of "science" or, as Boulding suggested was likely, opt for the descriptor "discipline" (Boulding 1956: 160–163) depends largely on the kinds of considerations raised by **Richards**. His idea that a cybersemiotic theatre laboratory that featured "the scheduling of a performance as an event among other events" (§2) might best facilitate fruitful experimentation is a practical suggestion worth considering. An event designed to "demonstrate the value of second-order cybernetic thinking" is likely to resemble something quite different from a typical evening at the theatre and may well benefit from built-in para-theatrical components to facilitate the type of second-order reflection desired. The development of such new paradigms is likely to require a Stafford-Beer-sized imagination. It is this reflection that brings me to my concluding thoughts.

15. In closing, I would like to address **Müller**'s suggestion that my project demonstrates a lack of "intellectual economy" and that, perhaps, I have been seduced by the beauty of a theory that does not survive Occam's Razor when applied in the manner that I am suggesting (§§4, 6) My response takes the form of two questions that I have found myself asking. What was the coalescence of cybernetics itself if not an exercise in stretching descriptions across domains and testing their elasticity, perhaps to the breaking point? And how has the field continued to grow and renew itself if not for the further stretching of "beautiful theories" beyond the margins of Occam's Razor from Stafford Beer's claim that an organization the size of a national government could be modeled upon the human nervous system to Luhmann's endlessly controversial extension of the theory of autopoiesis into the realm of social systems to the audacity of Glanville's claim that science itself is in fact a restricted subset of design that is, in turn, the embodiment of cybernetics? It no way is my intent to place myself, and my endeavor, on the same level as these giants of our field. It is, rather, to point out that if my proposal entails some rather radical, perhaps questionable, extensions of existing concepts, it is largely because I have been so profoundly inspired by the daring cyberneticians of the past, who have cleared the intellectual space for entirely new forms of thinking. Certainly, this flair for theoretical eccentricity has fueled the recurrent accusation that cybernetics is not a genuine science but rather an elaborate system of analogies (Medina 2011: 11); but, perhaps, we should simply follow **Richards**'s lead and happily surrender the scientific claim altogether. Either way, I am grateful that, in spite of his misgivings about its intellectual foundation, **Müller** feels that my "ongoing work could influence both theatre and research and might very well lead to lasting changes in concepts and theories as well" and that he "shall remain curious" (§7). To some, Beer's Cybersyn project might stand as a singularly spectacular quixotic failure. And yet many continue to sift through the detritus of its collapse for lessons that might still serve us well. In this spirit, I will endeavor to follow the admonition borrowed by Ranulph Glanville from Samuel Beckett: Try Again, Fail Again, Fail Better.

Combined References

Ashby W. R. (1956) An introduction to cybernetics. Chapman & Hall, London.

Auslander P. (1998) From acting to performance: Essays in modernism and postmodernism. Routledge, New York.

Ball D. (1983) Backwards and forwards: A technical manual for reading plays. Southern Illinois University Press, Carbondale and Edwardsville.

Bateson G. (1987) Cybernetic explanation. In: Bateson G., Steps to an ecology of mind: Collected essays in anthropology, psychiatry, evolution, and epistemology. Jason Aronson, Northvale: 406–416. Originally published in 1967 in American Behavioral Scientist 10(8): 29–32. http://cepa.info/2726

Bateson G. (1999) Steps to an ecology of mind: Collected essays in anthropology, psychiatry, evolution and epistemology. University of Chicago Press, Chicago IL. Originally published in 1972.

Blair R. (2000) The method and the computational theory of mind. In: Krasner D. (ed.) Method acting reconsidered. St. Martin's Press, New York: 201–218.

Bogart A. & Landau T. (2004) The viewpoints book: A practical guide to viewpoints and composition. Theatre Communications Group. New York.

Boulding K. (1956) The image: Knowledge in life and society. University of Michigan Press, Ann Arbor MI.

Brier S. (2008) Cybersemiotics: Why information is not enough! University of Toronto Press, Toronto.

Butler J. (2004) Undoing gender. Routledge, New York.

Chansky D. (2004) Composing ourselves: The little theatre movement and the American audience. Southern Illinois University Press, Carbondale.

Clarke B. (2001) Energy forms: Allegory and science in the era of classical thermodynamics. University of Michigan Press, Ann Arbor.

Clarke B. (2009) Heinz von Foerster's demons: The emergence of second-order systems theory. In: Clarke B. and Hansen M. (eds.) Emergence and embodiment: New essays in second-order systems theory. Duke University Press, Durham: 34–61.

Cohen R. (1978) Acting power: An introduction to acting. Mayfield Publishing Company, Paolo Alto.

Cole T. & Chinoy H. K. (eds.) (1976) Directors on directing: A sourcebook of the modern theater. Bobbs-Merrill Educational Publishing, Indianapolis.

Coleman D., Romao T., Villamen C., Sinnett S., Jakobsen T. & Kingstone A. (2013) Finding meaning in all the right places: A novel measurement of dramatic structure in film and television narratives. Projections: The Journal for Movies and Mind 7(2): 92–110.

Coleridge S. T. (1817) Biographia literaria. http://www.gutenberg.org/files/6081/6081-h/6081-h.htm

Dean A. & Carra L. (1989) Fundamentals of play directing. Holt, Rhinehart and Winston, Toronto.

Drinko C. D. (2013) Theatrical improvisation, consciousness, and cognition. Palgrave Pivot, London.

Elam K. (1988) The semiotics of theatre and drama. Routledge, London.

Esterhammer A. (2008) Romanticism and improvisation, 1750–1850. Cambridge University Press, Cambridge.

Fischer D. H. (1970) Historians' fallacies: Toward a logic of historical thought. Harper & Row, New York.

Foerster H. von (1960) On self-organizing systems and their environments. In: Yovits M. C. & Cameron S. (eds.) Self-organizing systems. Pergamon Press, London: 31–50. Reprinted in: Foerster H. von (2003) Understanding understanding: Essays on cybernetics and cognition. Springer, New York: 1–19. http://cepa.info/1593

Foerster H. von (1981) Objects: Tokens for (eigen-)behaviors. In: Foerster H. von, Observing systems. Intersystems Publications, Seaside CA: 258–271. Originally published in 1976. http://cepa.info/1270

Foerster H. von (2003a) Objects: Tokens for (eigen-)behaviours. In: Foerster H. von, Understanding understanding: Essays in cybernetics and cognition. Springer, New York: 261–271. Originally published in 1976 in: ASC Cybernetics Forum 8(3–4): 91–96. http://cepa.info/1270

Foerster H. von (2003b) On self-organizing systems and their environments. In: Foerster H. von, Understanding understanding. Springer, New York: 1–19. Originally published in 1960 in: Yovits M. C. & Cameron S. (eds.) Self-organizing systems. Pergamon Press, London: 31–50. http://cepa.info/1593

Foerster H. von (2003c) Understanding understanding: Essays on cybernetics and cognition. Springer, New York NY.

French D. (1988) Salt-water moon. Talonbooks, Vancouver.

Glanville R. (1982) Inside every white box there are two black boxes trying to get out. Behavioural Science 27(1): 1–11. Reprinted in: Glanville R. (2012) The Black B∞x. Volume 1: Cybernetic Circles. Edition Echoraum, Vienna: 439–453. http://cepa.info/2365

Glanville R. (2007) Try again. Fail again. Fail better: The cybernetics in design and the design in cybernetics. Kybernetes 36(9/10): 1173–1206. http://cepa.info/2464

Glanville R. (2009a) A (cybernetic) musing: Black boxes. Cybernetics and Human Knowing 16(1–2): 153–167.

Glanville R. (2009b) A (cybernetic) musing: Design and cybernetics. In: The black boox, volume III: 39 steps. Edition Echoraum, Vienna: 423–425.

Glanville R. (2009c) The black b∞x. Volume 3: 39 Steps. Edition Echoraum, Vienna.

Glanville R. (2012) The black b∞x. Volume 1: Cybernetic circles. Edition Echoraum, Vienna.

Glanville R. (2015) The sometimes uncomfortable marriages of design and research. In: Rogers P. A. & Yee J. (eds.) The Routledge companion to design research. Routledge, London: 9–22. http://cepa.info/2799

Glasersfeld E. von (1980) The concept of equilibration in a constructivist theory of knowledge. In: Benseler F., Hejl P. M. & Köck W. K. (eds.) Autopoiesis, communication, and society: The theory of autopoietic system in the social sciences. Campus Verlag, Frankfurt am Main: 75–85. http://cepa.info/1352

Glasersfeld E. von (1995) Radical constructivism: A way of knowing and learning. Falmer Press, London.

Glasersfeld E. von (2007) Key works in radical constructivism. Sense Publishers, Rotterdam.

Grinker R. (1958) Towards a unified theory of human behavior. Basic Books, New York.

Gumbrecht H. U. (2004) Production of presence: What meaning cannot convey. Stanford University Press, Stanford.

Hagen U. & Frankel H. (1991) Respect for acting. Wiley, New York. Originally published in 1973.

Hahn H. (1980) Superfluous entities, or Occam's Razor. In: Hahn H., Empiricism, logic, and mathematics: Philosophical papers. Edited by B. McGuinness. Reidel, Dordrecht: 1–19. German original published as: Hahn H. (1930) Überflüssige Wesenheiten (Occams Rasiermesser). Wolf, Vienna.

Henrich J., Heine S. J. & Norenzayan A. (2010) The weirdest people in the world? Behavioral and Brain Sciences 33(2–3): 61–83.

Heylighen F. (2013) Self-organization in communicating groups: The emergence of coordination, shared references and collective intelligence. In: Massip-Bonet A. & Bastardas-Boada A. (eds.) Complexity perspectives on language, communication, and society. Springer, New York: 117–150.

Hickok G. (2014) The myth of mirror neurons: The real neuroscience of communication and cognition. W. W. Norton, New York.

Hodge F. (1994) Play directing: Analysis, communication, and style. Prentice Hall, New Jersey.

Holzapfel A. (2014) art, vision and nineteenth century realist drama: Acts of seeing. Routledge, New York.
Houghton N. (1962) Moscow rehearsals: An account of methods of production in the soviet theatre. Octagon Books, New York.
Ibsen H. (2010) Hedda Gabler. Originally published in 1907. http://www.gutenberg.org/files/4093/4093-h/4093-h.htm
Jesse A. (1996) The playing is the thing: Learning the art of acting through games and exercises. Wolf Creek, Wisconsin.
Jonas W. (2007) Research through DESIGN through research: A cybernetic model of designing design foundations. Kybernetes 36(9/10): 1362–1380.
Jonas W. (2015) A cybernetic model of design research: Towards a trans-domain of knowing. In: Rogers P. A. & Yee J. (eds.) The Routledge companion to design research. Routledge, London: 23–37.
Kauffman L. H. (1987) Self-reference and recursive forms. Journal of Social and Biological Structures 10: 53–72. http://cepa.info/1816
Kauffman L. H. (2003) Eigenforms – Objects as tokens for eigenbehaviors. Cybernetics & Human Knowing 10(3–4): 73–90. http://cepa.info/1817
Kauffman L. H. (2005) EigenForm. Kybernetes 34(1/2): 129–150. http://cepa.info/1271
Kauffman L. H. (2015) Self-reference, biologic and the structure of reproduction. Progress in Biophysics and Molecular Biology 10(3): 382–409. http://cepa.info/2844
Kinser B. & Kleinman N. (1969) The dream that was no more a dream: A search for aesthetic reality in Germany, 1890–1945. Harper & Row, New York.
Knowles R. (2015) How theatre means. Palgrave Macmillan, London.
Landgraf E. (2009) Improvisation: Form and event – A Spencer-Brownian calculation. In: Clarke B. & Hansen M. B. N. (eds.) Emergence and embodiment: New essays on second-order systems theory. Duke University, Durham: 179–204.
Landgraf E. (2011) Improvisation as art: Conceptual challenges, historical perspectives. Continuum, New York.
Luhmann N. (1984) Soziale Systeme. Grundriß einer allgemeinen Theorie. Suhrkamp, Frankfurt am Main. English translation: Luhmann N. (1995) Social systems. Stanford University Press, Stanford CA.
Luhmann N. (1995) Social Systems. Stanford University Press, Stanford. German original published in 1985.
Luhmann N. (1997) Die Gesellschaft der Gesellschaft. Suhrkamp, Frankfurt am Main.
Luhmann N. (2000) Art as a social system. Stanford University Press, Stanford. Orignally published in German as: Luhmann N. (1995) Die Kunst der Gesellschaft, Shurkamp Verlag, Frankfurt am Main.
Mathews S. (2007) From agit-prop to free space: The architecture of Cedric Price. Black Dog, London.
Matuszek K. C. (2015) Ontology, reality and construction in Niklas Luhmann's theory. Constructivist Foundations 10(2): 203–210. http://constructivist.info/10/2/203
McConachie B. (2008) Engaging audiences: A cognitive approach to spectating in the theatre. Palgrave Macmillan, New York.
Medina E. (2011) Cybernetic revolutionaries: Technology and politics in Allende's Chile. MIT Press, Cambridge MA.
Merlin B. (2003) Konstantin Stanislavsky. Routledge, London.
Miller G., Galanter E. & Pribram K. (1960) Plans and the structure of behavior. Holt, Rinehart and Winston, London.
Mitchell K. (2009) The director's craft: A handbook for the theatre. Routledge, London.
Müller-Kampel B. (2003) Hanswurst, Bernardon, Kasperl. Spaßtheater im 18. Jahrhundert. Schöningh, Paderborn.

Nietzsche F. (2008) The birth of tragedy out of the spirit of music. Translated by Jan Johnston. Vancouver Island University, Nanaimo BC. German original published in in 1872. http://www.holybooks.com/wp-content/uploads/Nietzsche-The-Birth-of-Tragedy.pdf

Palm W. J. (2000) Modeling, analysis and control of dynamic systems. Wiley, New York.

Pask G. (1971) A comment, a case history and a plan. In: Reichardt J. (ed.) Cybernetics, art and ideas. Studio Vista, London: 76–99.

Pask G. (1980a) The limits of togetherness. In: Lavington S. (ed.) Proceedings IFIP World Congress in Tokyo and Melbourne. North Holland, Amsterdam: 999–1012.

Pask G. (1980b) Developments in conversation theory: Actual and potential applications. http://www.pangaro.com/pask-pdfs.html

Pask G. (2011) The meaning of cybernetics in the behavioral sciences. In: Scott B. (ed.) The cybernetics of self-organization, learning and evolution. Edition echoraum, Vienna: 511–536. Originally published in: Rose J. (ed.) (1969) Progress in cybernetics. Volume 1. Gordon and Breach, New York: 15–45.

Pavis P. (1998) Dictionary of the theatre: Terms, concepts, and analysis. University of Toronto Press, Toronto.

Pickering A. (2010) The cybernetic brain: Sketches of another future. University of Chicago Press, Chicago IL.

Pickering K. & Thompson J. (2013) Naturalism in theatre: Its development and legacy. Palgrave Macmillan, London.

Porr B. & Di Prodi P. (2014) Subsystem formation driven by double contingency. Constructivist Foundations 9(2): 199–211. http://constructivist.info/9/2/199

Porr B. & Wörgötter F. (2003) Learning a forward model of a reflex. In: Becker S., Thrun S. & Obermayer K. (eds.) Advances in neural information processing systems 15. MIT Press, Cambridge MA.

Porr B. & Wörgötter F. (2005) Inside embodiment – What means embodiment for radical constructivists? Kybernetes 34: 105–117.

Rebitzer J. (1995) Playing with feedback control systems: Thoughts on self-consciousness. In: Foerster von H. (ed.) Cybernetics of cybernetics. Future Systems, Minneapolis: 47–50.

Richards L. D. (2010) The anticommunication imperative. Cybernetics & Human Knowing 17(1–2): 11–24. http://cepa.info/925

Richards L. D. (2013) Difference-making from a cybernetic perspective: The role of listening and its circularities. Cybernetics & Human Knowing 20(1–2):59–68. http://cepa.info/924

Richards L. D. (2015) Designing academic conferences in the light of second-order cybernetics. Constructivist Foundations 11(1): 65–73. http://constructivist.info/11/1/065

Richards L. D. (2016) A history of the history of cybernetics: An agenda for an ever-changing present. Cybernetics & Human Knowing 23(1): 42–49. http://cepa.info/2781

Riegler A. & Müller K. H. (eds.) (2014) Second-order science. Special issue of Construcivist Foundations 10(1).

Rokotnitz N. (2011) Trusting performance: A cognitive approach to embodiment in drama. Palgrave and Macmillan, London.

Sawyer K. R. (2003) Improvised dialogues: Emergence and creativity in conversation. Ablex, Westport CT.

Scholte T. (2010) The Stanislavski game: Improvisation in the rehearsal of scripted plays. Canadian Theatre Review 143: 24–28.

Scholte T. (2015) Proto-cybernetics in the Stanislavski system of acting: Past foundations, present analyses and future prospects. Kybernetes 44 (8–9): 1371–1379.

Scott B. (2011) Conversation theory: A dialogic approach to educational technology. In: Scott B., Explorations in second-order cybernetics: Reflections on cybernetics, psychology and education. Edition Echoraum, Vienna: 304–328. Originally published in 2001 in Cybernetics & Human Knowing 8(4): 25–46. http://cepa.info/1803

Segal L. (2001) The dream of reality. Second edition. Springer, New York. Originally published in 1986.
Sievers W. D., Stiver Jr. H. E. & Kahan S. (1974) Directing for the theatre. W. C. Brown, Dubuque.
Spiller N. (ed.) (2002) Cyber reader: Critical writings for the digital era. Phaidon Press, London.
Stanislavski K. (2008) An actor's work: A student's diary. Routledge, New York.
Sweeting B. (2015) Conversation, design and ethics: The cybernetics of Ranulph Glanville. Cybernetics & Human Knowing 22(2–3): 99–105. http://cepa.info/2845
Sweeting B. (2016) Design research as a variety of second-order cybernetic practice. Constructivist Foundations 11(3): 572–579.
Zola E. (1893) The experimental novel. Originally published in French as: Zola E. (1880) Le roman expérimental. http://users.clas.ufl.edu/rogerbb/classes/readings/zola.pdf

Part II:
Reflecting on the Perspectives for a Fivefold Agenda of Second-Order Cybernetics

Remarks of a Philosopher of Mathematics and Science

Michèle Friend

Structural style

1. In the target article, Louis Kauffman discusses the notion of an eigenform. In technical mathematical terms: this is a fixed point for a transformation. In very general and woolly terms, it is a reflexive stability that is reached while something is still changing or moving. The image of a spinning top, before it destabilizes and collapses, can be counted as an image of an eigenform. There is movement (transformation) but also a fixed point (the material top, the axis of rotation and place on the surface of that rotation). The top turns upon itself. It is in this sense reflexive. The rate of spin changes as it dissipates energy. In this sense the "spinning top" (not just the material top at rest) but the top in its spinning state is affected by its own spin. It is, after all, like all material things, subject to the second law of thermodynamics. The energy transmitted to it, by the initial rotation of fingers that set it off, is dissipated in the movement.

2. Eigenforms are relative stabilities, or meta-stabilities, stabilities in motion, but more than that they affect, and draw on, the world around them. We are affected when we watch the spinning top, the physical world is affected since tops make a little noise and they affect small air currents. They draw on the world around them through their design, the aerodynamic, material and balancing properties, these are "put into" the top in its construction. The spin of the fingers that set it off transmits energy to the top, which is then dissipated in the motion.

3. The structural style of Kauffman's article is itself an eigenform. In the course of the article, Kauffman elegantly spins the ideas about themselves, widening the scope, or going into greater detail, returning to the main theme without repeating himself. The ideas develop in this reflexive circulating manner, and by the end of the article, we are affected by the information, and begin to recognize eigenforms and fixed points around us. We see the world differently. As people read the article, as commentators comment on it, the second-order properties of the eigenform manifest themselves. The recognition of the structure of eigenforms in objects and processes is part of cybernetics. When we reflect upon, or are affected by, such recognition, we move to the meta-level or the second-order level. This reflection iterates – we reflect upon the reflection itself, and it is *we* who do this, so the reflection is reflexive but transforming at the same time; we become incorporated into the

second-order eigenform of the style of the article. We in turn affect other people around us, maybe speaking to them a little differently since our repertoire of concepts has been widened, having been changed ever so slightly by the reading of the article, etc. This wider scope of the eigenform is its third-order transformative effect.

4. I cannot do justice to Kauffman's structural style by imitating it. I am afraid I shall have to be quite banal in my structural style: more piecemeal, since I want to discuss two other, quite different, themes: objectivity and sustainability.

Objectivity

5. In our daily lives, as we go about our routines, we have a simple account of objects and objectivity. From this perspective, there are physical objects in the world. They are there, and most are there independent of us. We can move some of them, we can alter them, but there are objective limitations to what we can do, since the objects have objective properties that we only rarely alter. The simple conception of objectivity is fine, since it is sufficient for us to navigate the world with some local success. We avoid bumping into hard objects, move out of the way of hard objects that are on a collision course with us, we can pick out groceries, we can estimate how much food to buy, we can put on our clothes fairly well in the morning and take them off at night. Each of these activities involves objects that we can move and alter. If we are interested in the survival of an individual person in the "modern" world, we can safely say that we bump around in the world fairly well. We do this because we can distinguish objects and their properties: hardness, nutritional value, the fit of an item of clothing (we do not wrap a scarf around our foot or try to stuff a sock into our ear). There are objective properties of objects in the world. If we get them wrong, the results can be fatal. Objectivity in this mundane sense is not all that mysterious, and the sense of objectivity is reinforced by our experience in the mundane world.

6. The mundane world, with its medium-sized dry goods and objective properties and relations is left behind in the outer reaches of mathematics and science. There, as we study increasingly complex systems, as we look at the very small, or at the very large, the objects and objectivity of mathematics and science become less tangible and more ethereal. In mathematics and science, the notion of "objectivity" is not as simple as many people suspect.

7. In mathematics we study infinite sets, and various infinite numbers; we study infinitesimals, irrational numbers, lowest upper bounds, self-referring systems, paradoxes and the formal relationship between formal mathematical theories that contradict one another. As we learn about the outer reaches of mathematics, we learn that mathematics, as a discipline, cannot be unified by one mathematical theory. There is no unique and consistent foundation for the whole of mathematics, at least at the present state of play in the discipline.

8. From the perspective of the mundane world that we navigate, this is cause for alarm. We expect to be able to transfer our experience of the everyday to the world of mathematics. We expect objects in mathematics to have properties and assume that these objects and properties are objective. There is an immovable, hard quality to them ("immovable" and "hard" are metaphors transferred from the world of medium-sized dry goods). The object "8" might not be a physical object, but it is objective in the sense that we cannot make up its properties. It does not change its properties, or so we think. The notion of

objectivity in mathematics starts to lose its grip when we think of mathematical properties that are true of an object in one theory but false of the "same" object in another theory. Even the innocent number 8 fails to have objective properties, *tout court*. What it does have are objective properties *relative to*, i.e., *that change with*, the theory one is working in. For example, the number 8 has an immediate successor if our domain is the natural numbers, but it does not have one if our domain is the real numbers. The context, or theory, changes the properties of "the" object 8.

9. From the cybernetics perspective, especially the second-order perspective, this is not cause for alarm. The "same" number 8 in the set of natural numbers *closely resembles* the number 8 in the set of reals. This is because the two 8s *share* many properties. They are both smaller numbers than 9, for example. Thus, objectivity is a metaphysical feature only found at a higher level of discussion, when we can discuss the object and its context, that is, within a theory. Moreover, we can shift from one theory to another, using a translation, knowing that we lose information in translation. 8 is not an object of mathematics in the sense of one stable well-defined entity such as "the moon." It is an eigenform in mathematics. "It" changes with mathematical theory development. "It" is defined differently in different mathematical theories, but there is still a recognizable "it" sitting relatively stable through the shifts in theory and perspective. This second-order notion of objectivity is conceptually rich and fruitful. It bears further development and exploration, and I encourage such development. The second-order notion of objectivity can also be found in the "hard" sciences.

10. In the "hard" sciences, the reproducibility of experiments is an indicator of objectivity. As Kauffman writes: "It is the repeatability that makes a successful experiment into an eigenform" (§59). The repeatability tells us that the result is robust or stable within the parameters of what is to count as "similar" experimental conditions. Water boils at 100 degrees centigrade, provided it is sufficiently pure, one is close to zero altitude, the pressure of the air is similar to that which we have on earth, and so on. "Water boils at 100 degrees centigrade" is relatively objective. It is a relatively stable fact. The stability and objectivity depend on, and vary with, the context, what is usually thought of as the experimental conditions. But it is richer than that.

11. The context not only includes the physical experimental conditions, but it also depends on our language and our theory. These in turn, depend on the underlying metaphysics (water is a compound substance), the mathematics we choose for our theory and the underlying logic of that mathematics and language. None of these is untouched, pure, independent of the other or independent of us. They are not simply objective. They feed off each other, inform each other, change each other with new discoveries and knowledge and with new participants in the theory, language, logic, and so on. Together, as a package, they have their second-order, or third-order, eigenforms. The eigenforms give us enough stability that we can do science, but when things become complicated, and we are surprised or puzzled, turning to the second- and third-order questions becomes important. The importance is more immediate in the "softer" sciences, where we are dealing with complex issues that are not as stable.

12. Kauffman's notion of eigenform as a mark of objectivity in mathematics and science is refreshing, and frees us from many sterile philosophical conundrums about identity conditions on objects, about the role of an observer, about knowing the truth of a theorem. I shall not give examples here. But in all of these areas of philosophical enquiry, the notion of eigenform can give us new insights.

Sustainability

13. "Sustainability" is a word that has come into vogue, but it is poorly defined if defined at all. Here is my suggestion for extending Kauffman's ideas, this time on the more political and cultural stage. We should look for the eigenforms of "sustainability." What would these look like? Why do I use the plural?

14. The word "sustainability" contains "sustain." "Sustain" suggests stability. But it is a stability in motion. It includes internal change, but change within stable parameters, on a first reading. It is also an "ability" that is, a potential, or a predisposition. With "sustainability" we want to preserve the ability of something or the ability to do something.

15. Here are two different conceptions of sustainability. We shall develop a more interesting third conception after exploiting the notion of eigenform. We might want to sustain the current state of affairs, what is commonly referred to as "business as usual." In this case, our economic, political and infrastructural institutions are sustainable in the very simple sense that we do not have to do anything. By leaving the institutions alone, and letting them change in the ways they have been changing, we sustain business as usual, and we might be tempted to think that this is almost *by definition* "sustainable." Here is why: what people usually mean when they favour "business as usual" is that they do not want government interference. They do not want a particular type of institution (such as a government or international agreement) to influence other institutions such as businesses or trade. Thus, to sustain the business-as-usual practice, to ensure that it is sustained, we need to do nothing new. Do nothing, and we sustain business as usual by definition. However, things are not so easy. There is a subtlety. Government institutions change, and this is also "usual." Thus, to properly sustain the business as usual model, we have to be clear about what counts as usual and what counts as unusual. In fact, we need to ensure against a government, or higher institution, interfering in business at a lower level. The problem with this conception of sustainability will not only come from government and international agreements.

16. Scientists tell us that this conception of sustainability is impossible in the long run, or that it is unwise, imprudent or immoral. Why? There are physical, biological and environmental limitations that we either have already passed or are about to reach. We live on a finite planet, with a finite amount of fertile land, a finite amount of ocean and a finite amount of biomass that cycles carbon into oxygen. Competing for these resources, aggravated by population growth, becomes increasingly costly (which is why it is an unwise and imprudent conception), and the competition will become ferocious, harming many people; hence the imprudence is immoral.

17. Of course, the situation is more complex than what I have described, thus, it requires more sophisticated conceptual tools to analyse. There are *rates* of replenishment of resources. A farmer can add more fertilizer, but the crops will still only grow at a certain rate. There are limits to the speed-up of plant growth through irrigation, genetic alteration, adding fertilizer, and so on. Carbon is converted into oxygen at a rate. Aquafer levels are replenished at a rate. Fossil fuels and coal are produced in the earth at a rate. When the rate of consumption of these resources and services (creating oxygen) is greater than the rate of replenishment, we are living *unsustainably* in a more scientific sense.

18. We have two eigenforms corresponding to the two conceptions of sustainability. One takes as its reflexive domain society and the economy within the context of neo-classical economic thinking. Neo-classical thinking is the fixed point allowing the transforma-

tion of business, society and politics. The second conception widens the reflexive domain to include the natural world as we know about it through science. It is us, in the world, shaping and changing the world that together have an eigenform. The relationship between us and the environment, the sustaining of our life *by*, and *within* the natural environment is the fixed point. The transformation is the cyclical nature of the seasons and replenishment of resources at, or above, the rate of consumption. That is how we live, or fail to live, sustainably under the second conception.

19. The first conception is first-order. We hardly perceive, or are not aware of, the context. We take neo-classical economics for granted and refuse to think of alternatives. The second conception is second-order. We have more obvious reflexivity, we are interested in a relationship between us and something else, we recognize that we influence the natural environment, and it influences us. Living in harmony with it is what affords stability to the whole system. This is the second eigenform of sustainability. This second form is unacceptable in the "more advanced" cultures.

20. Here is a third conception that is third-order. Today, only quite "primitive," "indigenous," "very poor" (living below the poverty line) or a few isolated commune cultures live sustainably under the second conception. Mankind lived in such a state, arguably, up to the industrial revolution. After that, we started to live unsustainably, according to the second conception. Those of us who enjoy the riches of the industrial revolution are often unwilling to give up those riches. The third conception, due to Kozo Mayumi, is that we should decide on a *culturally acceptable rate of entropy production per rate of consumption.*

21. That is, we accept that we are not living sustainably as per the second conception. We also recognise the science that warns us that we cannot live sustainably as per the first conception. We recognise, with the scientists, that we create entropy. That is, we use up resources at a rate that is *greater* than the rate of replenishment. Different cultures have different expectations about material and energy consumption. Each culture can make a *prima facie* decision independently of other cultures at what rate (above the rate of replenishment) they are willing to use up the resources.

22. Of course, we are talking of reflexive domains. The rates decided upon have their own momentum and direction. But cultures are rarely isolated. We influence one another. We exchange information. We try different lifestyles or read and hear about them. Thus, what is an acceptable rate of entropy production per unit of consumption has its own (meta-) rate of change. This is a third-order eigenform. The rate of entropy production per rate of consumption is the fixed point. The culture is the transformation. We have a new conception of sustainability that is quite abstract, and has the structure of an eigenform, and thus can be mathematically represented, and reasoned over rigorously, up to the standards of our best mathematics. As a culture seeking sustainability, we are after this third-order eigenform. We seek to sustain a rate of depletion because it is worthwhile. It is deemed to be worthwhile when we consciously allow ourselves to consume a certain amount of material and energy that will not be replenished. I think that awareness of this third level eigenform will help us understand what is at stake if we want to live "sustainably."

Conclusion

23. The conception of eigenform developed in an eigenform structural style by Kauffman has rich conceptual possibilities. It is another way of seeing things.

24. When dealing with complex questions, it is usually beneficial to spend time looking at the question from several perspectives. We find several different-looking solutions, and we are then faced with the also complicated task of making sense of the differences. This is all in the nature of enquiry into complex issues.

25. The notion of objectivity in mathematics and science, and the problem of defining and agreeing upon a notion of sustainability are examples of problems that benefit from the perspective suggested by Kauffman. Reading the article already starts us on the path of recognizing eigenforms around us. We then move on to develop our own, to bring the concepts to bear on other areas of enquiry. We form a community of people who share this perspective, and develop it further, creating our own eigenform of eigenforms.

The Past and the Future of Second-Order Cybernetics

Ronald R. Kline

1. I would like to thank the editors for the opportunity to comment on the fivefold research agenda for second-order cybernetics that is laid out in the special issue on "Second-Order Cybernetics" of *Constructivist Foundations* for July 2016, see http://constructivist.info/11/3.

2. As a historian of science and technology, I am struck by the role of the past in this book. History is important to the editors, Karl Müller and Alexander Riegler, and to the generation of scholars who studied under the founders of second-order cybernetics in the mid-1970s (Stuart Umpleby and Robert Martin, who did their PhDs with Heinz von Foerster at the University of Illinois; and Bernard Scott who did his with Gordon Pask at Brunel University). This is not surprising because they have all written participant histories of the field (Müller & Müller 2007; Umpleby 2003, 2005, 2007; Martin 2007; Scott 2004). But several members of the next generation also pay attention to history. These include Ben Sweeting, who did his PhD with Ranulph Glanville, a student of von Foerster and Pask, as well as scholars who do not have this academic lineage. The past is a resource for developing and critiquing the fivefold research agenda in the editorial, target articles, and commentaries.

3. At the very start of this volume, the editors employ their interpretation of how Robert Martin and I independently portray the history of second-order cybernetics (Martin 2015; Kline 2015) to construct the "Kline-Martin-Hypothesis," which states

> As a research program, second-order cybernetics was
> a. insufficiently developed,
> b. has had no sustainable consequences for other scientific disciplines in the past, and
> c. will remain mostly irrelevant in the future as well.

4. In regard to my work, they say I claim "that the move from first-order to second-order cybernetics was a dead end that did not produce long-lasting impacts for other disciplines. As such, it did not leave any significant traces" (Müller & Riegler 2016: §1). At the end of the editorial, which analyzes research in second-order cybernetics, Müller and Riegler confirm parts (a) and (b) of the hypothesis, reject part (c), and describe an innovative research agenda to promote the future of the field.

5. I am honored that the editors have attached my name to an hypothesis about second-order cybernetics. Nonetheless, I disagree with their interpretation of how I treated their field in my book, *The Cybernetics Moment*. In regard to part (a) of the hypothesis, I

did not comment on whether or not second-order cybernetics was sufficiently developed as a research program. I did quote Francisco Varela as saying in 1981 that von Foerster's framework for understanding cognition was "not so much a fully completed edifice, but rather a clearly shaped space, where the major building lines are established and its access clearly indicated" (Varela 1981: xi; Kline 2015: 197). But I also said the "work that [Humberto] Maturana and Varela had done to erect and fill-in that edifice became more widely known when they published *Autopoiesis and Cognition*" in 1980 (Kline 2015: 197). In regard to part (b), I said that second-order cybernetics was, at the time the book was published, a marginal field in the US with an institutional home in the American Society for Cybernetics and in the journal *Cybernetics & Human Knowing*, which partially supports that part of the hypothesis. But I also said that second-order cybernetics was "fruitful" in the US and Europe among several social scientists. I noted that it had more success in Europe through the work of sociologist Niklas Luhmann and in the area of socio-cybernetics (Kline 2015: 4f, 199, 201, 242f). In regard to part (c), I did not speculate on the future of second-order cybernetics.

6. Nevertheless, I can understand why the editors thought I had implied that second-order cybernetics was a "dead end." I argued that its emergence in the mid-1970s was one of two reinventions of cybernetics that occurred at the end of what I call the "cybernetics moment" in the US. I defined the end of that moment as the time when cybernetics was discredited among scientists (mainly because of its association with Soviet communism and so-called fringe groups such as dianetics), when it lost its status as a promising universal science, and when it lost institutional support from MIT and the University of Illinois. I argued that cybernetics emerged from that crisis by being reinvented as a science of social systems in its first-order form, and as second-order cybernetics (Kline 2015: chap. 7). To me, the marginality of both first-order and second-order cybernetics in the early 21st century did not signal a "dead end" for either one of those fields.

7. The past does a lot of work for Müller and Riegler. They use their interpretation of the past to construct parts (a) and (b) of the Kline-Martin-Hypothesis, from which they infer part (c). Then they test the hypothesis against research in second-order cybernetics to specify a new research agenda for the field. In that analysis, the past sanctions their two-part definition of second-order cybernetics: SOC_E or cybernetics in the endo-mode; and SOC_L or cybernetics at the second-order science level. They found that "SOC_E was based on the initial definition by von Foerster as the cybernetics of observing systems" in 1974, and that "SOC_L was linked to [Margaret] Mead and her self-referential desire to study cybernetic domains such as the American Society for Cybernetics with tools and methods from cybernetics", which she presented in a speech to the society in 1968 (Müller & Riegler 2016: §49; Foerster 1974: 1; Mead 1968).

8. This use of history has been common in science and engineering since the professionalization of these fields in Europe and the US in the late nineteenth century. When a new science is emerging, its promoters often write histories of their field in order to do what sociologists of science call "boundary work." That includes drawing closed or open boundaries around their field in order to exclude some legacies and include others, or to separate an emerging field from a competing one (Gieryn 1983, 1999).

9. Boundary work is evident in Müller and Riegler's editorial as well as in the articles and open-peer commentaries that follow it. I will discuss two examples that I, as an outside observer, find instructive.

10. The first one is the case of Luhmann, who is often cited as a prominent social scientist who has taken up second-order cybernetics (e.g., Wolfe 1995; Hayles 1999: chap. 6; Geyer 2001). While the older generation in this book have noted Luhmann's ties to second-order cybernetics (e.g., Müller 2007: 411; Umpleby 2005; Scott 2004), they do not embrace his work in the book. In their editorial, Müller and Riegler do not mention Luhmann. Umpleby (2016: §46) cites Luhmann as an example of someone doing cybernetics in the "exo-mode", i.e., not as second-order cybernetics. Scott (2016: §40) calls Luhmann's application of Maturana and Varela's concept of autopoiesis "controversial." He describes his proposal to unify psychology through Pask's conversation theory as an "alternative, cybernetics-based, concept of a social system to that developed by the sociologist Luhmann." Yet, the younger generation in the book is more receptive to the growing influence of Luhmann. Two of the articles draw on Luhmann for their analysis. In her article, Diana Gasparyan invokes his notion of autopoietic social systems to sanction her second-order-cybernetic theory of consciousness. Tom Scholte (§34) interprets the action in Shakespeare's play "Hamlet" as the "self-reorganization of an autopoietic social system (à la Luhmann) following a substantial perturbation." In addition, the open peer commentaries by Marcelo Arnold-Cathalifaud and Daniela Thumala-Dockendorff, Eva Buchinger, and Bruce Clarke and Dorothy Chansky engage with Luhmann's voluminous writings.

11. My second case involves the sharp boundary that is usually drawn between first-order and second-order cybernetics. The boundary is crucial to the identity of second-order cybernetics, a taken-for-granted reality rooted in the past and often expressed as a progress narrative (e.g., Umpleby 2016: §15). In contrast, Ben Sweeting, in his article on design research, draws on Glanville's interpretation of the past to argue against a progress narrative and for a porous boundary between the two orders of cybernetics. Sweeting proposes that the adjectives "'first' and 'second' should not, however, be understood as implying a sequence or the surpassing of one by the other. Rather, SOC is specifically the application of cybernetics to itself – 'the cybernetics of cybernetics'", as von Foerster titled Margaret Mead's (1968) paper. For Sweeting, the "terminology of 'first' and 'second' can obscure the continuity between SOC and earlier cybernetics" (Sweeting §18). Sweeting also refers to the diagrams drawn by Gregory Bateson in the interview Stewart Brand conducted with him and Mead in 1976 to show the continuity between the two orders of cybernetics (Sweeting 2016: §18; Brand 1976).

12. I will conclude by placing my brief account of present-day second-order cybernetics in the context of the boundary work performed in a related field, information theory. Many authoritative figures wrote "official histories" of information theory as it was being developed. In 1953, five years after Bell Labs mathematician Claude Shannon published "A Mathematical Theory of Communication", his classic paper that founded the field, Colin Cherry, an electrical engineer at Imperial College, London, presented a paper at the first London Symposium on Information Theory entitled "A History of the Theory of Information." While Shannon had limited the genealogy of information theory to a couple of researchers at Bell Labs and to Norbert Wiener at MIT, Cherry included a host of researchers in a much longer history of communications. That inclusive boundary work comported with the broad British interpretation of the term "information theory" (Cherry 1953; Shannon 1948; Kline 2004; Geoghegan 2008). In 1948, Wiener created an even smaller genealogy in his path-breaking book *Cybernetics*, by limiting information theory to the idea of equating information with entropy. He claimed that it was independently developed by statistician R. A. Fisher, Shannon, and himself (Wiener 1948: 18). As information theory

became established as a discipline in the US in the next two decades, its promoters did more boundary work by separating "Shannon theory" (coding) from Wiener's "statistical theory of communication" (prediction and filtering). Although some of this boundary work unfairly cut Wiener out of the genealogy of "Shannon theory", it marked off two robust research agendas that thrived in the Institute of Radio Engineers' Professional Group on Information Theory and in its successor society in the Institute of Electrical and Electronics Engineers (Kline 2015: chap. 4).

13. This example shows that exclusive boundary work can help an emerging scientific field prosper by marking off separate research agendas rather than allowing them to compete for the mantle of the discipline. Of course, exclusive boundary work can also hinder the development of an emerging field by shutting out promising lines of research.

14. I think that two trends in the present book are encouraging signs for the future of second-order cybernetics. First, the wealth of research cited in the articles and commentaries showed me that the field is in better shape than I thought it was when I wrote *Cybernetics Moment*. Second, the diversity of scholarship and viewpoint that the younger generation of scholars brought to bear on the proposed fivefold research agenda indicates to me that the editors' inclusive boundary work in selecting contributors for the volume will help second-order cybernetics thrive in the future.

References

Brand S. (1976) For god's sake, Margaret: Conversation with Gregory Bateson and Margaret Mead. CoEvolution Quarterly 10: 32-44.

Cherry C. (1953) A history of the theory of information. Transactions of the IRE Professional Group on Information Theory 1(1): 22–43.

Foerster H. von (ed.) (1974) Cybernetics of cybernetics. University of Illinois, Urbana.

Geoghegan B. D. (2008) The historiographic conceptualization of information: A critical survey. IEEE Annals of the History of Computing 30(1): 69–81.

Geyer F. (2001) Sociocybernetics. In: Smelser N. J. & Baltes P. B. (eds.) International encyclopedia of the social and behavioral sciences. Elsevier, Amsterdam: 14549–14554.

Gieryn T. F. (1983) Boundary work and the demarcation of science from non-science: Strains and interests in professional ideologies of scientists. American Sociological Review 48: 781–795.

Gieryn T. F. (1999) Boundaries of science. In: Jasanoff S., Markle G. E., Peterson J. C. & Pinch T. (eds.) Handbook of science and technology studies. Sage, London: 393–443.

Hayles N. K. (1999) How we became post-human: Virtual bodies in cybernetics, literature, and informatics. University of Chicago Press, Chicago.

Kline R. (2004) What is information theory a theory of? Boundary work among information theorists and information scientists in the United States and Britain during the Cold War. In: Rayward W. B. & Bowden M. E. (eds.) The history and heritage of scientific and technical information systems: Proceedings of the 2002 conference, Chemical Heritage Foundation. Information Today, Medford, NJ: 15-28.

Kline R. R. (2015) The cybernetics moment: Or why we call our age the information age. Johns Hopkins University Press, Baltimore.

Martin R. (2007) BCL and the heuristics seminars: A school for cybernetics. In: Müller A. & Müller K. H. (eds.) An unfinished revolution? Heinz von Foerster and the Biological Computer Laboratory (BCL), 1958–1976. Edition echoraum, Vienna: 117-129.

Martin R. (2015) Second-order cybernetics, radical constructivism, and the biology of cognition: Paradigms struggling to bring about change. Cybernetics & Human Knowing 22(2-3): 169-182.

Mead M. (1968) Cybernetics of cybernetics. In: H. von Foerster, et al. (eds.), Purposive Systems: Proceedings of the First Annual Symposium of the American Society for Cybernetics. Spartan Books, New York: 1–11.
Müller A. & Müller K. H. (eds.) (2007) An unfinished revolution? Heinz von Foerster and the Biological Computer Laboratory (BCL), 1958–1976. Edition echoraum, Vienna.
Müller K. H. (2007) The BCL – An unfinished revolution of an unfinished revolution. In: Müller A. & Müller K. H. (eds.) (2007) An unfinished revolution? Heinz von Foerster and the Biological Computer Laboratory (BCL), 1958–1976. Edition echoraum, Vienna: 407–466.
Scott B. (2004) Second-order cybernetics: An historical introduction. Kybernetes 33(9/10): 1365-1370.
Shannon C. E. (1948) A mathematical theory of communication. Bell System Technical Journal 27: 379–423, 623–656.
Umpleby S. A. (2003) Heinz von Foerster and the Mansfield Amendment. Cybernetics & Human Knowing 10(3-4): 161-163.
Umpleby S. A. (2005) A history of the cybernetics movement in the United States. Journal of the Washington Academy of Sciences 91(2): 54-66.
Umpleby S. A. (2007) Interview on Heinz von Foerster, the BCL, second-order cybernetics and the American Society for Cybernetics. In: Müller A. & Müller K. H. (eds.) (2007) An unfinished revolution? Heinz von Foerster and the Biological Computer Laboratory (BCL), 1958–1976. Edition echoraum, Vienna: 77-87.
Varela F. J. (1981) Introduction: The ages of Heinz von Foerster. In: Foerster, H. von. Observing systems. Intersystems, Seaside CA: xi–xvi.
Wiener N. (1948) Cybernetics: Or control and communication in the animal and the machine. John Wiley and Technology Press, New York and Cambridge MA.
Wolfe C. (1995) In search of post-humanist theory: The second-order cybernetics of Maturana and Varela. Cultural Critique 30(1): 33-70.

Embracing Realists Without Embracing Realism: The Future of Second-Order Cybernetics

Robert J. Martin

Introduction

1. In 2015 I wrote an ASC column for *Cybernetics and Human Knowing* (Martin 2015b), hoping to open a conversation about how to reenergize second-order cybernetics (SOC). SOC is currently working to overcome its being relegated to the sidelines of science (a point made by Kline 2015 and Martin 2015b). I am delighted that the present book more than fulfills my hopes for such a conversation. For this I am grateful to Alexander Riegler and Karl Müller. However, I would point out that the Kline-Martin hypothesis identified in their editorial is a misreading of the column I wrote. The 2015 paper does point out that SOC has failed "to achieve wide acceptance, particularly in science" (Martin 2015b: 169) not because of any lack of value in the ideas, but because of the realist tradition and structure of science. I appreciate the efforts of Müller and Riegler to challenge authors and readers to consider the three questions labeled the Kline-Martin hypothesis (Müller & Riegler §1) but I must point out that the Kline-Martin hypothesis is their invention and does not reflect my views, especially as regards the value of SOC. I appreciate their invitation to respond. I briefly summarize my arguments regarding the resistance to change of traditional science below. The remainder of the commentary clarifies and expands my position in light of this volume.

Resistance to change

2. The Martin (2015b) article makes the point that science has been resistant to considering the change in paradigm that came about with SOC, in part because scientists have no reason to see themselves as having problems that SOC can address. This can be attributed to three main factors:

a. Scientific disciplines have become silos that operate independently of the larger ideas that might otherwise influence their thinking;

b. In each discipline the cycles of proposal, funding, research, and publication are profoundly conservative, inhibiting changes in paradigms that are not rooted in the ongoing research of the discipline;
c. The structure of funding, research and publication inhibits or prevents scientists from adopting the paradigm and tools of SOC; essentially, scientists have no reason to see themselves as having problems that SOC can address.

Including realism within a constructivist approach

3. What constitutes the realism of science is that, as Louis Kauffman explains in his target article, "Cybernetics, Reflexivity and Second-Order Science" (see especially §10), physical science is a special case of producing theories that do not affect the phenomena they describe or explain – and they must be so by definition. If they do not conform to this criterion, they are not regarded as physical science. Note that the social sciences strive, but are not able, to emulate this feat because the phenomena they describe can be affected by the knowledge of the theories that describe and explain them. In other words, within the context of constructivism, the realism of science can be seen as a special case.

Including constructivism within realism

4. If realism can be included within constructivism and constructivism can be (and is) included within realism, how can this be? It is a matter of context. For example, using Jean Piaget's concept of adaptation through assimilation and accommodation (formalized by Heinz von Foerster in his 2003 paper on objects as tokens for eigen-behavior), realists can say that when learning has been successful, learners have constructed/invented an understanding of science concepts that reflects an accurate understanding of the universe. A constructivist understanding of learning science is that students must invent science concepts through interaction with the environment (including conversation) in order to understand science concepts accurately (Bybee & Sund 1982). This is a common paradigm in science education in the US, but science teachers do not necessarily reject science as a realist project; what they have learned through experience is that they cannot successfully teach science through lectures alone. Manipulation of objects plus reasoning through conversation is necessary for students to construct an understanding that coincides with accepted scientific explanations of the phenomenon.[1] Teachers who consider themselves constructivists may or may not accept that science itself is a true, accurate, and objective representation of the world.

---

1. See, for example, "A Private Universe," a film made by a Harvard University project on a constructivist approach to teaching an understanding of the phases of the moon and the change of seasons on earth. The film is available from the Harvard-Smithsonian Center for Astrophysics at http://www.learner.org/resources/series28.html

Context allows pragmatic bedfellows

5. If as argued above, realism can be contained within constructivism and constructivism can be contained within realism, there can also be a third context in which realists and constructivists can function together without difficultly. This is the context that Michael Lissack,[2] referring to Hans Vaihinger (2015), calls, "as if." A pragmatic approach allows both realists and constructivists to treat explanations that prove useful "as if" they were true statements without worrying about to what extent they are true. Vaihinger showed that any number of useful constructs can be shown to be fictions, but because of their usefulness, we treat them as if they were true.

6. Paring down Vaihinger, we do not need to concern ourselves with whether assumptions are fictions, but only whether or not the assumptions work to produce useful ideas – useful as measured by their ability to achieve a purpose. Here we cross paths with Ernst von Glasersfeld (1984), a parent of radical constructivism, who makes the same point in his metaphor of a lock and the keys that open it: ideas and actions are to be measured by whether they allow us to achieve our purpose. Just as there may be many different keys that can open a lock, reality is a lock that can be opened with many keys – without our ever knowing how the lock is constructed (ibid: 21). Constructivism is itself one these keys; realism is another – especially the realism of science that holds that our understanding is still, after all, only tentative and temporary; both constructivism and realism can provide ways of dealing effectively with one's environment; neither provides a way of understanding things "as they are." If this point happens to be lost on particular realists or constructivists, it makes no difference as long as we proceed "as if" a set of assumptions or findings were true. We stipulate that such and such is the case and we proceed to find out how far this stipulation will take us before it breaks down. We do not need to engage in a battle between realism and constructivism; it is sufficient that we agree that within a particular context, we investigate how and to what extent a particular idea works.

7. Constructivist concepts/tools do work and some realists will be willing to accept that they do work – but only within specific, limited, temporary contexts. I propose that this is an arrangement we can live with. We can live with this arrangement for four reasons:

a. because constructivism and realism are both keys that have allowed us to make progress in reaching various goals;
b. both can be regarded as "as if" approaches;
c. the truth of both is undecidable (using von Foerster's concept of undecidable questions) and therefore must be decided by each individual; and
d. there is no reason to require acceptance of either realism or constructivism as a prerequisite for working together.

2. Personal email received 14 December 2016, used with permission.

Embracing realists without embracing realism

8. A hint as to where to go next comes from von Foerster. Rather than speaking of constructivists and realists, von Foerster referred to inventors (those who believe the laws of science are invented) and discoverers (those who believe the laws of science are discovered).[3] As a way of expressing the possibility of bridging the gap between constructivists and realists, von Foerster, no doubt with tongue in cheek, suggested that inventors might invent discoverers and discoverers might discover inventors. That is, constructivists and realists might choose to work with one another. My own experience has been that most of my colleagues are realists. Their willingness to defend their positions has helped me to clarify my own position while also becoming convinced of the futility of trying to convince my academic colleagues of the viability of the constructivist approach as an alternative way of thinking. Nevertheless, as soon as we begin to discuss specific empirical research projects, we find ourselves on common ground. From this I learn that our difficulties lie in differences in epistemology, not at the level of accepting empirical research on perception and cognition or of working together on how to approach problems and how to analyze research findings – including those research findings that point to ways in which individuals construct/modify/filter their present experience to fit previous experience and existing biases. In other words, for the purpose of solving many problems, there is little to prevent cooperation between realists who accept traditional science and constructivists who wish to incorporate traditional science into a larger framework.

9. We need to engage realists because most professionals, including scientists, are realists and that's where the action is. In the words of Lissack, "If we cannot find a way to deal with…. [pragmatic] realism and with those whose worldview it comprises, we are irrelevant" (personal email). By pragmatic realists, Lissack means those who "believe there is a singular truth but that we can only approximate it in our representations" (ibid). We have an opportunity to develop more robust tools to understand, describe, and cope with the problems faced by humanity. The commonalities between realist and constructivist approaches make this possible. Constructivists and realists who accept traditional science (though not necessarily the truth claims of traditional science) also accept:

a. *Empirical evidence*. Empirical evidence is defined in terms of observation and experience. While constructivists maintain that experience is constructed and realists, to one extent or another, believe in the possibility of representing reality, both agree that science relies on observation and experience.
b. *Temporary truth*. Both realists and constructivists accept scientific findings as temporary.
c. *Human bias in perception*. Both realists and constructivists accept social science findings that show perception and decision-making in human beings to be biased, often in systematic ways that cannot be easily overcome even when the perceiver is aware that they exist. Constructivists believe that this limitation, in principle, cannot be overcome. Realists believe that science, in principle, can overcome this limitation, but this difference does not preclude working together.

---

3. This and the following sentences are based on my personal recollection of hearing von Foerster say this a number of times during the time I took classes with him and attended conferences at which he spoke. Unfortunately, I have not been able to find this anywhere in print.

Concepts/tools for participant-observers

10. Just as science has accepted that "what the frog's eye tells the frog's brain," is not an objective reality but a fit between the frog and its environment, science can accept that what the human's perception tells the human is not an objective reality but a fit between the human being and her environment. Going forward, what we have learned and are learning about human perception and cognition may be moving realists toward accepting that human perception (and thinking) has systematic biases that can be investigated and reflected upon so that decisions and actions can be based on more comprehensive understanding. In other words, realists are moving toward constructivist evidence and conclusions, just not toward constructivist epistemology.

11. Professionals in a variety of disciplines – law, heath care, psychotherapy, government, and business – are participant-observers in ways that traditional scientists are not. They have constituencies to whom they bring expertise (often based, at least in part, on a scientific knowledge base) and to whom they are responsible for the consequences of their advice, decisions, and actions. They are faced with a need to expand the concepts and tools of their practice. Such a list of tools would include the following four sets of tools:

a. Tools to describe and think about complex processes that are not adequately described by linear causality are needed to more fully describe processes such as learning and other processes investigated by the social sciences as well as the biological processes studied by medical research. Circularity provides a set of concepts/tools that can be used to address this need.
b. Tools to examine professional practice and its effects are needed. Reflexivity is a set of concepts/tools that can be used to address this need. An example of reflexivity in practice – an example that is familiar to many – is Donald Schön's reflective practitioner (1983).
c. Tools to place an emphasis on behavior as a way of understanding, teaching, and making decisions in professional contexts. The concept/tool of objects-as-tokens-of-eigen-behavior can be useful in this endeavor.
d. Tools to address ethical questions that come to light as a result of addressing the first three needs. The concept of ethics as developed by SOC can be useful in this undertaking.

12. The four concepts/tools I have mentioned form the core of SOC: circularity, reflexivity, the understanding of objects as arising from (Eigen-)behavior (Foerster 2003), and an expanded ethics. Other concepts/tools of SOC could certainly be included in the list. While others outside the SOC community may be developing these concepts/tools within their disciplines, the practitioners of SOC are already deeply engaged in developing these four concepts/tools as the core of SOC – and opportunities for working with those outside the SOC can increase as practitioners of SOC turn to work with interested individuals and disciplines on problems they wish to cope with.

Contributions past and present

13. SOC has already contributed to disciplines outside SOC; contributing to other disciplines is part of what SOC does. This book contains many examples. Kauffman's target article addresses the question "How then, do physical and natural science manage to obtain their apparently objective results? The answer lies in circularity and eigenform" (§10) – which he then develops. Ben Sweeting's target article "Design Research as a Variety of Second-Order Cybernetic Practice" develops Ranulph Glanville's take on design as a cybernetics activity that encompasses science (rather than the other way around). Tom Scholte's article "Black Box Theatre: Second-Order Cybernetics and Naturalism in Rehearsal and Performance" develops a theory and a way of describing performance and rehearsal of theater that is leading to new ways of rehearsing. Frank Galuszka (2009) shows how cybernetic ideas can be used to describe and understand visual art, especially painting. Klaus Krippendorff (2000) shows that the stance of objectivity taken by sociology privileges scientists in that area vis-à-vis those they describe – and that this is an ethical problem. There are examples of cybernetics applied to teaching and learning that show the value of strategies that are part of, or that developed from, SOC (e.g., see Baron 2016; Martin 2016). There are many more examples, many of them going back years or even decades; the point is that engaging others who are not part of the SOC community has potential for engaging what Umpleby, in his article "Second-Order Cybernetics as a Fundamental Revolution in Science", calls "actor based disciplines" (§§70-73) and what I have referred to here as professionals, that is, practitioners of specific disciplines.

14. Engaging professionals, including social scientists, can be seen as an extension of ASC past projects, such as those described by Umpleby (§§33-62), especially in his discussion of "doing science from within" and his example of the work of the Institute of Cultural Affairs, an example that goes back to the 1950s. We know that individuals in health care, medicine, law, education, government, business, and the arts, as well as individuals within the traditional sciences, have shown interest in the website, publications, discussion groups, and conferences of the American Society for Cybernetics. In particular, the current leadership of the ASC has moved in the direction of creating conversation-centered conferences that focus on topics of interest to those outside the ASC community (see Richards 2015; Martin 2015a; Herr 2015; Hohl 2015; Lombardi 2015; Baron & Griffiths 2015).

15. I want to be clear that in no way am I implying that second-order cybernetics should focus on becoming compatible with realist epistemologies. Rather I am pointing out that in the investigation of practical problems, the concepts/strategies of second-order cybernetics have much to offer professionals including those in science, especially social science – whether they are realists or constructivists.

Recommendations

16. Because this commentary is concerned with how to increase the role of SOC within the sciences and other professions, I offer the following as recommendations:

17. Identify topics central to SOC that are of interest to non-SOC communities and create opportunities to bring members of these communities together. The ASC is working to do this through conversational conferences (see Herr 2015; Lombardi 2015; Martin 2015a; Richards 2015; Scholte 2015; Schroeder 2015). We are learning how to do these

conferences through experimenting with topics, speakers, opportunities for publication, and conference design and books such as this volume, which are designed to provide both a broader context and multiple viewpoints from a variety of fields.

18. Use the language of the individuals we wish to reach. Note that while in learning a discipline, a student learns the language of that discipline, in consulting and in doing therapy, the consultant uses the language of the client. SOC can retain its own language but also use the language of individuals, groups, or disciplines interested in its concepts.

19. When seeking funding or collaboration with other organizations such as the AAAS and NSF, adopt the language of those organizations.

20. Encourage others to use the concepts/strategies associated with SOC within their own disciplines and using their own language. For example, many individuals within science education find the constructivism of Piaget and Lev Vygotsky both convincing and helpful. Whether they hold realist or constructivist epistemologies is irrelevant; within the context of understanding and teaching science, they understand and operate as constructivists.

21. Adopt a deliberate stance that welcomes all who are interested in ways of dealing with complexity in their own disciplines and practice – without regard to their epistemologies. Even ardent realists can appreciate tools and research that come from SOC that help them understand and cope with problems and contexts in their areas.

22. Present ideas through conversation rather than teaching or instruction. Providing arguments and evidence for adopting a constructivist epistemology does not work. Change comes through behavior, and although listening is a behavior, it is not as effective as conversation and collaboration.

23. Avoid privileging constructivism; it drives non-constructivists away. This may irritate some constructivists, but as a radical constructivist I have learned in dealing with colleagues, students, and clients, that I can share all my insights without pushing a constructivist agenda – no matter how much I would like to.

24. Work toward strategic change. Focus on projects that establish cybernetics and SOC as relevant contributors to the sciences and those professions that use science as a knowledge base.

25. These recommendations are not answers; they suggest a direction – a direction that will need to be pursued through practical work such as giving conferences, making webpages, publishing, encouraging learning communities, and having conversations.

Conclusion

26. Originally cybernetics was seen as being within the paradigm of the objective observer of the real world. Acceptance of cybernetics was easy when the descriptions of systems were descriptions of physical machines (Kline 2015). The research that followed, especially that of Maturana, Varela, von Foerster, and others, led to the logical dissolution of the paradigm of realism and the objective observer – a change that solidified into what became SOC (Martin 2004). The sciences and those professions that use the sciences as a knowledge base have not changed to a more constructivist epistemology. I now see that the expectation that this would happen was unreasonable, given what cognitive science tells us about the tendency of human beings – including scientists – to persist in ways of perceiving and thinking that serve them well in their environments. The most important change

that took place in the transition from cybernetics to second-order cybernetics was a change from the accepted science paradigm of the objective observer describing an objective reality to the paradigm of a participant-observer who is a closed circular system constructing a world as a result of behavior. As already noted, fifty years of effort have shown that we are not going to shift traditional science to the latter paradigm in the foreseeable future. The point is to not be defeated by this situation.

27. Every profession, including every science, is a practice. Through practice, everything that has a name becomes an object. This includes the concepts and models that form the basis of every discipline, including every science. The concepts and models become objects that their users see as real objects – objects that cannot be questioned because they appear as real as a car on the street. Only behaving differently enables us to see differently. It is only through behavior that realists and constructivists can come to see the world differently. And it is in practice that we learn both to act and see differently (Martin 2015c, 2011). It is through engaging others in projects and in conversation, both verbal and written, that, regardless of our epistemologies, we change.

Acknowledgements

Thanks to Dr. Suzanne Wildhagen Martin for comments and editorial suggestions.

References

Baron P. (2016) A cybernetic approach to contextual teaching and learning. Constructivist Foundations 12(1): 91–100.

Baron P. & Griffiths D. (2015) The tensions between second-order cybernetics and traditional academic conferences. Constructivists Foundations 11(1): 86-88.

Bybee R. W. & Sund R. B. (1982) Piaget for educators. Second edition. Waveland Press, Prospect Heights IL.

Foerster H. von (2003) Objects: Tokens for (eigen-)behaviors. In: Understanding understanding: Essays on cybernetics and cognition. Springer, New York: 261–271. Originally published in 1976.

Galuszka F. R. (2009) Towards a cybernetic-constructivist understanding of painting. Constructivist Foundations 5(1): 1-18

Glasersfeld E. von (1984) An introduction to radical constructivism. In: Watzlawick P. (ed.) The invented reality: How do we know what we believe we know? Contributions to constructivism. W. W. Norton & Co., New York: 17-40.

Herr C. M. (2015) Can conversations be designed? Constructivist Foundations 11(1) 74-75.

Hohl M. (2015) Desires, constraints, and designing second-order cybernetics conferences. Constructivist Foundations 11(1) 84-85.

Kline R. R. (2015) The cybernetics moment: Or why we call our age the information age. The Johns Hopkins University Press, Baltimore.

Krippendorff K. (2000) Ecological narratives: Reclaiming the voice of theorized others. In: Ciprut J. V. (ed.) The art of the feud: Reconceptualizing international relations. Praeger, Westport CT: 1–26.

Lombardi J. Cybernetics (2015) Conversation and consensus: Designing academic conferences. Constructivist Foundations 11(1) 79-81.

Martin R. J. (2004) The once and future: thoughts and notes. Cybernetics and Human Knowing 11(2): 71-76.
Martin R. J. (2011) Education as recursive cycles of learning to see through acting and learning to act through seeing: the influence of Heinz von Foerster. Cybernetics and Human Knowing 18(3-4) 123-128.
Martin R. J. (2015a) Connections of conversation-based conferences to the foundations of radical constructivism. Constructivist Foundations 11(1): 88-90.
Martin R. J. (2015b) Second-order cybernetics, radical constructivism and the biology of cognition: paradigms struggling to bring about change. Cybernetics and Human Knowing 22(2-3): 169-182.
Martin R. J. (2015c) The role of experience in the ASC's commitment to engage those outside the cybernetics community in learning cybernetics. Kybernetes 44(89) 1331-1340.
Martin R. J. (2016) How change happens with difficulty. Constructivist Foundations 12(1): 109-110.
Richards L. D. (2015) Designing academic conferences in the light of second-order cybernetics. Constructivist Foundations 11(1): 65-73.
Scholte T. (2015) Embed and unzip. Constructivist Foundations 11(1): 76-77.
Schön D. A. (1983) The reflective practitioner: How professionals think in action. Basic Books, New York NY.
Schroeder P. C. (2015) Nurturing conversation through innovative conference design. Constructivist Foundations 11(1): 77–79.
Vaihinger H. (2015) The philosophy of as if: A system of the theoretical, practical and religious fictions of mankind. Translated by C. K. Ogden. Harcourt Brace and Company Well-formatted version of the 1924 original by Harcourt, Brace and Company. Createspace.com Random Shack, Lexington, Kentucky. German original published in 1911.

Some Implications of Second-Order Cybernetics

Anthony Hodgson

From emergence to methodology

1. The overview that Karl Müller and Alexander Riegler draw of the actual and future possible field of SOC makes it easier to see the distinctions between first-order and second-order. Without those distinctions it is unlikely that second-order cybernetics (or science) will be able break out of the challenge of irrelevance asserted in the Kline-Martin Hypothesis. It seems to me that the time is ripe for a major thrust to take SOC into further domains not only for its deeper theoretical understanding but also because of its potential relevance to social transformation.

2. The history, momentum and successes of first-order science, coupled with enormous global investment in its institutions, communities and funding makes it no easy task to get the voice of second-order thinking heard. When this is coupled with the conclusions, which the review here confirms, that SOC has been little developed over the last fifty years and that there is very little evidence in the form of sustainable consequences for other domains of science, then the difficulty is exacerbated.

3. The emphasis Müller and Riegler place on research and development agendas for the future of SOC is a timely affirmation that the potential of SOC has hardly been tapped and developed. For example, the potential of the great insights of Heinz von Foerster, Gregory Bateson and Ernst von Glasersfeld coupled with the championing by Ranulph Glanville and Bernard Scott stands out as a rich field for exploration. I will introduce an additional way of framing the essential features of SOC, which I propose should also be extended to second-order science to broaden out the relevance. This framework emerged from the interdisciplinary workshop "Second-order Science and Policy" (SOSP, http://www.decisionintegrity.co.uk/page44.html) held early in 2016 to explore the potential relevance of second-order science to policy formation. I will then use this framework as a way of noting salient points from the various contributory target articles in an attempt to indicate the inclusive relevance of the rich diversity of perspectives represented.

4. Michael Lissack makes the case (§1) that the label "cybernetics" has, in use, lost contact with its essential meaning and needs new forms of expression. I see the switch to the term "second-order science" as one way of contributing to this task, but only if the concept is enlarged and made more comprehensive.

5. It seems fairly typical of paradigm clash that the incumbent marginalises the new that questions its assumptions. This leads to a kind of subterranean collusion to not even acknowledge there is a challenge. This leads further to the issue being hidden in plain sight:

> It seems that we have a blind spot for the fact that experience is the most basic and unavoidable medium of our being. Not only do we normally not notice how all our beliefs about ourselves and the world constitute experience; we do not notice that we do not notice. (Kordeš §5)

6. The development of a flourishing community of second-order researchers is likely to be the most powerful way to open this up in discourse on the nature of science. Thomas Flanagan (§3) points out that the very nature of the interactions in a community of researchers changes expectations of what is considered to be a scientific finding. He summarised that a finding is a complex function of:

> (a) the observational and communication dynamics within the system under study
> (b) the conjoint sense-making methodology selected for use by the researcher and fellow actors
> (c) the focus on boundary conditions of the inquiry specified by the researcher
> (d) the adequacy of the reporting narrative.

7. The construction of second-order methodology could fall into an inconsistency trap unless it is recognised that a community of scientists have made choices as to what second-order methodology is. Approaching second-order methodology with first-order assumptions is unlikely to go very far. First-order methodology is limited by being based on linear causality (Martin §11), which is inconsistent with the importance of circularity.

8. This trap is avoided if the approach to methodology is itself second-order. The positive angle on this perspective is the scope for the conscious design of methodology. Lissack (§16) summarises seven insights that cybernetics has given rise to that need to be taken into account:

- the role of the observer
- the law of requisite variety
- the importance of the observer in cognition
- the use of black boxes
- the idea that all action is in some ways a conversation
- the importance of recognising that true models differ from descriptive representations
- the importance of narratives

9. Another useful view of methodology is provided by John Warfield's domain of science model (Flanagan §16) in which there is a continuing cycle between foundation, theory, methodology and application, which is portrayed basically as a learning cycle.

10. Müller (2016) outlines a seven-stage methodology that incorporates the principles of second-order investigation searching for systemic universality. In parallel to this I developed a similar approach in my doctoral research (Hodgson 2016).

Second-order science as designed practice

We need an inclusive big tent rather than a divisive faction fight. (Cariani §14)

11. An excursion was made by SOSP (see §3), an interdisciplinary workshop to enlarge the concept of second-order science building on the core work of second-order cybernetics by "sweeping in," to use Charles West Churchman's (1979) expression, a broader range of approaches to science than is usually discussed in second-order cybernetics. All of these shared the initial starting point of the presence of the observer as critical in some way or another. The perspective that emerged from interdisciplinary dialogue across diverse fields of science, policy and practice provided a richer picture as to what the hallmarks of a second-order science practice might be. The enquiry was also facilitated to identify possible links between this emergent pattern and the challenges facing contemporary policy development – what the possible relevance and use of a second-order discipline might be. I will use this distillation of principles as a platform to explore second-order science methodology.

12. A key "so what?" question is "so how do we go about science differently if we intentionally adopt a second-order approach?" This can be treated as a design task affecting how we might carry out both theoretical and empirical investigation. Grandon Gill (§16) affirms:

> To be effective in highly complex environments, research designs need to be highly localised and need to shed some of the formalisms of the traditional scientific method, such as the hypothesis test (intended to support or refute stable, general propositions). In place of these approaches, the researcher needs to become highly aware of the interactions affecting the local context and must also become expert in the art of observation and the construction of models that reflected the local reality.

13. The concept used from here on in this review can be regarded as a meta-model to prompt the attention and awareness of the second-order practitioner. It is basically cast in what Müller and Riegler define as the endo mode, science from within.

14. The concept comprises seven distinct areas. However, they are also systemically connected and overlapping so, rather than a check-list they are better represented as a nonlinear visual pattern as in Figure 1.

1. The triadic network: Observer, language, society

> Triadic networks in science can be built between (1) actors or researchers, (2) an environment or domain of investigation and (3) a common language, grammar, rule system or, more generally, a knowledge base. (Müller & Riegler §21)

15. The co-presence of observer, language and society (of scientists) is placed as central and strengthens the principle that the observer role is critical. The triadic relation is essential to von Foerster's (2003) viewpoint. He characteristically summarised this as analogous to the relationship between the chicken, the egg and the rooster. "You cannot say who was first and who was last. You need all three to have all three." The presence of the observer in the observation is proposed as a fundamental condition in second-order science. In this sense, all scientific knowledge is some form of intersubjective consensus

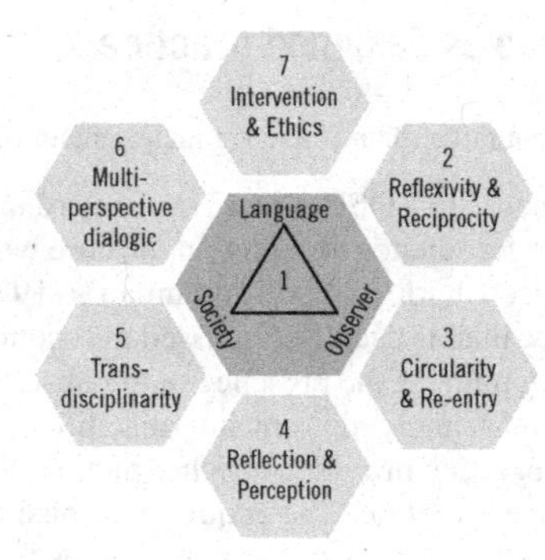

Figure 1. Seven domains constituting second-order science.

amongst a community of scientists. Where those scientists are ignorant of their assumptions about knowing, they are restricted by second-order blindness to the implications of their position. However, the observer is also a decider and actor and, in that sense, imposes forms of policy by the very nature of the way she frames observation. This goes further than a sociology of science as developed by Kuhn (Becerra §6).

16. Terms, symbols or images are situated; they acquire meaning through collective use in actual situations. This triadic relationship is also dialogic and has emergent properties of a living language. Flanagan points out:

> A collective sense-making methodology for second-order cybernetics must include provisions for languaging because people use language that is uniquely coded for expressing only certain parts of their immediate needs [...] The meanings behind statements need to be decoded and clarified within a consensual linguistic divide so the parties engage in collective sense making can accurately share understandings. (Flanagan §11)

17. Bryony Pierce (§2) sees second-order science as grounded in enaction within a community. This view also resonates with Konstantin Pavlov-Pinus's (§1) positioning of second-order science as bridging between phenomenological and analytical styles of research.

2. Reflexivity and reciprocity

> For the science system in general, the reflexive turn to a mode from within, or an endo-mode, can yield at least four groups of new opportunities, ... (Umpleby §48)

18. For Stuart Umpleby (§7) the essence of second-order science (in so far is it might be captured in one phrase) is "the science of reflexivity." Second-order science is able better to take into account the way that scientific ideas and findings entering the awareness of society change the nature of society and in turn the nature of the science that created the ideas. However, in the absence of recognition of second-order understanding this largely goes unnoticed. This leads to the idea of the study of observing systems. In observing the observed is changed but there is also the feedback of that change to changing the observer.

19. In reflexive systems, observation and intervention are not one-way streets. There is reciprocity between the observer and the world observed. The observer is *participating* in the system and *there are consequences*. Making the observation may not leave the observed in a constant condition. Intervention often creates new conditions (sometimes referred to as unintended consequences), for example in social systems, by provoking new ways of gaming the system. A second-order viewpoint would pay much more attention to this effect and as a result would have to go beyond the conventional categories of first-order thinking. Louis Kauffman (§3) sums up the situation thus:

> A *reflexive domain* is an abstract description of a conversational domain in which cybernetics can occur. Each participant in the reflexive domain is also an actor who transforms that domain. In full reflexivity, each participant is entirely determined by how he or she acts in the domain, and the domain is entirely determined by its participants.

3. Circularity and re-entry

> In the face of the circularity of context and observer it is still possible to explore and come to agreements that have every appearance of being scientific facts. (Kauffman §1)

20. In second-order thinking any stable properties of "the world" are not fixed things but eigenforms. Art Collings (§9) emphasises that circular processes have the property of generating eigenforms that are defined as fixed points in a transformation. From the process perspective, things or objects are eigenforms generated by the circularity of that process. In observing, the observer makes a distinction. The distinction in turn reflects back on the observer. Second-order circularity implies that the condition of the observer is changed by the feedback of the observation. Tatjana Schönwalder-Kunz (§5) also sees science as a self-referential structure that is not simply between observer and observation but in relation to the disciplinary context. The principle of re-entry proposes that any field can be applied to itself as, for example, theory of theory, method of methods, and cybernetics of cybernetics. From a second-order perspective the observer is continuously bringing forth a world and responding to and learning from that world. This stance supports the view that comprehensiveness is impossible. Knowledge is not some static object "out there" but is constantly reforming through the engagement of the knower; and the knower is changed by the encounter with knowledge.

21. Kauffman sees all cybernetics, not just second-order, as inherently circular: "Cybernetics is the study of systems and processes that interact with themselves and produce themselves from themselves" (Kauffman §38). In this sense, all cybernetics is second order. He takes the view that there is no definition of cybernetics that is not circular. Indeed, he takes this further and asserts:

> … all attempts to find stable knowledge of the world are attempts to find theories accompanied by eigenforms in the actual reflexivity of the world into which one is thrown. The world itself is affected by the actions of its participants at all levels. (Kauffman §89)

Such forms are discovered and then codified to become the objective results of that domain of science. A wider perspective on the situation reveals that the larger landscape of the reflexive domain has been significantly influenced by the theories it has given rise to. He concludes that circularity is both legitimate and unavoidable.

4. Reflection and perception

> The subject–object dualism has inherent insoluble contradictions, which make it impossible to come up with an adequate idea about reflection. (Gasparayan §33)

22. It is interesting to me that the development of a science of qualities has not yet entered into the main discourse of second-order cybernetics. This may well relate to the tendency of even this study to be locked into its own roots and language and inadvertently miss out on other parallel explorations that use seemingly different starting points and language. However, there seem to be at least two common elements. One is the attempt to privilege qualities to be as important as quantities and thus challenge the limitations of how measurement is conceived. The other is the presence of the observer in the observation and intervention. Although this is not touched on directly in the work under review there are notable linkage points in the section introduced by Diana Gasparayan on consciousness. She makes the point that consciousness should be present in the study of consciousness and therefore a first-order approach is self-defeating.

23. Perhaps a linking area around this question is that of qualia as the characteristics of all and any conscious experience (Pickering §1) including the practice of second-order cybernetics. A science of qualities treats the self-experiencing mind as the primary conscious instrument of the science, prior to the tools of investigation and measurement – microscopes, telescopes, computer modelling and so on. This is essentially the inclusion of the question "how does the scientist/decision maker construct his or her reality?" (Becerra §4).

5. Transdisciplinarity

> A mere call to interdisciplinarity is not enough. What is needed is a new methodology, explicit lines of work, and new tools and techniques easing such integration. (Becerra §15)

24. In the founding of the systems sciences the aspiration was for a universal language of similarities that recurred in many fields. Mathematics is clearly one form of language that has become supplemented by other forms of systems modelling. This has never sat comfortably with the ingrained paradigms of academic institutions. A major necessity for contemporary complexity and uncertainty is the tackling of challenges that cut across traditional disciplines. The fields open for investigation are much broader than most conventional research. Indeed, a key question is: what constitutes a field for investigation?

25. Answers to this question are blown wide open by second-order science, which is more congruent with so-called wicked problems that surface wide fields of connectedness requiring understanding beyond knowledge and data. The perspectives of understanding become critical.

26. First-order science has built its structure of knowledge through specialised disciplines. Transdisciplinarity is an attempt to go under and beyond these distinctions and seek other forms of insight. From the perspective of Jean Piaget's genetic epistemology, Rolando Garcia (Becerra §14) suggests that improvement of approaches to complex problems requires integration of different aspects of knowledge and to construct the study object among multi-disciplinary team members.

6. Multi-perspectival dialogic

Interdisciplinary perspective is not a matter of theoretical reconciliations, but rather it is a co-construction of new theory through the reconfiguration of meaning drawn from joint consideration of primary observations. (Flanagan §3)

27. A key component of second-order method is its accommodation of emergent processes and emergent findings, some of which may be created in the process of application. An interesting illustration of this is provided in Tom Scholte's incorporation of Gordon Pask's conversation theory as a key component in the process in theatre research of Active Analysis. Conversations regarding circumstances and objectives of the selected drama are followed by active improvisation to be followed by a second conversation from the observer/participant perspective. This in turn re-enters and reframes the original conversation. It could well be that this approach generated in the context of the theatre is also applicable in the theatres of action of other sciences. In some ways it reflects more congruently the live experience of research. Eva Buchinger (§5) also proposes that Pask integrated with Niklas Luhmann as a way of conceptual integration between disciplines. A comparison with dialogical design science would also shed further light on this (Bausch & Flanagan 2013).

28. In dealing with complex situations that do not yield to a single discipline it is valuable to take several perspectives. The process of dialogue around a question from a number of disciplinary or stakeholder perspectives enables a creative emergence.

29. Allenna Leonard (§4) emphasises that these approaches provided a field for the emergence of second-order cybernetics allowing for multiple constructs of stakeholder positions in different cultural and social contexts. Such background must lead to the challenging of existing power relationships in both the scientific communities and their managerial communities.

7. Intervention and ethics

The associated ethics and responsibilities that arise out of second-order cybernetics may be overwhelming. This is an unsettling no-man's land for many scholars and students, who in turn opt out of this challenging reflexive epistemological domain. (Baron §8)

30. The observer is not merely an observer. We can substitute terms like actor, decider, and intervenor. From an enactive second-order perspective, in a world that is highly structurally coupled, there can be no such thing as a totally detached observer. Any position (even that of a decision not to observe) is an intervention. Assumptions are being made based on values and judgements as to what is "in" and what is "out" of consideration. Yet these judgements are often invisible and remain unquestioned, leaving research as a methodological game played on a field where the game itself is taken as objective and unquestioned. An implication for policy is that the use of "objective evidence" is at risk of being interpreted and used as an argument for political ends without making clear the value assumptions behind its supposed objectivity. A complementary second-order discipline would seek to make clear the position assumed by the objectivity of the research. It is interesting that the primary reference for the place of ethics is still von Foerster's (1997) often-cited work. This has not been very well developed and is of paramount relevance to the relationship between social responsibility and science, a relationship the first-order

culture of science has great difficulty in making sense of. Some explorations of ethics and second-order cybernetics in decision making were explored by Hodgson (2010).

31. Making a distinction is making an intervention. This act defines content but, as Lissack (§3) points out, context always matters. Although not a contributor to this book there is much that could be brought to bear here from Gerald Midgley's (Midgley & Ochoa-Arias 2001) work on systemic intervention and boundary critique. The very nature of second-order science implies that consideration of ethical dimensions is essential to a full methodology.

> Clearly, there are ethical repercussions from seeking to understand a second-order science that includes the observer and, by extension, the environments to which the observer is structurally coupled. (Forsythe §18)

Towards a conscious methodology

32. Can an operation of re-entry here contribute useful perspectives to research in practice? I am thinking of two fields of practical relevance. The first area is early career researchers, PhD students who are attracted to explore the second-order paradigm and wish to conduct a piece of research that is intentionally and deliberately second-order, and significantly, explicitly so in general terms beyond any specialist discipline that might be its field of research. Many will recognise the typical actual experience reported by Michael Hohl:

> Told as a story, my research appears pretty straightforward and top-down. In fact it was bottom-up and came together step-by-step over three years. The research process was a constant learning process. (Hohl §5)

33. The second area is that of policy making where there is increasing questioning of the limits of evidence-based policy, where the evidence is based on first-order science and yet the application is in the world of complexity and uncertainty of the society and its politics. Consider applying to policy development as well as design the following statement by Jose dos Santos Cabral Filho:

> A significant advance in design towards a second-order level will come when designers embrace an all-encompassing systemic approach that will necessarily have the inclusion of the observer, at all possible levels, as its pivotal point. (Cabral §11)

34. The elements of both design practice (Scholte-Halprin §12) and naturalist theatre (Scholte §3) can provide suggestions as to how innovative second-order ways of going about policy in a complex world with emergent properties might be tackled.

35. The facilitation of shared exploratory thinking in groups, especially using visual thinking, has direct parallels with digital design (van Stralen §4) and the summary points that Ben Sweeting (§22) highlights, namely:

- Reflective conversation
- Forward looking research
- Use of drawings and models as a part of thinking
- Importance of participation
- Circular reflective process
- Reciprocal relationship

The implications for human affairs

> The governance of the contemporary world and the interconnections among governance, democracy and knowledge are far more complex than most observers recognize. No single level is decisive in shaping the world in which we live. (Stingl §10)

36. The predominant view of scientists in society is that first-order research is paramount and that if implementation was not going well it was not a scientific problem and so a problem for scientists. Umpleby strongly points out that this is no longer tenable:

> The present time is characterized by an abundance of societal and environmental problems locally, nationally, and globally, where a high accumulation of theoretical scientific knowledge is accompanied by a deep deficiency in extension, implementation, or transformational knowledge. (Umpleby §56)

37. He sees the emergence of second-order science as one contribution to redressing this problem in creating better channels between science and practical human affairs especially in the social and environmental domains.[1]

38. Substitute the word "policy" for "design" in Sweeting's statement (§8) "…design involves the creation of new situations, design questions cannot be fully formulated in advance but shift and change as they are explored and as proposals are enacted" this description would most likely be agreed by policy researchers.

39. Some further areas for research that the progressive policy community might consider are:

- Recursion of many levels; character and eigenvalue (Clark & Chansky §9)
- Outside and inside (Landgraf §7)
- Theatre is uniquely positioned to provide methods and tools to understand consequences of differing configurations forming perception and conception. (Christy §18)
- Caution in transferring from one context to another (Müller §4)
- Cybernetic theatre for exploring policy (moving on from Sweeting [A Theatre for Exploring the Cybernetic] §5; Richards [The Many Varieties of Experimentation in Second-Order Cybernetics] §4)

40. Of course, there are what may amount to intractable problems in all this for the status quo culture. For example, recognising circularity in learning, cognition, problem solving, etc., does not by itself change a rootedness in linear causality (Martin §11). In my view, mainstream deterministic science is a special case of second order in which the community of scientists (and their political sponsors) have implicitly agreed to remain ignorant (or to suppress) that the experiencer is "hidden in plain sight." Two tricky areas are:

- SOC challenges the dominant paradigm of power and control (Martin §21) where evidence-based policy is limited to that of first-order science

1. A second-order approach to research in socio-ecological transformation is being pursued by Ioan Fazey, Professor of the Social Dimensions of Environmental Change at the University of Dundee. This includes a series of conferences on "Transformation in Practice" http://www.transformations2017.org/30/12/16

- Aspects of the theories that are held become engaged with and engaged by power and shape the subterranean assumptions that drive policy behaviour (Bohinc §4).

41. Nevertheless, in conclusion, I suggest that there is potential value in research into the relationship between second-order science and policy development, which could contribute to a number of areas including:

- Providing a common language of engagement for collaboration between "hard" and "soft" sciences and policy development that includes explicit ethics
- Providing a meta-framing for exchange across the disciplines of sciences, humanities and the arts, and design more congruent with wicked problems
- Providing some possible underpinnings for the limitations and possibilities of any discipline's contribution to societal transformation more clearly with transparency of assumptions
- Doing all this in a manner that renders the ethics of human activity transparent

References

Bausch K. C. & Flanagan T. R. (2013) A confluence of third-phase science and dialogic design science. Systems Research and Behavioural Science 30: 414–429.

Churchman C. W. (1979) The systems approach and its enemies. Basic Books, New York.

Foerster H. von (1995) Ethics and second-order cybernetics. Stanford Humanities Review 4(2): 308–319.

Foerster H. von (2003) Understanding understanding: Essays on cybernetics and cognition. New York, Springer.

Hodgson A. (2010) Decision integrity and second order cybernetics. In: Wallis S. (ed.) Cybernetics and systems theory in management: Tools, views and advancements. IGI Global, Hershey: 52–74.

Hodgson A. (2016) Time, pattern, perception: Integrating systems and futures thinking. PhD Thesis, University of Hull.

Midgley G. & Ochoa-Arias A. E. (2001) Unfolding a theory of systemic intervention. Systemic Practice and Action Research 14(5): 615–649.

Müller K. H. (2016) Methodologies for second-order science. Chapter 8 in: Second-order science: The revolution of scientific structures. Edition Echoraum, Vienna.

New Directions in Second-Order Cybernetics

Larry Richards

What is new?

1. Chapter 1 of Ross Ashby's classic, *An Introduction to Cybernetics* (1956), is titled "What is New", without the question mark. With the fiftieth anniversary of Margaret Mead's 1968 paper "Cybernetics of Cybernetics" approaching, it might be instructive to ask that question of SOC – not from the perspective of a cybernetician or of first-order cybernetics, but from the perspective of an interested bystander who happens on the term and asks about its relevance today, not in the early 1970s. Different people who associate themselves with cybernetics will respond to this question in different ways, and there does not appear to be even general agreement on a response, as the various contributions to this volume attest. However, there are some common threads that can be pulled through most of the responses, and it is precisely in the variety of possible responses that some new directions can be discerned. I have my own perspective on what is new and will try to reconcile it with at least some of the ideas of the other contributors.

2. Another consideration in responding to this question is that any claim of newness can be challenged by those familiar with similar ideas in fields of inquiry other than cybernetics. So, it is important to be careful about what we claim in order to avoid being dismissed as uninformed or as charlatans. For example, for SOC to claim that it is the only or the first field of inquiry to advocate for including the observer of phenomena in the phenomena being observed ignores the participant-observer approach to research, and its ethnographical methods, widely used in anthropological and management studies, among others. Separate from cybernetics, Paul Feyerabend (1975) developed a strong case for scientists to build their own values and motivations into their research approach and findings, and then to take responsibility for the consequences. Reflexivity,[1] a topic of special interest in SOC, has been of interest in communication theory, psychotherapy and other pursuits as well. This is not new.

---

1. I use the word "reflexive" here to speak of a "turning back on to." I will use the word "recursive" to speak of a "returning to." The word "recursive" can be used to describe many forms of circularity. Reflexivity is specifically about actions taken (or utterances made) that turn back on the actor (or speaker).

3. I claim that cybernetics (that is, SOC) offers a vocabulary that is still relatively new (and continues to evolve) and useful for talking and thinking[2] about the dynamics of relations and behavior in a way that accounts for the dynamics of relations and behavior of the observer/listener (participant) who is doing the talking and thinking. That is, it offers a way of thinking about ways of thinking, making the way of thinking (pattern of thought) we might employ in a particular situation a choice rather than a default, without awareness, to the prevailing way of thinking. What is new from SOC is that this way of thinking is a way of thinking about itself, turning cybernetics into a process of conceptualization rather than a set of relatively stable concepts like those that characterize other fields of inquiry. There is certainly still a role for temporarily stable concepts in cybernetics, as a conceptualization process generates concepts, and those that are useful will be retained while they remain useful. However, there is no claim to truth, only to the desirability of the process, which itself can change. So, SOC has the potential to change the approaches we take to human inquiry itself.

4. After a discussion of the concept of success and what success would be for SOC, I present a case for cybernetics as a way of thinking about ways of thinking in the form of six conceptual problems that it could address. I claim that these six conceptual problems, among others, represent current constraints on the possibilities for human sustainability and associated societal transformation. These constraints are constraints on thinking, and SOC can help to overcome them. I recognize that any presentation of a new way of thinking must rely on current language and the way of thinking that is embedded in its logic. My choice of using the formulation of six problems as a presentation device is an attempt to throw light on both the challenge that SOC represents and the hope that it offers. I am under no illusion that it is sufficient as argument or that any mode of presentation at this time could be. My desire is that it generate and sustain some conversations.

What would success look like?

5. I will not go into a lengthy definition or history of cybernetics or SOC. I assume the reader is familiar with the terms or can locate reasonable expositions elsewhere. I will only say that I regard SOC and cybernetics to be the same. While there may still be occasions when it is useful to make the distinction between first and second order, the original motivation for doing so has changed. Any scientist, engineer, technologist, designer, artist or craftsperson who does not account for themselves and their desires and intentions in the actions they take or the statements they make is not doing cybernetics, even if they expropriate the vocabulary to justify what they do. As the person credited with naming

2. I use the word "thinking" here to speak of an awareness, in a language, of a set of concepts/ideas and the connections among them. Metaphors for the experience of thinking might include: a sequential unfolding or a sustained churning up of the set of concepts and their connections. I use the term "way of thinking" to speak of a particular pattern of connecting, unfolding or churning, with or without awareness. Logic is a common way of connecting, although not the only way, and there are many possible logics.

modern cybernetics, Norbert Wiener, advocated (1954), doing cybernetics includes taking responsibility[3] for the consequences of our actions.

6. Of course, there are those who will point to technologies like automatic control, artificial intelligence, robotics, virtual reality and bionics as examples of the success of cybernetics. Some would qualify these as successes of first-order cybernetics. Unless these technologies were and are employed with awareness of their consequences for humankind, and not mindlessly for their commercial or military value, I would not call them successful or cybernetic. So, what might have once been considered a success of cybernetics no longer qualifies under the transformation to SOC. In other words, I want the use of the cybernetic label to imply an accounting for my desires and a taking of responsibility for the consequences of my actions, as required of the cybernetics I now have. To the extent that technology, any technology, is used with awareness of the desirability of the consequences for humanity, I can accept it as potentially cybernetic. However, there is no guarantee of desirable outcomes, and hence of success in the usual sense. We need new ways of thinking about desire, intention, humanness and consequences, and success itself.

7. If I know what I want and I know it is possible to achieve it, I do not need cybernetics – I just go and do what I need to do to achieve the outcome. However, when I only have a vague idea about what I want or do not want and I do not know how to pursue or avoid it in the current society, the vocabulary of cybernetics can be useful.[4] Cybernetics is not about success and the achievement of goals; it is about the reconfiguration of constraints (resources) in order to make possible what was not previously possible, including the avoidance of what was previously inevitable. When desires are treated as constraints, they become subject to reconfiguration as well. So, to talk about the success of cybernetics as though it is a tool that could help people who use it become wealthy and famous by solving current problems dismisses what it has to offer. SOC is distinguished more by the new questions it asks than by the answers it might provide to current questions, and its value is more in the new systems it imagines than in the rehabilitation of current systems. The dilemma of SOC is that recognizing the value of this way of thinking may require simultaneously assimilating it; its value is realized by doing it, but likely only by the people doing it. This is consistent with its reflexive character.

8. I propose, then, that, rather than call on criteria from external and mainstream sources, we set our own criteria for realizing the value of SOC: namely, as long as the vocabulary of cybernetics is contributive in maintaining interest in ongoing conversations[5]

---

3. I use the word “responsibility” here to speak of an awareness of my desires with respect to the consequences of my actions.
4. A claim could be made that the role of applied philosophy in society might be similar to that of cybernetics as described here, and in fact philosophy might contribute in addressing the same situation. However, I wish to distinguish cybernetics from philosophy, even though cybernetics may inform philosophy and vice versa. Cybernetics depends on no specific philosophical tradition. Its formulation starts with the idea of difference or change, and the set of concepts and their vocabulary are consistent with the simple act of drawing a distinction. It does not seek truth in the philosophical sense; rather, its value lies in its usefulness as an alternative to the prevailing linear, hierarchical, goal-oriented and especially theistic way of thinking about how the world does and should work. Philosophers may disagree; but then, I am not a philosopher.
5. I use the word “conversation” here to speak of a particular dynamics of interaction among two or more participants, in a language (verbal, graphical, audial, kinesthetic, gestural, etc.), such that the dynamics begins with an asynchronicity (a conflict, disagreement, friction, contradiction, be-

on nothing short of the transformation of society to a more just and equitable one, it has value. These conversations can happen in any current arena: for example, science, design, the arts, government, business, education, health care or everyday life, with the latter deserving special attention. I speak of a just and equitable society as:

a. one that supports every individual in their quest for significance through the unique contributions each has to offer; and
b. one without violence, or at least where violence is an alternative of last resort.

This implies that the basic needs of all humans be met unconditionally (or there will be violence), and that, once met, everyone can participate in the conversations on the transformation and design of the society in which they are a member. Conversation (Pask 1976) becomes a thread that can be pulled through all new directions proposed for SOC, and one of consequence for virtually all human beings.

Second-order cybernetics and six conceptual problems

9. I have chosen six conceptual problems for which SOC can provide some assistance. They all draw on SOC as a way of thinking for that assistance. The result in each case is a slightly different way of thinking about each problem. These problems cannot be resolved or dissolved as long as currently prevailing ways of thinking dominate, and these current ways are not likely to change easily, as they are essential to maintaining the status quo. The status quo both benefits those who are rewarded by it and operates without awareness by most others. That is, we depend on these prevailing ways of thinking to make sense of the current world and cannot imagine as desirable alternatives that would alter that sense-making. Yet, the problems persist. The six articles in this volume identified as target articles each correspond roughly to one of these problems, although with significant overlaps. My discussion below does not draw directly from those articles or the critique of those articles. Rather, I discuss each problem from the directions implied by SOC, as I see them.

Paradigms and the problem of the two cultures: Science and art

10. In his classic *The Two Cultures and the Scientific Revolution* (1961), C.P. Snow identified a rift between the ways of thinking in the sciences and related technical fields and those of the arts and humanities. He presents this rift in part as a warning and in part as an opportunity. If these ways of thinking do not talk to one another (metaphorically speaking), they may drift further apart, exacerbating the socioeconomic stratification and extreme inequality that characterize the human predicament world-wide. If, on the other hand, they could talk with one another, new ideas and approaches for addressing the disastrous consequences that could result if the situation is not addressed might emerge. In a class I had

---

ing on different planes, being out of phase, i.e., out of sync, etc.) and moves toward synchronicity (agreement, including the agreement to disagree, i.e., a mutual understanding). A conversation is sustained by a mutual preference for recurrent interaction and can stop when other preferences (including the avoidance of boredom) are given priority. A person can have a conversation with herself, where the participants are different roles, perspectives, positions, etc., adopted by that person. A conversation with oneself is the process that generates thinking.

with Russell Ackoff, he would refer to the approach of science as searching for similarities among phenomena that appear different and that of the arts as searching for differences among phenomena that appear the same. While this is an obvious oversimplification, it does point to the incommensurability of the two cultures and to the potential value of both when embraced, without bias, simultaneously.

11. In *The Structure of Scientific Revolutions* (1970), Thomas Kuhn popularized the word "paradigm" to speak of the prevailing way of thinking or pattern of thought embedded in the language and behaviors of a culture, and that more often than not operates without the awareness of the thinkers using it. The differences in paradigm become particularly noticeable when one crosses from one culture to another and tries to live in it. As a way of thinking about ways of thinking, SOC could be characterized as a paradigm of paradigms of which it – cybernetics (SOC) – is one. This turns cybernetics into a process, a way of thinking that is continually changing and must change whenever the set of ways of thinking changes. It can also raise awareness of the way of thinking being employed in a situation, rendering the way of thinking a choice.

12. In his target article, Stuart Umpleby promotes SOC as providing a framework for a fundamental revolution in science, and in their editorial Karl Müller and Alexander Riegler discuss such a framework and call it second-order science. Cybernetics has been treated as a science since its modern origins as a science of control and communication (Wiener 1961), and indeed its concepts have been of particular interest to mathematicians, biologists, psychologists and social scientists, including management scientists, over the years. As an important human enterprise, new directions in science as suggested by Umpleby, Müller and Riegler deserve to be supported, and I would regard them as directions compatible with SOC under four conditions:

a. that the new direction be treated not only as representing a new paradigm for scientific inquiry but also as recognizing that multiple paradigms can provide simultaneously useful insights;
b. that the scientist be aware of the paradigm(s) being applied and make it(them) a choice based on the desirability of its(their) consequences;
c. that SOC, as the overriding paradigm, be a process and itself subject to change as a consequence of the doing of scientific inquiry; and,
d. that science not be given superior status relative to other modes of inquiry, particularly the arts.[6]

13. SOC has the potential to unstick us from the trap of a single paradigm and the ethical dilemmas accompanying such a trap.

Disciplines and the problem of logical types

14. Bertrand Russell's theory of logical types addresses the self-referential paradox of sets that include themselves as a member. The problem of logical typing arises when

6. Some cyberneticians have recognized the mutual value of science and art in cybernetics. Ashby referred to cybernetics as "the art of steersmanship." Pask talked about cybernetics as "the art and science of manipulating defensible metaphors." Maturana defined cybernetics as "the science and art of understanding." For von Glasersfeld, cybernetics was "the art of creating equilibrium in a world of possibilities and constraints" (Richards 2009: 101f).

I have a situation involving categories (sets that do not include themselves as members) that, if mixed, would produce a paradoxical formulation – the vicious circle or "damned if you do, damned if you don't" syndrome. This is the problem of observers who include themselves in their formulations of the phenomena they are observing. The standard way out of this dilemma, within the logic that produced it, is to create a hierarchy of logical types that exclude all sets that include themselves as members – that is, paradox is not allowed. Connections among logical types are then dealt with at the next higher level of the hierarchy (as generalizations that maintain consistency) rather than at the lower level where interactions among the logical types would result in inconsistency (hence the term *logical* type). Logical typing is still the predominant way to simplify a complex situation and may be a sufficient way to deal with a particular phenomenon, or it may not be. Paul Watzlawick, John Weakland and Richard Fisch (1974) in their book, *Change: Principles of Problem Formation and Problem Resolution*, demonstrate how an awareness of these paradoxes in therapeutic situations can be dealt with by a therapist who operates at a higher level. It does not account for the situation when the therapist is a part of the problem – this would require a supra-therapist.

15. I claim (without argument here) that academic disciplines arise as a way to deal not only with the complexity of what would otherwise be the interconnectedness of all phenomena but more significantly with the inconsistencies that would rear their heads when prevailing logic is applied to this interconnectedness. The prevailing logic does not include time (and has been referred to as "time-less") and therefore dismisses the role that dynamics[7] might play in the behaviors of certain phenomena. Disciplines are logical types within which attempts are made to maintain consistency. Humberto Maturana and Francisco Varela (1992: 207-212) introduced the term "non-intersecting phenomenal domains" to speak of domains of concepts that, if mixed, would create inconsistencies not allowed in scientific explanations. Disciplines deal with different phenomenal domains, creating problems of inconsistency when they attempt to interact with one another. Cybernetics has been referred to as transdisciplinary, offering a vocabulary that could assist individuals from different disciplines when they interact with each other – a language they can agree on as they attempt to learn something new about their respective disciplines or the relations between them (Müller 2012). Cybernetics has also been referred to as an anti-discipline, providing the motivation to break down disciplinary walls by embracing paradox and developing alternative logics (e.g., Spencer Brown 1972; Varela 1975; Kauffman 1987) and to organize instead around problems or projects where all currently best available knowledge can be applied. What is learned is then no longer owned by a discipline.

16. For now, we live in knowledge structures that are, given our current language and its logic, disciplinary. What we can do, and what I think SOC suggests, is to work within our disciplines to:

---

7. I use the word "dynamics" to speak of a pattern of changes, where change, rather than object or entity, is fundamental. Dynamics, in its pure sense, cannot be articulated and is therefore of little use in explanations of phenomena. Attempts to include dynamics in explanations involve extreme simplifications, typically linear and kinematic, as measured by an external and standard clock. This actually removes the dynamics from what was experienced by the observer/listener/participant prior to the explanation. Dynamics can be appreciated, just as our experiences can be appreciated. Dynamics can be approached through art forms by manipulating language, as in poetry (verbal), dance (kinesthetic), theatre (verbal and gestural), music (audial), drawing and sculpture (visual), etc.

a. raise awareness of the limits of our domains of inquiry and their logics;
b. use differences in language and logic to compose asynchronicities as triggers for conversations across disciplines;
c. learn the languages and logics of other disciplines to generate new ideas in our own;
d. explore alternative logics for reformulating our disciplines and the domains within them, particularly those logics that offer alternatives to hierarchical structuring; and
e. reformulate our disciplines to account for time and dynamics.

17. Maturana (1978, 1980), for example, has reformulated biology as a relational science rather than a physical science. He explicitly recognizes the non-intersecting domains of relations (explanations) and dynamics (experience), and creates the dialectical pair: structure (a changing pattern of relations) and organization (an invariant pattern of dynamics) in the formulation. By taking this dialectical approach, he both accounts for dynamics and introduces a non-hierarchical way of thinking into science. A dialectical way of thinking[8] generates processes rather than hierarchies. In his target article, Bernard Scott has proposed reformulating the discipline of psychology using an SOC framework. I look forward to a reformulation of psychology that is as useful to me as the reformulation of biology has been, and perhaps even to a breakdown of disciplines and the hierarchies that support them, as incompatible or opposing phenomenal domains bump into one another in non-disciplinary conversations.

Desires and the problem of conscious purpose

18. The idea that systems are purposeful has been around from well before the advent of modern cybernetics. In a paper regarded as seminal in the lead-up to cybernetics, Arturo Rosenbleuth, Norbert Wiener and Julian Bigelow (1943) demonstrate how the idea of purpose arises from simple feedback loops that maintain the value of one or more variables within limits, i.e., the system's constraints. Any system with constraints appears to have a purpose as there are outcomes precluded from the set of possibilities. The problem in SOC arises when the observers/designers specify the purpose of their designs, giving conscious intent to their actions. Gregory Bateson (1972a, 1972b) warned of the dysfunctions of conscious purpose when the actions taken do not and cannot account for all the ecological circularities of the situation and the unanticipated consequences inherent in taking such actions. Yet, humans have needs, desires, preferences and values; we are self-aware of our actions and alternatives; and, we can act with intent to satisfy our needs and desires. To act without self-awareness of our desires and the possible consequences of our actions would be irresponsible. SOC may point to a different way of thinking about desires, intentions and consciousness that addresses Bateson's concerns.

19. Conscious purpose has come to be associated with the specification of desires as goals or objectives to be achieved. The word goal is used as a point concept, a target to

8. I use the term "dialectical thinking" to speak of a process of connecting concepts such that: for every idea, create at least one incompatible and/or opposing idea, let these ideas interact to generate new ideas and their incompatible and opposing ideas, and so on. Some ideas will emerge as sufficiently desirable to try out. Others will serve to further the conversations that support this way of thinking.

pursue; a goal is a future end state separate from the means used to achieve it. The word objective is often used as a vector concept to distinguish it from goal, i.e., a direction to be pursued (as in maximize or minimize); an objective, like goal, is also future-oriented and separated from the means used to accomplish it. I have proposed that desires be treated as a set of constraints (Richards 1991, 2007). Specifying a set of constraints treats desires as a spatial concept, focusing attention on the states we wish to exclude from happening, leaving open a space of possible outcomes deemed currently acceptable. This approach is present-oriented, merging ends and means: the set of constraints that represent our desires and the actions we take to avoid what we do not want are here and now, and our evaluation of possible consequences is based on current best available knowledge. Our desires, actions and evaluations can change as we experiment, learn and change, making it important to be careful about excluding outcomes that could become useful as circumstances change. Treating desires as constraints and intention as an awareness of desires as constraints opens the door for an alternative to the consciousness of purpose about which Bateson was concerned. I have called this alternative a "consciousness of presence" (Richards 2013a, 2013b), similar to the treatment offered by Eugen Herrigel (1953).

20. Treating desires as a set of constraints and shifting consciousness from the achievement of goals in the future to the presentation of self in the here and now have significant implications for science, design and society. While I think it is useful to continue to try to apply SOC ideas to an explanation of human consciousness (see Diana Gasparyan's article), I also think it is important to recognize that SOC suggests that the important questions of science are undecidable questions (Foerster 2003b: 293) – questions that only we can decide, questions of desire. Under SOC, scientists would be explicit about their desires and values, as advocated by Feyerabend (1975), and build them into their research designs. There may be no better example of an undecidable question than the question of consciousness. We have the opportunity to explain human consciousness as a desirable attribute. Designing also involves desires; Thomas Fischer and I (2017) discuss the implications of constraint-oriented design for designers and society. I have no doubt that a shift to a consciousness of presence implies a society quite different from what we have now. Technologies, behaviors and attitudes toward one another would be different; conversation would be an activity of choice and the society would be fully participative and dialogic.

Craft and the problem of time

21. The prevalence of a time-less logic in commonly accepted paradigms, the unchallenged assumption of an external standard clock[9] in science and other human endeavors, and the prevailing reliance on linear, unidirectional time in the application of conscious purpose all point to the problem of time – namely, time is a human invention created to resolve the paradoxes of self-reference and allow us to move and act with volition (Richards 2016). Norbert Wiener recognized an aspect of the problem when he titled the first chapter of his book "Newtonian and Bergsonian Time" (1961). A recent book (Canales 2015) details a debate on the subject of time between Bergson and Einstein. The standard clock is, of course, useful to scientists who wish to replicate the results of experiments performed by

9. I use the word "clock" to speak of a way of sampling a dynamics. The idea of a clock is often used as a surrogate for the idea of time. The standard clock discussed here is one that samples in seconds, minutes, hours, days, years, etc., as synchronized with the regular rotation of the earth on its axis and the regular revolution of the earth around the sun.

other scientists, an aspect of science regarded as essential. Clocks also serve a social function for humans; we synchronize our clocks in order to get to the same place at the same time as others with whom we wish to have conversations and to exhibit the kind of caring for each other specific to humans. Different cultures employ different clocks and concepts of time. The arts play with time and dynamics, even if artists do not express what they do that way. In his target article, Tom Scholte proposes using the theatre as a playground to experiment with the dynamics of human interaction, and therefore with time.

22. When we accept unwittingly an external and standard clock, we also accept that it can become a tool of oppression – for example, the time sheet/card at work, or the continual reminder that "time is money." Our lives become regulated by the clock, stable reward-oriented hierarchies arise to implement this mode of regulation, and goal-achievement becomes that to which we devote our every moment. A society of oppression is not a just and equitable one. However, if time is a human invention, it stands to reason that it can be manipulated. SOC, as a way of thinking that recognizes many possible concepts of time, can raise awareness of alternatives. In my response to Scholte's target article, I speak of the cybernetician as a potential craftsperson in and with time. If cybernetics is realized in the doing of it, the cybernetician draws on both science and the arts, equally and without bias, and then adds the craft of cybernetics. The idea of craft merges art and science and adds action. Being a craftsperson in and with time is different from being an artisan, where one works with physical media – although I claim that all art manipulates time. Time is not a medium in the same way that sound and paint are. Knowing when to say or do something as an intervention, how loudly or softly to speak, how fast or slowly to move, what rhythm to use, how to turn a flow into an event, when to emphasize or not – all of these involve a kind of craftsmanship in and with time.

23. The cybernetician's craft gets enacted in conversation. The idea of a fully participative and dialogic society implies that time take on a different significance than is manifest in most current societies. Ernst von Glasersfeld (2007), Herbert Brün (2004) and Maturana[10] have written about time in a way that reflects SOC. Luigi Boscolo and Paolo Bertrando's *The Times of Time* (1993) also deserves consideration. The idea that we can manipulate time while still remaining social and without going crazy is new and needs SOC. For example, if all technology mediates human interaction in some way or another, new technologies that support a participative-dialogic society are indicated. When a society adopts an ever-changing present approach to time and life and when clocks are treated as a human choice in the moment, a revival of analog computing may be required in order to accommodate and facilitate the non-linear dynamics represented by a conversation – that is, the intensity and complexity of conversation may exceed the capability of what can be accommodated or facilitated by the digital simulation of that dynamics. In a conversation, multiple clocks can be operating simultaneously. Gordon Pask (1979) advocated for the kinetic design of computing devices, as an alternative to the current kinematic design approach. Designing and building alternative devices in support of conversation is another role, a collaborative one, for the craftsperson in and with time.

---

10. Unpublished manuscript "The Nature of Time" at http: //www.slideshare.net/Longsthride/the-nature-of-time-humberto-maturana-1995

Design and the problem of participation

24. I regard participation in a society, and in the decisions that alter that society and our participation in it, as a human need. I use the word "participation" to speak of my awareness that what I say or do makes a difference. To feel my participation restricted and my ability to make a difference in the world non-existent is a classic description of oppression and to accept it as inevitable is to give up at least some of my humanity. I claim that this feeling of inevitability and alienation is supported by prevailing ways of thinking about power and causality and by the social structures that emanate from these ways of thinking – namely, reward-oriented hierarchies (Marianne Brün 2004). SOC may offer a way of thinking (about ways of thinking) that could provide an alternative to thinking that the only way to make a difference in the world is to have the power to cause things to happen, and therefore to be in a sufficiently high position in the relevant hierarchy to be granted that power. The challenge for the practice of design is that we need not only a process that creates technologies and social systems that encourage and facilitate participation but also one that is itself participative. It is design that can satisfy the human need for participation. Ranulph Glanville's (2007) insistence that the design process be embedded in SOC is instructive for new directions in both design and SOC.

25. I have written elsewhere (Richards 2013b) that thinking in terms of how dynamics works as opposed to how causality works may be useful as an alternative to thinking in terms of power as the only way to make a difference in the world. The mechanism through which this kind of dynamics can work is conversation. Conversation undoes power (Krippendorff 1995). Speaking metaphorically, the smallest of perturbations can trigger a transformation of the fabric of dynamics world-wide: think in terms of a ripple that radiates immediately throughout that fabric. There is no causal link that can be traced back to the perturbation, and no evidence that can be brought to bear on the direct effects of actions taken. Change just happens. There may be constraints on what transformations can happen that may be worth studying prior to taking action – this is the cybernetician's craft. In a conversation, it is the asynchronicities introduced that can trigger new ideas and/or new conversations. Hence, the Anticommunication Imperative: if you seek the new, compose asynchronicity (Richards 2010).

26. Hierarchy is a way of thinking about simplifying a phenomenon that would otherwise be too complex to control or manage by breaking it into subsystems or categories that are more manageable. When applied to social structures, an intent is to keep people who make up a unit of the social system from interacting or having to interact with people in another unit. Responsibility for coordinating the units resides with a manager at the next level of the hierarchy. So, hierarchical social structures, by intent, suppress conversation (Richards 2013b). To get people to want to move up to managerial positions, rewards are offered – both monetary rewards and increased authority and power. These systems are then driven by goal achievement. I have discussed elsewhere the design of alternatives to hierarchically organized academic conferences (Richards 2015a) and the design for greater participation in these conferences (Richards 2015b).

27. The challenge of introducing non-hierarchical social structures into current society, like Stafford Beer's icosahedral (1994) or other polyhedral structures, is that hierarchical thinking is ubiquitous. Government structures are set up to accommodate hierarchical organizations; they have legal status; and, technologies support and reinforce this way of thinking. In almost everything we do day-to-day, we deal with social hierarchies in some

form or another. To realize a fully participative and dialogic society, we need a change of system, a new order of things, not just a change in the current system. I look to a collaboration between design and SOC, like that proposed by Ben Sweeting in his target article, to imagine a new society and how to bring it about, without violence. I would suggest that a precondition for such a society is that all basic needs (food, fresh water, shelter, sanitation, health care, education, etc.) be met unconditionally and not used as rewards. Until all basic needs are met, however they are defined, the uniquely human need for participation will be secondary and dismissed as a luxury, thus reinforcing non-participative processes and non-dialogic structures.

Ethics and the problem of human nature

28. Von Foerster (2003b: 291) argued that, in a desirable society, ethics should operate under the surface, like an underground river that we never see but that affects our behavior in every moment. He also offered the Ethical Imperative: act always to increase the number of choices (ibid: 303). Brün (2003: 118) formulated the concept of ethics as: dilute power, increase freedom. The problem is that leaving ethics unlegislated or even unspoken opens the door for what is sometimes called the dark side of human nature. Human nature is then used to justify systems of oppression and control. Hierarchical structures and the control of resources by the few is regarded as necessary to suppress the instincts referred to as human nature. Without a system of incentives and sanctions, human society runs amuck.

29. Greed, cruelty, abuse of power, in fact all of the so-called deadly sins, are often considered aspects of this dark version of human nature, and as such constitute behaviors to be addressed and controlled in the design of a society. I claim that these aspects of human nature are constructs of the current society and used to place blame on individual humans rather than on the systems that reinforce, even require, those behaviors – competition, rewards, self-interest, family. In a participative-dialogic society, humans exhibit kindness, generosity, compassion and creativity, with systems that are compatible with and support those qualities. What would constitute these alternative systems? Louis Kauffman's characterization of SOC as a reflexive domain (see his target article) might be helpful as a way of thinking about such systems. In a domain where self-reflection and self-other reference are inherent, perhaps von Foerster's statement of desire, "A is better off when B is better off" (Foerster 2003a: 209), could be extended reflexively to "A is better off when B is better off, and B is better off when A is better off," and we could then work together towards its realization. Umpleby's explicit incorporation of ethics as a requirement in an SOC-based science (§§60-62) represents a hopeful direction.

New directions and questions

30. Conversation is a common theme running through all six of these conceptual problems. Conversation is enacted in cybernetics, and its SOC version is of potential interest to all humans world-wide. There may be many ways to talk and think about SOC and conversation, of which the six problems is only one. Personally, I think a new approach to science is needed, I think we need more from the arts, and I think the craft of the cybernetician needs substantial development. The challenge is that the only way we have of thinking and talking about new ways of thinking is with the language and logic we cur-

rently have, which reinforce many of the behaviors we may wish to avoid with new logics and languages. I express the six problems in current language and thinking, yet they point toward ways of thinking that suggest alternative language. Cybernetics provides a helpful vocabulary, but we may also need new grammatical and syntactical structures. In the meantime, we can move ahead on addressing the six problems and pursuing the directions implied by the six target articles, taking into consideration the questions raised by the open peer commentaries.

31. There are three trends in SOC that I detect in some cybernetic articles and occasionally at cybernetic conferences that I think deserve questioning if not outright avoiding.

32. First, I regard any vision of using SOC to unify all human knowledge to be both futile and an affront to what it has to offer. The idea of unification is hierarchical, pointing ultimately to a supreme being. SOC is about change and processes of change, including change of SOC itself. The value of SOC is as a process that supports dialogue and participation. Conversation is a dialectical process of generating the new, of continuing to retard the decay of variety that is an inevitable consequence of cybernetic thinking. That variety decays should not be taken as a negative aspect of cybernetics; on the contrary, if variety did not decay, we humans would have nothing to do.

33. Second, and similarly, SOC is not a candidate for a new ideology. SOC is not asking to be believed; it is asking to be accepted as useful in understanding the world and ourselves, designing new systems, and living our day-to-day lives in a way that gives some hope for overcoming seemingly insurmountable obstacles to realizing our desires, and beyond our desires the desirable. While our work in SOC may decay towards unification or ideology, we must be vigilant in seeking ways to retard that decay and embracing new ideas that challenge that direction.

34. Third, and most recently, cybernetics (and SOC) is being described as a "science of context" (Lissack §§16-19). I am willing to exhibit a little more patience with this idea, given its recent appearance, as it might still be useful as a trigger for some new questions – that is, it qualifies as a potential asynchronicity. However, I don't know where to go with it in order to have the conversation. I understand what science is, and I understand the word "context", but I have yet to draw any significance from the combination: a science of context. The question at this point is: How can we understand the word context so that a science devoted to it is useful? If the word "context" is used in its traditional sense, as whatever a receiver of a message needs to know to understand the intent of the sender or to formulate their own intent, then we are talking about a science of communication, which is not new. If the word "context" is used in its broad sense, as any and all circumstances that might affect an observer's looking, listening and acting in a particular situation – that is, language, culture, education, family history, experiences, etc. – then we may be talking about a science of everything, and that is not particularly useful. If the word "context" is used to talk specifically about the ways of thinking, including the desires and values, brought to bear on a situation, then we have a science of context-making, or better: the art, science and craft of context-making. This is a description of design and an argument for the merger of cybernetics and design (Glanville 2007; Fischer & Richards 2017).

References

Ashby W. R. (1956) An introduction to cybernetics. Chapman & Hall, London.
Bateson G. (1972a) Conscious purpose versus nature. In: Bateson G. Steps to an Ecology of Mind. University of Chicago Press, Chicago IL: 432-445.
Bateson G. (1972b) Effects of conscious purpose on human adaption. In: Bateson G. Steps to an Ecology of Mind. University of Chicago Press, Chicago IL: 446-453.
Beer S. (1994) Beyond dispute: The invention of team syntegrity. John Wiley, Chichester, UK.
Boscolo L. & Bertrando P. (1993) The times of time: A new perspective in systemic therapy and consultation. W. W. Norton, New York.
Brün H. (2003) Irresistible observations. Non Sequitur Press, Champaign IL.
Brün H. (2004) Teaching the function of time in art. In: Chandra A. (ed.) When music resists meaning: The major writings of Herbert Brün. Wesleyan University Press, Middletown CT: 3-5.
Brün M. (2004) Paradigms: The inertia of language. In: Chandra A. (ed.) When music resists meaning: The major writings of Herbert Brün. Wesleyan University Press, Middletown CT: 292-300.
Canales J. (2015) The physicist & the philosopher: Einstein, Bergson and the debate that changed our understanding of time. Princeton University Press, Princeton, NJ.
Feyerabend P. (1975) Against method: Outline of an anarchistic theory of knowledge. Verso, London.
Fischer T. & Richards L. D. (2017) From goal-oriented to constraint-oriented design: The cybernetic intersection of design theory and systems theory. Leonardo 50(1): 36-41
Foerster H. von (2003a) Perception of the future and the future of perception. In: Foerster H. von, Understanding understanding: Essays on cybernetics and cognition. Springer-Verlag, New York: 199-200.
Foerster H. von (2003b) Ethics and second-order cybernetics. In: Foerster H. von, Understanding understanding: Essays on cybernetics and cognition. Springer-Verlag, New York: 287-304.
Glanville R. (2007) Try again. Fail again. Fail better: The cybernetics in design and the design in cybernetics. Kybernetes 36(9/10): 1173-1206.
Glasersfeld E. von (2007) The conceptual construction of time. In: Glasersfeld E. von, Key works in radical constructivism. Edited by Marie Larochelle. Sense Publications, Rotterdam: 225–230.
Herrigel E. (1953) Zen and the art of archery. Vintage Books, New York.
Kauffman L. H. (1987) Self-reference and recursive forms. Journal of Social and Biological Structures 10(1): 53-72.
Krippendorff K. (1995) Undoing power. Critical Studies in Mass Communication 12(2): 101-132.
Kuhn T. S. (1970) The structure of scientific revolutions. Second edition. University of Chicago Press, Chicago.
Maturana H. R. (1978) Biology of language: The epistemology of reality. In: Miller G. A. & Lenneberg E. (eds.) Psychology and biology of language and thought. Academic Press, New York: 27-63.
Maturana H. R. (1980) The biology of cognition. In: Maturana H. R. & Varela F. J. Autopoiesis and cognition. D. Reidel, Dordrecht, Holland: 1-58.
Maturana H. R. & Varela F. J. (1992) The tree of knowledge: The biological roots of human understanding. Revised edition. Shambala, Boston.
Mead M. (1968) Cybernetics of cybernetics. In: Foerster H. von, White J., Peterson L. & Russel J. (eds.) Purposive systems. Spartan Books, New York: 1-11.
Müller A. (2012) What does it mean to be a trans-, inter- or meta-disciplinary subject? In: Glanville R. (ed.) Trojan horses. Edition echoraum, Vienna: 15-17.
Pask G. (1976) Conversation theory: Applications in education and epistemology. Elsevier, Amsterdam.

Pask G. (1979). An essay on the kinetics of language, behavior and thought. Proceedings of the silver anniversary meeting of the Society for General Systems Research. SGSR, Washington DC: 111-128.

Richards L. D. (1991) Beyond planning: Technological support for a desirable society. Systemica 8(2): 113-124.

Richards L. D. (2007) Connecting radical constructivism to social transformation and design. Constructivist Foundations 2(2/3): 129–135.

Richards L. (2009) Craft and constraint, clocks and conversation. Edited by J. Marrero. Online book, https://pearlheatherforge.wordpress.com/2009/05/20/4

Richards L. D. (2010) The anticommunication imperative. Cybernetics & Human Knowing, 17(1/2): 11-24.

Richards L. (2013a) Idea avoidance: Reflections of a conference and its language. Kybernetes 42(9/10): 1464–1470.

Richards L. (2013b) Difference-making from a cybernetic perspective: The role of listening and its circularities. Cybernetics & Human Knowing 20(1/2): 59-68.

Richards L. D. (2015a) Designing academic conferences in the light of second-order cybernetics. Constructivist Foundations 11(1): 65-73.

Richards L. D. (2015b) Design for participation: Culture, structure, facilitation. Constructivist Foundations 11(1): 93-97.

Richards L. (2016) A history of the history of cybernetics: An agenda for an ever-changing present. Cybernetics & Human Knowing 23(1): 42-49.

Rosenbleuth A., Wiener N. & Bigelow J. (1943) Behavior, purpose, and teleology. Philosophy of Science. 10: 18-24.

Snow C. P. (1961) The two cultures and the scientific revolution. Cambridge University Press, New York.

Spencer Brown G. (1972) Laws of form. The Julian Press, New York.

Varela F. J. (1975) A calculus for self-reference. International Journal of General Systems 2: 5–24.

Watzlawick P., Weakland J. H. & Fisch R. (1974) Change: Principles of problem formation and problem resolution. W. W. Norton, New York.

Wiener N. (1961) Cybernetics: Or control and communication in the animal and the machine. Second edition. The MIT Press, Cambridge, MA.

Wiener N. (1954) The human use of human beings: Cybernetics and society. Avon Books, New York.

Epilogue

Possible Futures for Cybernetics

Karl H. Müller, Stuart A. Umpleby & Alexander Riegler

1. In the field of science studies the term "boundary work" is used to describe writings by scientists that distinguish their field from other fields (see also Kline §8). This volume presents the boundary work for second-order cybernetics (SOC) and describes what we believe will be its future scope and dimensions.

2. During the Macy conferences in the late 1940s and early 1950s a group of academics from a wide range of fields came together to discuss the development of a general theory of control and communication, of information and regulation, of learning, adaptation and understanding, as a complement to the general theory of matter and energy in the natural sciences.

3. In the 1950s and 1960s the field of cybernetics attracted widespread interest. People from many social science and engineering disciplines read and were influenced by the publications from the Macy group. However, the commitments of most readers remained to their home disciplines. They took ideas like feedback and the importance of communication and decision-making and used them in their home disciplines. Since most researchers were employed teaching traditional courses, after a few years of reading about cybernetics, they chose to work in the traditional disciplines.

4. Fortunately, a few people remained attracted to the new field, even if they continued to teach in their traditional disciplines. As a result, cybernetics continued to evolve, and as it did, new ideas were invented:

- self-organizing systems in the 1960s,
- the biology of cognition, management cybernetics and autopoiesis in the 1970s,
- reflexivity and its connection to ethics and macro-economics in the 1980s,
- design in the 1990s, and
- a fruitful critique of science in the 2000s.

5. In the 1950s and 1960s centers and institutes dedicated to systems and cybernetics were established on several campuses in the US. But in the 1980s and 1990s these centers and institutes began to close as their founders, often immigrants from Europe, retired. Universities had degree programs and tenure-track positions in the traditional disciplines but not in systems and cybernetics.

6. Nevertheless, the ideas in cybernetics continued to attract enthusiastic multi-disciplinary groups on several campuses. While from the late 1970s to about 2010 journal articles by North American authors declined steadily (see Figure 1), in the last 15 years an increasing number of books about the achievements of cybernetics have been published and earlier books have been reissued (Figure 2). There is widespread recognition that break-through, game-changing research results from multi-disciplinary teams (Umpleby,

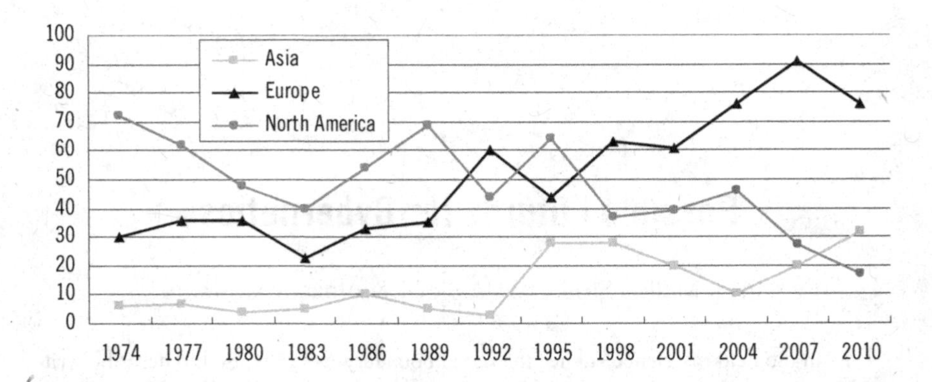

Figure 1. Number of articles per year by region over time in the three journals *Cybernetics and Systems, Kybernetes* and *Systems Research and Behavioral Science* (adapted from Umpleby, Wu & Hughes 2017).

Anbari & Müller 2007), even though universities continue to operate with reward systems that favor narrow specialization.

7. Although there are currently no courses or academic positions in the theoretical and philosophical aspects of cybernetics, we feel the future of the field is bright for several reasons: Cybernetics is a transdisciplinary field that has influenced and has been influenced by many fields, including neurophysiology (Maturana 1975), psychology (Watzlawick 1983), engineering (Sage 1992), management (Beer 1972; Ackoff 1981; Schwaninger 2008), mathematics (Wiener 1948; Kauffman 2016), political science (Deutsch 1966), sociology (Buckley 1968), economics (Soros 1987), anthropology (Bateson 1972; Mead 1964), philosophy (Abraham 2016) and design (Glanville 2015). Cybernetics conferences attract people from all of these fields and the conference participants communicate easily with one another because of shared assumptions, principles, and models.

8. Cybernetics and its sister field of systems science provide transdisciplinary ideas that make it easier for people from diverse fields to work together. The problems of the future will increasingly require the efforts of people from several disciplines. Although Americans, with their focus on practical problem-solving, tend to neglect theory and philosophy, other countries, such as Russia and China, in addition to Europe, are taking an interest in cybernetics.

9. Today professional people are spending several hours a day in cyberspace, cybersecurity is a major concern domestically and internationally, and cyber infrastructure is spreading around the world, but most people, including academics, have not known that a science of cybernetics exists. Some people have thought that the field of cybernetics ended in the mid-1970s, at about the same time that SOC was invented (Pickering 2010; Kline 2015).

10. In its contemporary form SOC continues to focus on circularities and their inherent paradoxes as well as on solutions to operate in and with these circularities. SOC is tied to circularities like Prometheus to the stone. In Greek mythology Prometheus is a great benefactor who steals fire from Mount Olympus and presents it to mankind. Similarly,

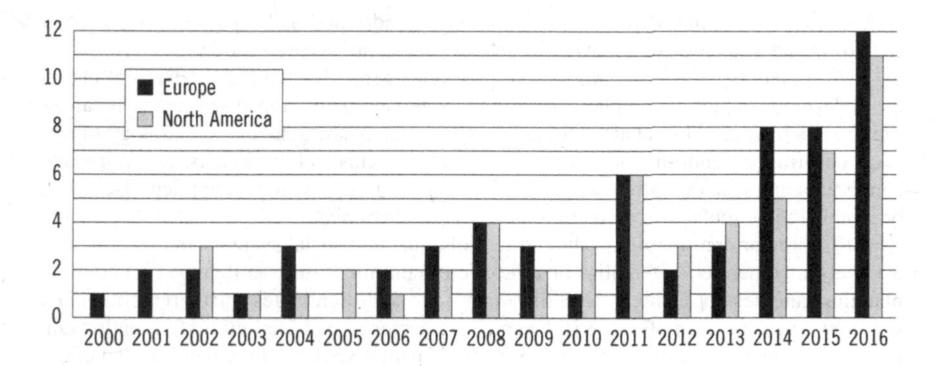

Figure 2. Number of recent books about cybernetics (adapted from Umpleby, Wu & Hughes 2017).

SOC steals circularities from their paradoxical and self-destructive past and offers them as a great gift to science.

11. SOC is to cybernetics what quantum mechanics is to physics – new ideas and a new direction. SOC has fundamentally reinterpreted the field of cybernetics. Assuming that cybernetics is another term for computer technology fundamentally misunderstands the field because it misses completely the theoretical and philosophical implications of cybernetics for our understanding of cognition, society and ethics.

12. SOC starts with circular configurations and their paradoxes and transforms them into new horizons for scientific research or reflective practices in applied or artistic fields. This cybernetic core function can be classified as the cybernetic attractor or as the eigenform for cybernetics itself. Eigenforms, following Heinz von Foerster (1976), reproduce themselves through circular configurations regardless of their starting points.

13. One of these circular configurations was presented at the first conference of the American Society for Cybernetics when Margaret Mead gave a lecture with the title "Cybernetics of Cybernetics" and spoke of the challenging task of exploring cybernetics with the methods and tools of cybernetics itself (Mead 1968). The circularities inherent in Mead's lecture have been expanded and generalized to a circular configuration within science where first-order science is the science of exploring the world and second-order science becomes the science of reflecting on these explorations. In this configuration SOC offers new perspectives for science and motivates a *second-order science* (Riegler & Müller 2014) as a vast and largely unexplored science frontier with a high innovation potential and multiple methods for quality improvements in science.

14. Another circular configuration deals with observers and their domains of research. It can be classified as an observer-circularity following von Foerster's distinction between first-order cybernetics as the cybernetics of observed systems and second-order cybernetics as the cybernetics of observing systems (Foerster 1974).

15. The fifty-six contributions in this book deal with the observer-circularity and generalize and transform it to different epistemic modes of science, i.e., explorations from

without (exo-mode) and explorations from within (endo-mode). They present the current state of the art with respect to new horizons for science in the endo-mode.

16. As can be seen in the present volume, the agenda for innovative explorations in the endo-mode is highly complex, diversified, challenging and transdisciplinary. It can be separated into several broad areas focusing on building a new general methodology of science, constructing endo-methodologies for scientific fields and disciplines, reframing and contextualizing research problems, searching circular practices within applied disciplines or establishing circular reflexive approaches in artistic domains.

17. Circular configurations abound in biological and social systems and include also a variety of other self-referential or reflexive circularities, such as therapy, negotiation, planning, and design. Cybernetics calls attention to and illuminates such circularities in a way new to science. In all these instances SOC performs a crucial midwife function and converts circular configurations and their inherent paradoxes into new scientific challenges and innovative perspectives for the respective state of the scientific art. This function has remained the stable core of SOC since its early days and will continue as such in the future even if SOC disappears from the map of scientific landscapes.

18. The present book shows that cybernetics and second-order cybernetics are again enjoying a widespread interest that is maintained by a vital and living organization and shared by a large community of researchers, designers or artists in a variety of academic disciplines, applied domains or artistic niches around the world.

References

Abraham T. (2016) Rebel genius: Warren S. McCulloch's transdisciplinary life in science. MIT Press, Cambridge MA.

Ackoff R. L. (1981) Creating the corporate future: plan or be planned for. Wiley, New York.

Bateson G. (1972) Steps to an ecology of mind. Chandler Publishing, San Francisco.

Beer S. (1972) Brain of the firm: A development in management cybernetics. Herder and Herder, New York.

Buckley W. (ed.) (1968) Modern systems research for the behavioral scientist: A sourcebook. Aldine Publishing, Chicago.

Deutsch K. (1966) The nerves of government: Models of political communication and control. Free Press, New York.

Foerster H. von (ed.) (1974) Cybernetics of cybernetics, or the control of control and the communication of communication. BCL Report 73.38. Biological Computer Laboratory, Urbana IL. Republished in 1995 by Future Systems, Minneapolis MN.

Foerster H. von (1976) Objects: Tokens for (eigen-)behaviors. ASC Cybernetics Forum 8(3–4): 91–96. Reprinted in: Foerster H. von (2003) Understanding understanding. Springer, New York: 261–271. http://cepa.info/1270

Glanville R. (2014) The black box. Edition Echoraum, Vienna.

Kauffman L. (2016) Cybernetics, reflexivity and second-order science. Constructivist Foundations 11(3): 489-504. http://constructivist.info/11/3/489

Kline R. (2015) The cybernetics moment or why we call our age the information age. Johns Hopkins University Press, Baltimore.

Maturana H. R. (1975) The organization of the living: A theory of the living organization. International Journal of Man-Machine Studies. 7(3): 313–332

Mead M. (1964) Continuities in cultural evolution. Yale University Press, New Haven.

Mead M. (1968) Cybernetics of cybernetics. In: Foerster H. von, White J. D., Peterson L. J. & Russell J. K. (eds.) Purposive systems. Spartan Books, New York: 1–11. http://cepa.info/2634

Pickering A. (2010) The cybernetic brain: Sketches of another future. University of Chicago Press, Chicago IL.

Riegler A. & Müller K. H. (eds.) (2014) Special issue "Second-order science." Constructivist Foundations 10(1). http://constructivist.info/10/1

Sage A. (1992) Systems engineering. Wiley, New York.

Schwaninger M. (2008) Intelligent organizations: Powerful models for systemic management. Second edition. Springer, Berlin.

Soros G. (1987) The alchemy of finance: Reading the mind of the market. Simon and Schuster, New York.

Umpleby S. A., Anbari F. T. & Müller K. H. (2007) Highly innovative research teams: The case of the biological computer laboratory (BCL). In: Müller A. & Müller K. H. (eds.) An unfinished revolution? Heinz von Foerster and the Biological Computer Laboratory 1958-1976. Echoraum, Vienna: 189-209. http://cepa.info/4094

Umpleby S., Wu X.-H. & Hughes E. (2017) Advances in cybernetics provide a foundation for the future. Special issue on "The Future of Systems." International Journal of Systems and Society 4(1): 29–36.

Watzlawick P. (1983) How real is real? Confusion, disinformation, communication. Souvenir Press, London.

Wiener N. (1948) Cybernetics, or control and communication in the animal and the machine. MIT Press, Cambridge MA.

Author Biographies

Marcelo Arnold-Cathalifaud is a professor at the Department of Anthropology at the University of Chile. His research is focused on the development of sociopoiesis, constructivist epistemology, organizational systems, emerging complexities of contemporary society such as social effects of biosciences on societies, organisms and environment, changes in social collaboration and solidarity; politics, social inequity and exclusion; development of regional social sciences; and, lately, the impacts of population ageing. marnold@uchile.cl

Yochai Ataria has written a number of articles on various topics relating to altered states of consciousness, mainly concerning the relationship between the sense of self, the sense of time and the sense of body during traumatic experiences. He has also published a number of articles regarding the meditative experience. yochai.ataria@gmail.com

Philip Baron currently works at the University of Johannesburg and teaches post-graduate studies in electrical engineering. His main interests are in interdisciplinary works, having qualified in the fields of psychology, engineering, philosophy, and religious studies. The University of Johannesburg presented him with its highest award for teaching and learning excellence in 2015 (Vice-Chancellor's Distinguished Awards: Teaching Excellence). He currently serves as an associate editor for *Kybernetes* journal. pbaron@uj.ac.za

Gastón Becerra is an assistant professor at the Universidad de Buenos Aires. His research focuses on the emergence of "complexity" as an object of study for the social sciences and the constructivists traditions, with special attention to Rolando García's and Niklas Luhmann's ouvres. gastonbecerra@sociales.uba.ar

Eva Buchinger specializes in systems theory and innovation policy. Her tasks include research, teaching, research management and policy consulting. She is, among other roles, board member of the Research Committee on Sociocybernetics (RC51) of the International Sociological Association (ISA) and consultant for the Austrian government and EU institutions (ERAC, Commission). eva.buchinger@ait.ac.at

Jose dos Santos Cabral Filho is an architect and Professor at the School of Architecture at UFMG (Brazil). He holds a master's and a PhD from Sheffield University and has been a visiting scholar at McGill University, NTNU and the RCA. His interests range from cybernetics to architectural performances, electronic music, play and games. jcabral@arq.ufmg.br

Peter Cariani has worked in theoretical biology, biological cybernetics, and neuroscience. His doctoral work developed a semiotics of percept–action systems, formulated a taxonomy of self-constructing adaptive systems, and explored epistemic implications of evolutionary robotics. He teaches courses at Harvard and Boston Conservatory related to the psychology of music and to the neural and psychological basis of conscious awareness. His current research investigates temporal codes and neural timing nets for pitch and rhythm perception. He is also working on a general theory of brain function based on complex temporal codes and timing nets. cariani@bu.edu

Dorothy Chansky is Director of the Humanities Center at Texas Tech University, where she also teaches history, theory, and criticism in the School of Theatre and Dance. Her publications include *Kitchen Sink Realisms: Domestic Labor, Dining, and Drama in American Theatre* (Iowa 2015) and *Food and Theatre on the World Stage* (Routledge 2015), co-edited with Ann Folino White. She writes criticism for New York Theatre Wire and edits *Theatre Annual: A Journal of Theatre and Performance of the Americas*. dorothychansky@gmail.com

Lowell F. Christy is the director of Evolutionary Intelligence, LLC. He is an experimental epistemologist trained in how to probe living systems, particularly cultural systems, identify the axis of social change and conduct experiments in order to scale transformation. He has worked at major think tanks in the US, including Stanford Research Institute and the Economic Strategy Institute; was Chairman of Micro LSI in Silicon Valley; and created large demonstration projects in cybernetic technology/human systems with governmental agencies. He studied with and was mentored by Heinz von Forester, Gregory Bateson and Douglas Engelbart. lowell.christyphd@gmail.com

Bruce Clarke is Paul Whitfield Horn Professor of Literature and Science in the Department of English at Texas Tech University. His research focuses on systems theory and narrative theory. Clarke edits the book series *Meaning Systems*, published by Fordham University Press. Recent books include *Neocybernetics and Narrative* (Minnesota 2014) and *Posthuman Metamorphosis: Narrative and Systems* (Fordham 2008). brunoclarke@gmail.com

Art Collings is Vice President for Land Conservation for Dutchess Land Conservancy, a non-profit conservation land trust in New York's Hudson Valley. He is a cartographer and conservation planner by profession, and a many-valued mathematical logician by avocation. He lives in Red Hook, NY, and is currently the Treasurer of the American Society for Cybernetics. otter@mac.com

Thomas Flanagan is a research neuroscientist and collaboration consultant who serves as President of the Board for the Institute for 21st Century Agoras, an international research and education non-profit promoting systems approaches to democratic processes. tom@globalagoras.org

Kathleen Forsythe has the title Doctor of Knowledge Architecture on the impetus of Louis Kauffman, through an unusual jury of peers. Her work has not followed the traditional paths of academia and has been deeply influenced through conversations in the communities of second-order cybernetics over the past 35 years. She is an artist, poet, novelist and educator. She is currently the Executive Director of the SelfDesign Learning Foundation in Vancouver, BC Canada. Her current work is focused on more deeply understanding languaging in non-verbal children diagnosed with autism. kathleenforsythe@selfdesign.org

Michèle Friend is Associate Professor in the department of Philosophy at George Washington University. Her specialty is pluralism in mathematics, but she also works in philosophy of logic, philosophy of relativity theory and philosophy of chemistry. Quite separately, she does some work in philosophy of the environment, ecological economics and policy analysis. In general, she holds a principled skeptical position in philosophy, where rigor of argument, clarity of thought and metaphysical honesty are prized. The skepticism is "principled" in the sense of not being global, i.e., she is not skeptical for the sake of it, and she is not skeptical come-what-may. It is simply a natural position arising from sensitivity to our epistemic, linguistic and other representational limitations. michele@gwu.edu

Diana Gasparyan has held fellowships in the Department of Philosophy at M. V. Lomonosov Moscow State University. Currently she works at the National Research University Higher School of Economics in Moscow, Russia, where she is an Associate Professor of Philosophy. In 2009–2010, she was a visiting Professor at Clark University (Massachusetts) within the Fulbright Program. Her webpages can be found at: http://www.hse.ru/en/org/persons/66551 and https://suhse.academia.edu/DianaGasparyan anaid6@ya.ru

T. Grandon Gill is a professor in the information systems and decision sciences department of the University of South Florida, where he also serves as academic director of the Doctor of Business Administration program, which helps working executives learn to apply research methods to their business problems. His research focus is the transdisciplinary field known as informing science. grandon@usf.edu

David Griffiths, commonly known as Dai, is a Professor of Education at the University of Bolton. He holds a PhD from Universitat Pompeu Fabra, Barcelona, and has worked as a teacher at many levels of the educational system. For the past twenty-five years, he has focused on understanding the systemic implications of the use of technology in education. In the course of this work, he has participated in and led a large number of European projects. He was co-chair of the American Society for Cybernetics (ASC) 2013 conference in Bolton, and was a co-editor of the proceedings for ASC conferences in 2012, 2013, and 2014. d.e.griffiths@bolton.ac.uk

Christiane M. Herr is an architectural researcher and educator focusing on the areas of cross-disciplinary and cross-cultural studies in design and education as well as digitally supported design. In her PhD, she explored cellular automata as a means to establish architectural design support, which led to her strong interest in diagrams, designerly ways of seeing, and radical constructivism. Christiane is a board member of the American Society for Cybernetics and CAADRIA (Computer-Aided Architectural Design Research in Asia).
christiane.herr@xjtlu.edu.cn

Anthony Hodgson is Honorary Research Fellow at the University of Dundee attached to the Centre for Environmental Change and Human Resilience (CECHR https://www.dundee.ac.uk/cechr). He is also a founder member of the International Futures Forum (http://www.internationalfuturesforum.com) where he facilitates the forum and leads research. His research includes a doctoral thesis from the Hull University Centre for Systems Studies titled *Time, Pattern, Perception: The integration of systems and futures thinking,* which develops the concept of second-order anticipatory systems. He has published a number of papers on applied systems thinking and foresight methods as well as consulted to a variety of corporate and public organisations. tony@decisionintegrity.co.uk

Michael Hohl is a designer, design researcher and educator. He is interested in how design researchers can be educated to develop the well-balanced skills, empathy, rigor and intuition necessary to conduct their research. Central to his research are "designing for all senses," for designers to consider qualities beyond mere usability or visual appeal, taking into account all our sensorial modalities, considering embodied cognition. Related to this are interests in the epistemology of data-visualizations and mentoring students in making use of appropriate research methods in the context of their research. mh@hohlwelt.com

Andrea Jelić holds a PhD in architecture from Sapienza University of Rome for her dissertation "Architecture and Neurophenomenology: Rethinking the Pre-reflective Dimension of Architectural Experience" (2015). As an architect and researcher, her current work focuses on the intersection of architecture, neuroscience, and phenomenology, aimed at investigating issues of experience, meaning, and pre-reflective place-making in architectural and urban environments. jelic.andrea@gmail.com

Louis H. Kauffman is Professor of Mathematics at the University of Illinois in Chicago. He has a PhD in Mathematics from Princeton University (1972) and a BS in Mathematics from MIT (1966). He has taught at the University of Illinois since 1971 and has been a visiting professor and researcher at numerous universities and institutes. Kauffman is the founding editor and editor in chief of the *Journal of Knot Theory and Its Ramifications* and the editor of the World Scientific Book Series on Knots and Everything. Kauffman is the author of numerous books and research papers on form, knots, and related disciplines. He is a past President of the American Society for Cybernetics, author of the column Virtual Logic for the journal Cybernetics and Human Knowing, and a 1993 recipient of the Warren McCulloch award of the American Society for Cybernetics. kauffman@uic.edu

Vincent Kenny is the Director of the Accademia Costruttivista di Terapia Sistemica in Rome. His current activities involve applying psychology and philosophy to three areas of human conflict: firstly to individuals' interpersonal difficulties between themselves and others (known as "psychotherapy"); secondly to problems in organisational communications between and within networks of conversations (known as "organisational consulting"); and thirdly to problems of professional tennis players who run into difficulties of self-interruption in the tennis courts of the ATP/WTA circuits around the world (known as "tennis psychology"). kenny@acts-psicologia.it

Ronald Kline is Bovay Professor of History and Ethics of Engineering at Cornell University, where he holds a joint appointment with the Science and Technology Studies Department and the School of Electrical and Computer Engineering. He is past president of the Society for the History of Technology, which awarded him its Leonardo da Vinci Medal in 2016. He has published numerous articles on the intellectual, cultural, and social history of technology and three books: *Steinmetz: Engineer and Socialist* (1992); *Consumers in the Country: Technology and Social Change in Rural America* (2000) and *The Cybernetics Moment, Or Why We Call Our Age the Information Age* (2015), all with Johns Hopkins University Press. rrk1@cornell.edu

Urban Kordeš is professor of cognitive science and first-person research at the University of Ljubljana, where he is currently heading the cognitive science program. He holds a bachelor's degree in mathematical physics, and a doctorate in philosophy and cognitive science. His research interests include in-depth first-person inquiry, neurophenomenology, methodological issues in the research of non-trivial systems and collaborative knowledge creation. urban.kordes@pef.uni-lj.si

Edgar Landgraf is Professor of German at Bowling Green State University (Ohio). Recent publications include articles on Goethe, Kant, Kleist, Nietzsche, DeLillo, and Niklas Luhmann. His book Improvisation as Art. Conceptual Challenges, Historical Perspectives was published in 2011 with Continuum (reissued as paperback by Bloomsbury in 2014). He is currently working on a monograph on Nietzsche's Posthumanism. elandgr@bgsu.edu

Allenna Leonard has been involved in the cybernetics community since the 1980s. She worked with Stafford Beer's viable system model and team syntegrity process and on exploring using cybernetics to advance approaches to sustainability and improving the reliability of auditing soft information. She is a director of the Cwarel Isaf Institute, an organization affiliated with Malik Management to share and develop the work of Stafford Beer. She earned her doctorate with a thesis comparing perspectives on broadcast regulation. leonard.allenna@gmail.com

Sergei Levin is a lecturer in philosophy and logic at the Higher School of Economics Saint Petersburg. In 2015 he earned a PhD from Saint Petersburg State University. His research fields include metaethics, philosophy of mind and free will debates. serg.m.levin@gmail.com

Michael Lissack is the Executive Director of the Institute for the Study of Coherence and Emergence, President of the American Society for Cybernetics, and Professor of Design and Innovation at Tongji University. His academic work is accessible at http://lissack.com and at http://remedy101.com lissack@isce.edu

Robert J. Martin is a composer, psychologist, and professor emeritus at Truman State University. He completed a doctorate in educational psychology at the University of Illinois at Urbana-Champaign with an interdisciplinary thesis guided by Heinz von Foerster and Herbert Brün. He has a life-long interest in composition, creativity, learning, psychotherapy, constructivism, and cybernetics/systems science. He has written two books, numerous articles, and composed music for a variety of solo instruments and ensembles.
rmartin@truman.edu

Albert Müller is historian and works at the Department of Contemporary History, University of Vienna. Among other things, he is responsible for archival collections concerning Warren M. Brodey, Heinz von Foerster, Ranulph Glanville, Richard Jung, Gordon Pask and Stuart Umpleby. albert.mueller@univie.ac.at

Karl H. Müller is Director of the Steinbeis Transfer Centre New Cybernetics, senior researcher at the University of Ljubljana and Member of the International Academy for Systems and Cybernetic Sciences (IASCYS). His recent publications include *Second-Order Science: The Revolution of Scientific Structures* (2016), *Surveys and Reflexivity: A Second-Order Study of the European Social Survey (ESS)* (with Brina Malnar, 2015), and *Post-Disciplinary Cybernetics: The Science of Reflecting on Reflections* (forthcoming). khm@chello.at

Konstantin Pavlov-Pinus works at the Institute of Philosophy in Moscow. His areas of interest include the boundaries of logic, philosophy of consciousness, and the logic of historical processes. He is the editor-in-chief of the electronic philosophical journal *Vox*, http://vox-journal.org pavlov-koal@ya.ru

Following degrees from the Universities of Edinburgh and Sussex in the UK and post-doctoral fellowships in the US, at Rochester and Stanford, **John Pickering** has worked at Warwick University in the UK, where he lectures on psychology, philosophy and environmental issues. His principal research interests are consciousness, process thought, ecological psychology and biosemiotics. j.a.pickering@warwick.ac.uk

Bryony Pierce is a Research Associate at the University of Bristol. Her doctoral thesis was on "The Role of Consciousness in Action" and she has published papers on consciousness, philosophy of action, rationality and experimental philosophy. She is a Founder Member of Experimental Philosophy Group UK, and a former member of the European Science Foundation CNCC "CONTACT" research group. bryony.pierce@bristol.ac.uk

Bernd Porr has degrees in physics and journalism. He did his PhD thesis in computational neuroscience, where he developed closed loop learning algorithms. As a lecturer at the School of Engineering at the University of Glasgow, he has worked on social robotics, bipedal walking, deep brain stimulation, depression/schizophrenia models and auditory closed loop processing. He is also a filmmaker who produces fiction and educational videos for the flipped classroom. bernd.porr@glasgow.ac.uk

Larry Richards is Professor Emeritus of Management and Informatics at Indiana University East. He most recently served as Interim Vice Chancellor and Dean of the Indiana University – Purdue University Columbus campus and Executive Vice Chancellor and Interim Chancellor for IU East. He is a past-president of the American Society for Cybernetics (ASC) and the American Society for Engineering Management (ASEM). He received the ASC Norbert Wiener Medal in 2007 and was elected an ASEM Fellow in 2002 and an Academician in the International Academy of Systems and Cybernetic Sciences in 2010. His interests include policy support systems; the arts, technology and society; and social transformation and design. laudrich@iue.edu

Alexander Riegler obtained a PhD in artificial intelligence and cognitive science from Vienna University of Technology in 1995 with a dissertation on artificial life. Riegler's interdisciplinary work include diverse areas such as knowledge representation and anticipation in cognitive science, post-Darwinian approaches in evolutionary theory, and constructivist and computational approaches to epistemology, see http://tinyurl.com/riegler. Since 2005 he has been the editor-in-chief of *Constructivist Foundations*. ariegler@vub.ac.be

Adriana Schetz is an assistant professor at the Institute of Philosophy, University of Szczecin, Poland. Her interests include philosophy of mind, cognitive science, philosophy of psychology, and especially the problem of perception, consciousness, and animal cognition. She is the author of *Biological Externalism in the Theories of Perception*, published in Polish in 2014. She is currently working on a new book on animal learning theories.
adriana.schetz@gmail.com

Tom Scholte is an actor/director/writer for theatre and film whose work has been seen at such film festivals as Sundance, TIFF, Rotterdam and the Berlinale. His research focuses on cybernetics in the Stanislavski System of Acting and narrative drama as a modeling facility for the study of complex social systems. tom.scholte@ubc.ca

Tatjana Schönwälder-Kuntze is Professor of Philosophy at the LMU Munich. Her research is focusing more and more on post-structuralist practical philosophy, especially with reference to theory building, its grounding and the question of complete inclusion. Historically, she is interested in modern philosophy subsequent to Kant, with "excursions" to St. Aquinas and Descartes. Her books include a study on Sartre's moral philosophy, *Authentische Freiheit* (2001), a commentary to *George Spencer Brown's Laws of Form* (2nd edition, 2009), a study on Kant's theory building, *Freiheit als Norm?* (2010) and recently, *Philosophische Methoden zur Einführung* (Second edition, 2016). t.schoenwaelder@lmu.de

Bernard Scott graduated from Brunel University, UK, in 1968 with a first-class honours degree in psychology. He completed a PhD in cybernetics from the same university in 1976. His supervisor was Gordon Pask, with whom he worked between 1967 and 1978. Bernard is former Reader in Cybernetics, Cranfield University, UK. He is now Gordon Pask Professor of Sociocybernetics with the International Center for Sociocybernetics Studies, Bonn. Bernard is a Fellow of the UK's Cybernetics Society. He is an Associate Fellow of the British Psychological Society, a Fellow of the American Society for Cybernetics and an Academician of the International Academy of Systems and Cybernetics Sciences. Bernard is Past President of Research Committee 51 (on Sociocybernetics) of the International Sociological Association. bernces1@gmail.com

Mateus de Sousa van Stralen received a master's degree in architecture from the Federal University of Minas Gerais (UFMG) in 2009 and is currently a doctoral student at the same university. He is the director of KUBUS4D (an architecture office focused on digital design) and a researcher at LAGEAR (Graphic Laboratory for the Experience of Architecture, School of Architecture at UFMG). Based on his experience in architecture and urbanism, in particular planning and building projects, he focuses on the following topics: architecture and digital technologies, design process and digital fabrication.
mateus-stralen@ufmg.br

Tilia Stingl de Vasconcelos Guedes is an autonomous systemic and business consultant and member of the European Society for Education and Communication. She has a bachelor's degree and a master's degree in computer science and business administration from the Vienna University of Technology, a post-graduate qualification in magazine journalism and has earned her PhD in communication science from the University of Vienna. Her research interests are in the field of organizational communication and systemic approaches.
comunic@tiliastingl.com

Lea Šugman Bohinc is an assistant professor at the University of Ljubljana, Faculty of Social Work and Faculty of Education, and at the Sigmund Freud University in Ljubljana, Faculty of Psychotherapy Science. Her research interests include the epistemology of help, post-modern collaborative approaches to help, common factors in successful psychotherapy and social work, and transdisciplinary sciences of complexity, such as cybernetics and synergetics. lea.sugmanbohinc@fsd.uni-lj.si

Ben Sweeting is a senior lecturer at the University of Brighton where he is the course leader for the undergraduate degree in architecture. He studied architecture at the University of Cambridge and the Bartlett, UCL. In his PhD, supervised by Neil Spiller and Ranulph Glanville and funded by the Arts and Humanities Research Council, Ben explored epistemological and ethical questions in relation to architecture and design, drawing on second-order cybernetics and radical constructivism as well as on practice-based research methods. He is a member of the American Society for Cybernetics and was awarded the Heinz von Foerster Award in 2014. r.b.sweeting@brighton.ac.uk

Daniela Thumala Dockendorff is a clinical psychologist and a full-time assistant professor at the Department of Psychology at the University of Chile. She has participated in various research programs and publications in the eld of aging from a constructivist systemic perspective. Currently, she is a member of the research team FONDAP-Geroscience Center for BrainHealth and Metabolism. She has worked in several academic programs in the eld of psy- chology and social sciences at several universities and has contributed to the establishment of the clinical field of psychogerontology in Chile. This work was supported by CONICYT/FONDAP/15150012 and CONICYT/ FONDECYT Initiation 11150355.

dthumala@u.uchile.cl

Stuart Umpleby is Professor Emeritus in the Department of Management and Director of the Research Program in Social and Organizational Learning (http://www.gwu.edu/~rpsol) in the School of Business at The George Washington University in Washington, DC. The courses he has taught include operations research, organizational behavior, process improvement, systems thinking, and philosophy of science. Umpleby has published many papers in the fields of cybernetics and systems science. He is a past president of the American Society for Cybernetics and Associate Editor of the journal Cybernetics and Systems. Website: http://www.gwu.edu/~umpleby umpleby@gmail.com

SERIES ON KNOTS AND EVERYTHING

ISSN: 0219-9769

Editor-in-charge: Louis H. Kauffman *(Univ. of Illinois, Chicago)*

The Series on Knots and Everything: is a book series polarized around the theory of knots. Volume 1 in the series is Louis H Kauffman's Knots and Physics.

One purpose of this series is to continue the exploration of many of the themes indicated in Volume 1. These themes reach out beyond knot theory into physics, mathematics, logic, linguistics, philosophy, biology and practical experience. All of these outreaches have relations with knot theory when knot theory is regarded as a pivot or meeting place for apparently separate ideas. Knots act as such a pivotal place. We do not fully understand why this is so. The series represents stages in the exploration of this nexus.

Details of the titles in this series to date give a picture of the enterprise.

Published:

Vol. 60: *New Horizons for Second-Order Cybernetics*
edited by A. Riegler, K. H. Müller & S. A. Umpleby

Vol. 59: *Geometry, Language and Strategy: The Dynamics of Decision Processes — Vol. 2*
by G. H. Thomas

Vol. 58: *Representing 3-Manifolds by Filling Dehn Surfaces*
by R. Vigara & A. Lozano-Rojo

Vol. 57: *ADEX Theory: How the ADE Coxeter Graphs Unify Mathematics and Physics*
by S.-P. Sirag

Vol. 56: *New Ideas in Low Dimensional Topology*
edited by L. H. Kauffman & V. O. Manturov

Vol. 55: *Knots, Braids and Möbius Strips*
by J. Avrin

Vol. 54: *Scientific Essays in Honor of H Pierre Noyes on the Occasion of His 90th Birthday*
edited by J. C. Amson & L. H. Kauffman

Vol. 53: *Knots and Physics (Fourth Edition)*
by L. H. Kauffman

Vol. 52: *Algebraic Invariants of Links (Second Edition)*
by J. Hillman

Vol. 51: *Virtual Knots: The State of the Art*
by V. O. Manturov & D. P. Ilyutko

More information on this series can also be found at http://www.worldscientific.com/series/skae

Vol. 50: *Topological Library*
Part 3: Spectral Sequences in Topology
edited by S. P. Novikov & I. A. Taimanov

Vol. 49: *Hopf Algebras*
by D. E. Radford

Vol. 48: *An Excursion in Diagrammatic Algebra: Turning a Sphere from Red to Blue*
by J. S. Carter

Vol. 47: *Introduction to the Anisotropic Geometrodynamics*
by S. Siparov

Vol. 46: *Introductory Lectures on Knot Theory*
edited by L. H. Kauffman, S. Lambropoulou, S. Jablan & J. H. Przytycki

Vol. 45: *Orbiting the Moons of Pluto*
Complex Solutions to the Einstein, Maxwell, Schrödinger and Dirac Equations
by E. A. Rauscher & R. L. Amoroso

Vol. 44: *Topological Library*
Part 2: Characteristic Classes and Smooth Structures on Manifolds
edited by S. P. Novikov & I. A. Taimanov

Vol. 43: *The Holographic Anthropic Multiverse*
by R. L. Amoroso & E. A. Ranscher

Vol. 42: *The Origin of Discrete Particles*
by T. Bastin & C. Kilmister

Vol. 41: *Zero to Infinity: The Fountations of Physics*
by P. Rowlands

Vol. 40: *Intelligence of Low Dimensional Topology 2006*
edited by J. Scott Carter et al.

Vol. 39: *Topological Library*
Part 1: Cobordisms and Their Applications
edited by S. P. Novikov & I. A. Taimanov

Vol. 38: *Current Developments in Mathematical Biology*
edited by K. Mahdavi, R. Culshaw & J. Boucher

Vol. 37: *Geometry, Language, and Strategy*
by G. H. Thomas

Vol. 36: *Physical and Numerical Models in Knot Theory*
edited by J. A. Calvo et al.

Vol. 35: *BIOS — A Study of Creation*
by H. Sabelli

Vol. 34: *Woods Hole Mathematics — Perspectives in Mathematics and Physics*
edited by N. Tongring & R. C. Penner

Vol. 33: *Energy of Knots and Conformal Geometry*
by J. O'Hara

Vol. 32: *Algebraic Invariants of Links*
by J. A. Hillman

Vol. 31: *Mindsteps to the Cosmos*
by G. S. Hawkins

Vol. 30: *Symmetry, Ornament and Modularity*
by S. V. Jablan

Vol. 29: *Quantum Invariants — A Study of Knots, 3-Manifolds, and Their Sets*
by T. Ohtsuki

Vol. 28: *Beyond Measure: A Guided Tour Through Nature, Myth, and Number*
by J. Kappraff

Vol. 27: *Bit-String Physics: A Finite and Discrete Approach to Natural Philosophy*
by H. Pierre Noyes; edited by J. C. van den Berg

Vol. 26: *Functorial Knot Theory — Categories of Tangles, Coherence, Categorical Deformations, and Topological Invariants*
by David N. Yetter

Vol. 25: *Connections — The Geometric Bridge between Art and Science (Second Edition)*
by J. Kappraff

Vol. 24: *Knots in HELLAS '98 — Proceedings of the International Conference on Knot Theory and Its Ramifications*
edited by C. McA Gordon, V. F. R. Jones, L. Kauffman, S. Lambropoulou & J. H. Przytycki

Vol. 23: *Diamond: A Paradox Logic (Second Edition)*
by N. S. Hellerstein

Vol. 22: *The Mathematics of Harmony — From Euclid to Contemporary Mathematics and Computer Science*
by A. Stakhov (assisted by S. Olsen)

Vol. 21: *LINKNOT: Knot Theory by Computer*
by S. Jablan & R. Sazdanovic

Vol. 20: *The Mystery of Knots — Computer Programming for Knot Tabulation*
by C. N. Aneziris

Vol. 19: *Ideal Knots*
by A. Stasiak, V. Katritch & L. H. Kauffman

Vol. 18: *The Self-Evolving Cosmos: A Phenomenological Approach to Nature's Unity-in-Diversity*
by S. M. Rosen

Vol. 17: *Hypercomplex Iterations — Distance Estimation and Higher Dimensional Fractals*
by Y. Dang, L. H. Kauffman & D. Sandin

Vol. 16: *Delta — A Paradox Logic*
by N. S. Hellerstein

Vol. 15: *Lectures at KNOTS '96*
by S. Suzuki

Vol. 14: *Diamond — A Paradox Logic*
by N. S. Hellerstein

Vol. 13: *Entropic Spacetime Theory*
by J. Armel

Vol. 12: *Relativistic Reality: A Modern View*
edited by J. D. Edmonds, Jr.

Vol. 11: *History and Science of Knots*
edited by J. C. Turner & P. van de Griend

Vol. 10: *Nonstandard Logics and Nonstandard Metrics in Physics*
by W. M. Honig

Vol. 9: *Combinatorial Physics*
by T. Bastin & C. W. Kilmister

Vol. 8: *Symmetric Bends: How to Join Two Lengths of Cord*
by R. E. Miles

Vol. 7: *Random Knotting and Linking*
edited by K. C. Millett & D. W. Sumners

Vol. 6: *Knots and Applications*
edited by L. H. Kauffman

Vol. 5: *Gems, Computers and Attractors for 3-Manifolds*
by S. Lins

Vol. 4: *Gauge Fields, Knots and Gravity*
by J. Baez & J. P. Muniain

Vol. 3: *Quantum Topology*
edited by L. H. Kauffman & R. A. Baadhio

Vol. 2: *How Surfaces Intersect in Space — An Introduction to Topology (Second Edition)*
by J. S. Carter

Vol. 1: *Knots and Physics (Third Edition)*
by L. H. Kauffman

Printed in the United States
By Bookmasters